Recent Advances in Smart Materials for Sensor and Actuator Applications

Recent Advances in Smart Materials for Sensor and Actuator Applications

Guest Editors

Jung Woo Sohn
Heung Soo Kim

MDPI

Basel • Beijing • Wuhan • Barcelona • Belgrade • Novi Sad • Cluj • Manchester

Guest Editors

Jung Woo Sohn
School of Mechanical Engineering
Kumoh National Institute of Technology
Gumi
Republic of Korea

Heung Soo Kim
Department of Mechanical, Robotics and Energy Engineering
Dongguk University
Seoul
Republic of Korea

Editorial Office
MDPI AG
Grosspeteranlage 5
4052 Basel, Switzerland

This is a reprint of the Special Issue, published open access by the journal *Sensors* (ISSN 1424-8220), freely accessible at: https://www.mdpi.com/journal/sensors/special_issues/smart_materials_sensor_actuator.

For citation purposes, cite each article independently as indicated on the article page online and as indicated below:

Lastname, A.A.; Lastname, B.B. Article Title. *Journal Name* **Year**, *Volume Number*, Page Range.

ISBN 978-3-7258-6484-3 (Hbk)
ISBN 978-3-7258-6485-0 (PDF)
https://doi.org/10.3390/books978-3-7258-6485-0

Contents

About the Editors

Jung Woo Sohn

Jung Woo Sohn is a Professor in the School of Mechanical Engineering at Kumoh National Institute of Technology, Gumi, Republic of Korea. Prof. Sohn's research interests include AI-based robot control, including reinforcement learning and imitation learning. He is also interested in prognostics and health management (PHM) of composite structures.

Heung Soo Kim

Heung Soo Kim is a Professor in the Department of Mechanical, Robotics, and Energy Engineering at Dongguk University, Seoul, Republic of Korea. Prof. Kim's research interests include prognostics and health management (PHM) of composite structures, power plants, industrial robots, and mobility batteries. He is also interested in biomimetic actuators, adaptive structures, and structural analysis. He serves as director of the BK21 AIMS Center and the DGU Global Intelligent Robot Center, leading significant research initiatives. He has published 245 SCIE journal papers, achieving an H-index of 46 with over 9,000 citations. His outstanding research contributions have been recognized through numerous Academic Achievement Awards from various academic societies in the PHM field.

Preface

The field of smart materials represents one of the most dynamic and rapidly evolving frontiers in modern engineering and materials science. These materials, defined by their unique ability to sense and respond intelligently to external stimuli—such as magnetic fields, electric fields, temperature, stress, or chemical changes—have transitioned from scientific curiosities to foundational components of next-generation technologies.

Among their most impactful applications, the integration of smart materials into sensors and actuators has yielded transformative results. They serve as the active "nerves" and "muscles" of intelligent systems, enabling capabilities that were once thought impossible. From high-sensitivity detection and autonomous structural health monitoring to precise motion control and adaptive vibration damping, smart materials are pushing the boundaries of what machines can perceive and accomplish.

This reprint, "Recent Advances in Smart Materials for Sensor and Actuator Applications," originates from a highly successful Special Issue published in the journal *Sensors*. As editors, we witnessed an overwhelming influx of high-quality research that underscored the sheer pace of innovation in this domain. The field is moving quickly, and the proliferation of new materials, novel device architectures, and sophisticated control strategies highlighted the need for a consolidated, comprehensive text that captures this exciting moment.

Our primary goal for this reprint is to provide a definitive snapshot of the state of the art. The collection of works herein spans the full interdisciplinary spectrum of the field, bringing together contributions from leading experts in materials science, mechanical engineering, robotics, and control theory.

This volume is intended for a diverse audience. It serves as an essential reference for graduate students, academic researchers, and industrial engineers seeking to understand the latest trends and technical challenges in the field. We hope it will also function as a catalyst, inspiring new ideas, fostering interdisciplinary collaboration, and paving the way for the next generation of smart material-based systems.

We extend our deepest gratitude to all the contributing authors. Their exceptional work and dedication are the cornerstone of this preprint. It is their research that is actively shaping the future, and we are honored to present their findings to the wider scientific community.

Jung Woo Sohn and Heung Soo Kim

Guest Editors

sensors

MDPI

Article

Monitoring the Early Strength Development of Cement Mortar with Piezoelectric Transducers Based on Eigenfrequency Analysis Method

Guocheng Wang [1], Wenying Qiu [1], Dongkai Wang [1], Huimin Chen [2], Xiaohao Wang [1,2] and Min Zhang [2,*]

[1] Tsinghua-Berkeley Shenzhen Institute, Tsinghua University, Shenzhen 518055, China; wgc19@mails.tsinghua.edu.cn (G.W.); qwy17@mails.tsinghua.edu.cn (W.Q.); wdk18@mails.tsinghua.edu.cn (D.W.); wang.xiaohao@sz.tsinghua.edu.cn (X.W.)
[2] Shenzhen International Graduate School, Tsinghua University, Shenzhen 518055, China; chenhm19@mails.tsinghua.edu.cn
* Correspondence: zhang.min@sz.tsinghua.edu.cn

Abstract: Monitoring the early strength formation process of cement is of great importance for structural construction management and safety. In this study, we investigated the relationship between the eigenfrequency and the early strength development of cement mortar. Embedded piezoceramic-based smart aggregates recorded the early strength of cement mortar. An eigenfrequency analysis model demonstrated the relationship between strength and frequency. Experiments were performed by using piezoelectric transducers to monitor the early strength formation process during the testing period. Three types of specimens with different strength grades were tested, and the early strength formation processes were recorded. The experimental results demonstrate that cement mortar strength has a good linear relationship with the resonance frequency, and the average square of the correlation coefficient is greater than 0.98. The results show that structural health monitoring technology is a feasible method of assessing structural safety conditions and has a broad market in the structural construction industry.

Keywords: cement strength; structural health monitoring; piezoelectric transducers; smart aggregates; eigenfrequency

1. Introduction

Cement is widely used in buildings, bridges, dams, and other buildings, as a civil engineering material [1–5]. As an inorganic cementitious material in powder form, cement is also an essential raw material for making concrete. Cement is mixed with water to form a slurry which hardens in air or water, binding the various aggregates together. Commonly used cement is Portland cement, including ordinary Portland cement, slag Portland cement, volcanic ash Portland cement, fly ash Portland cement, etc. [6–14]. The strength of cement is gradually formed during hydration, and the final strength grade of cement is an essential factor affecting building strength. Monitoring cement strength is useful when studying the formation of building strength and ensuring the safety of building construction. The cement strength is related to maturity, which is the index of evaluating cement and its mixture formation and depends on temperature and time [15]. A cement mortar block is a test block made of mixed and hardened sand, cement, and water, with a standard proportion, often used to test cement strength [16–18]. As a standard test block with a fixed size, the study of its modal and eigenfrequency changes does have specific significance to cement strength formation.

Structural health monitoring (SHM) and damage identification are technologies that have received much attention in recent years. It is an important driving force in civil engineering construction to develop informatization and intellectualization. By using sensing technology to monitor the mechanical properties of a building structure, SHM can diagnose the damage, aging, and corrosion degree of the structure and evaluate its safety, durability, applicability, and bearing capacity in real-time [19,20]. Moreover, SHM technology is a way to minimize manual intervention in the building structure with automatic, real-time, continuous online monitoring and to make reliable estimates of the health of the building. It is a viable method to acquire and process data from a structure in an operational state and assess the key performance indicators of the structure with a nondestructive method [21]. There are many sensors used for SHM, including cement-based sensors [22–26], fiber optic sensors [27–30], and piezoelectric ceramic sensors. Compared with other methods, piezoceramic has some unique advantages based on the characteristics of piezoelectric materials, which can be used not only as sensors but also as actuators to provide vibrational excitation for active SHM.

The basic principle of the piezoelectric wave propagation method is to attach a pair of piezoceramics, called smart aggregates (SA), to the surface of the structure or to embed them into the structure. One piezoelectric transducer transmits stress waves and the other receives stress waves. By analyzing the changes between the transmitted and received signals, such as the attenuation of amplitude and the delay of the propagation time caused by strength formation, the mechanical properties and damage condition of the structure can be identified. When the structure is significant, embedded piezoelectric transducers are better suited to detecting internal structure factors of the concrete structure than surface-bonded transducers. In addition, this approach can overcome the disadvantage of lead zirconate titanate (PZT) vulnerability and provide protection for the core of PZTs for better structural health monitoring. To provide electrical safety and mechanical security, respectively, for the PZT patches, a waterproof layer and protective layer were used to wrap them [31].

Piezoelectric SA is a piezoelectric piece or piezoelectric functional element embedded into a structure in various ways. Compared with external paste transducers, this mounting method will not affect structure shape, thus reducing external interference factors and improving the reliability of the test results. However, the stability and reliability of the structure could be affected by an embedded piezoelectric element to a certain degree. This method can effectively extend the longevity of piezoelectric material and provide long-term, dynamic and real-time health monitoring and damage identification. In 2006, Song et al. [32] used an innovative piezoelectric-based approach to conduct early-age strength monitoring of concrete structures. Piezoelectric transducers in the form of SA were embedded into concrete structures as actuators and sensors during casting for strength monitoring purposes. In 2012, Hou et al. [33] used a PZT-based SA for seismic compressive stress measurement. The sensing system for dynamic compressive stress included proposed SAs and a commercially available charge amplifier. With a lower limit of frequency response stress amplitude, the sensing system is also capable of monitoring seismic pressures of low- and middle-rise buildings subject to moderate earthquakes. Meng et al. [34] achieved SHM of reinforced concrete columns under a static loading procedure by using PZT smart aggregates as sensors and actuators based on an active sensing system. In 2014, a combination of SA for monitoring the overall structural condition and surface-bonded PZT transducers for monitoring the crucial sections of construction was proposed as an effective SHM method for large infrastructures [35]. In 2017, Kong et al. [36,37] finished developing a novel embeddable spherical smart aggregate (SSA) for health monitoring of concrete structures. This novel SA offers omnidirectional actuating and sensing capabilities. In 2019, Zou et al. [38] provided a comprehensive examination regarding temperature effects on SA-based active monitoring methods for concrete structures. Zhou et al. [39] used a series of piezoceramic transducers that enabled the time-reversal method to identify the structural damage mechanism of basalt fiber reinforced plastic (BFRP) bar-reinforced concrete beams. Four SA transducers were bonded to the surface of a concrete beam to detect its damage in the bend-shear zone, the pure bending zone, and the overall zone. Gao et al. [40] made a novel tubular smart aggregate designed to be more compatible with

2-D structures. As seen by the above research, piezoelectric smart aggregates have been widely used in the research and application of structural health monitoring. However, their potential for strength monitoring of specific shapes and sizes, as well as the change of cement strength during curing by eigenfrequency analysis, has not yet been discovered.

In this paper, a novel method of monitoring the early strength development of cement mortars using embedded SA is proposed. Cement mortars were cast as standard specimens and analyzed by eigenfrequency analysis. The change of cement strength affected the maximum amplitude and resonant frequency of the received signal through wave propagation. The experimental results suggest that the actual strength of the cement can be reflected in the wave propagation characteristics. It proves that eigenfrequency is correlated with the strength of cement through maturity, and an exact functional connection was established.

2. Wave Propagation Monitoring and Eigenfrequency Analysis Method

The piezoelectric wave propagation method uses a pair of SAs embedded in the structure. One SA is used for transmitting stress waves and the other serves as the receiver, as shown in Figure 1. Stress wave parameters such as frequency, amplitude, and waveform, are directly related to the mechanical properties of the monitored structure, including the Young's modulus, Poisson's ratio, and eigenfrequency. When the stress wave propagates, it carries real-time information about the changes in the mechanical parameters of the structural material. The structural strength is identified by analyzing the signals received by the sensor under different conditions, such as the attenuation of the signal amplitude, the change of mode, and the delay of the propagation time caused by the strength. This method has a particular inhibitory effect on environmental noise and interference, which ensures the accuracy and reliability of the monitoring process and results.

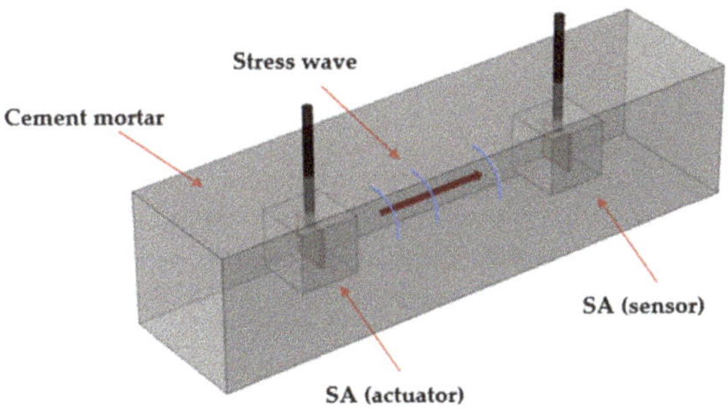

Figure 1. The configuration of the piezoelectric sensing system in the cement mortar test block.

The eigenfrequency or natural frequency is the discrete frequency at which a system tends to vibrate. Resonance occurs when the frequency of the external force is the same as the eigenfrequency of the system. In a resonant state, the vibration amplitude of the system will be maximized and will be significantly different from that in a non-resonant state. Based on the self-excitation characteristic of SA, the external force of frequency change can be applied to the system by applying a sweep signal to SA. The eigenfrequency of the system can be found by analyzing changes in the received signal.

The change of signal is related to the shift of structure eigenfrequency. Eigenfrequency is often related to the stiffness and mass of the structure itself [41–44]. In theory, the Young's modulus changes regularly during the early strength development of concrete or cement mortar [32]. Structure stiffness and the Young's modulus are equivalent to some extent, so the eigenfrequency analysis method could indicate the early strength formation process of cement mortar.

3. Experimental Method

3.1. Design and Fabrication of Smart Aggregates

Piezoceramic patches (PZT-5H, Baoding Hongsheng Acoustic Electronic Equipment Co., Ltd., Baoding, Hebei, China) were selected according to piezoelectric response coefficient and mechanical properties. As embedded transducers in concrete and cement mortar, the strength of piezoelectric ceramics should not be lower than the high strength and stiffness of concrete and cement mortar after forming. Additionally, the embedded transducers must have certain dimensional compatibility. The specific indicators are given in Table 1. SA has dual functions of excitation and reception as both the actuator and the sensor. Moreover, considering the sensitivity under the action of transmitting power and high frequency, PZT-5H was selected.

Table 1. Performance parameters of PZT-5H.

Parameter	Numerical Value	Unit
Density	7.45	10^3 kg/m^3
Young's modulus	46	GPa
Mechanical quality factor	70	/
Poisson ratio	0.33	/
Planar electromechanical coupling factor	0.65	/
d_{31}, d_{32}	−186	10^{-12} C/N
d_{33}	670	10^{-12} C/N

As shown in Figure 2, the SA consisted of a PZT patch with dimensions of 10 mm × 10 mm × 0.3 mm and an epoxy resin waterproof layer of 12 mm × 12 mm × 1 mm in the center of a cubic cement block of 20 mm × 20 mm × 20 mm with a protective layer of similar dimensions. A multimeter measured the resistance of each PZT patch before welding the wire to it. The infinite proved the whole monitoring system was packaged well, and the signal was transmitted through the stress wave instead of the current in the concrete. After the PZT patch was welded, a layer of epoxy resin potting adhesive was evenly applied to its surface using a polyvinyl alcohol (PVA) mold. After curing for 24 h, the epoxy resin-covered PZT patch (Figure 3a) was demolded by dissolving the PVA mold in water. The rigidity of cured epoxy resin is similar to that of cement mortar specimens, hence it had little influence on the regular operation of the PZT transducers, ensuring exemplary acoustic coupling. In addition, epoxy resin avoids short circuits while using PZT patches and direct contact between PZT patches and cement mortar as a waterproof layer.

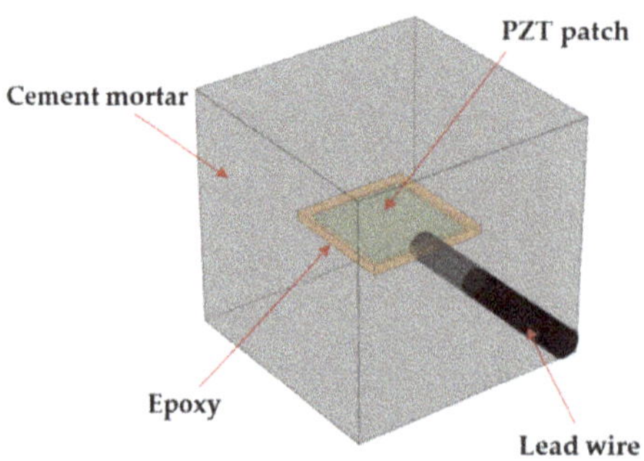

Figure 2. The structure of smart aggregate.

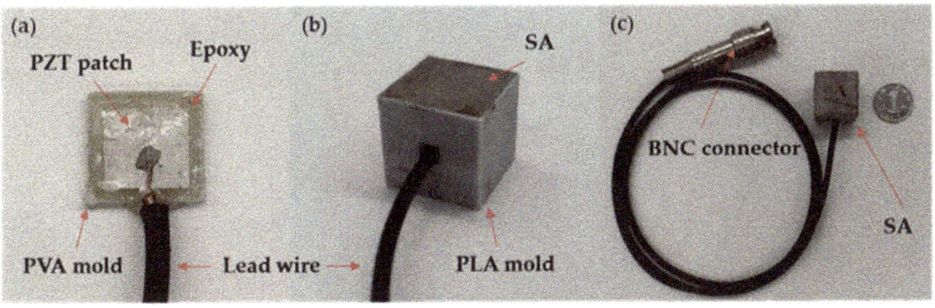

Figure 3. (**a**) PZT patch covered with an epoxy resin layer in a PVA mold; (**b**) Cement mortar with the epoxy resin-wrapped PZT patch cast in a PLA mold; (**c**) A photo of a smart aggregate.

The epoxy resin-covered PZT patch was then put into a new PLA mold, and the mixed cement mortar was cast into the mold (Figure 3b). To avoid errors caused by the difference in material properties, the same cement material as the specimens was used. The cast cement mortar was vibrated using a vibratory table and then compacted and smoothed. Finally, demolding was carried out after 48 h of curing, and the completed SA is shown in Figure 3c. SAs were used for early strength development monitoring of cement mortar specimens after 14 days. The mass proportion in cement mortar specimens, subsequent hydration heat, and strength variation of the cement mortar protective layer are tiny, hence the early strength development monitoring of cement mortar specimens was affected very little. The cement mortar protective layer of the SA had the same mechanical properties as the monitored cement mortar concrete to ensure the exemplary mechanical coupling and interface coupling of SA and monitoring structure. The cement mortar protective layer also bore a certain load to protect the PZT patch. The mass proportion of the SA was tiny and the mechanical properties of the SA are similar to the cement mortar specimens. No other sensors or impurities were between the two SAs for monitoring. Therefore, embedded SAs monitored the strength of cement mortar without affecting the strength development of cement mortar specimens.

3.2. Preparation of the Cement Mortar Test Specimen

Three types of cement, i.e., Portland slag cement 32.5 (P.S 32.5), ordinary Portland cement 42.5 (P.O 42.5), and ordinary Portland cement 52.5 (P.O 52.5), were selected for cement mortar specimens of 40 mm × 40 mm × 160 mm. The SAs were fixed at both ends of the specimen mold at a distance of 80 mm before casting and each SA was 40 mm away from one end of the cement mortar specimen. As shown in Figure 4a, three groups of specimens (S1, S2, and S3) were fabricated with different types of cement and ingredients (Table 2). The primary raw materials of ordinary Portland cement are silicate, gypsum, and some mixed materials. In addition to gypsum and silicate, Portland slag cement contains granulated blast furnace slag and other mineral materials. Because the composition of ordinary Portland cement and Portland slag cement is different, the hydration of ordinary Portland cement is faster than Portland slag cement. In terms of particle size, ordinary Portland cement is finer than Portland slag cement, so the mixture will be more uniform. Cement mortar specimens harden gradually at 20 °C curing temperature and 90% curing humidity after they are fabricated. The process is called setting or hardening. Fresh cement mortar or concrete gains strength most rapidly during the first few days. The strength of cement mortar or concrete will become stable after 28 days. The design of cement mortar or concrete structures is generally based on the 28th day strength. The strength of cement mortar or concrete before 28 days is called early strength.

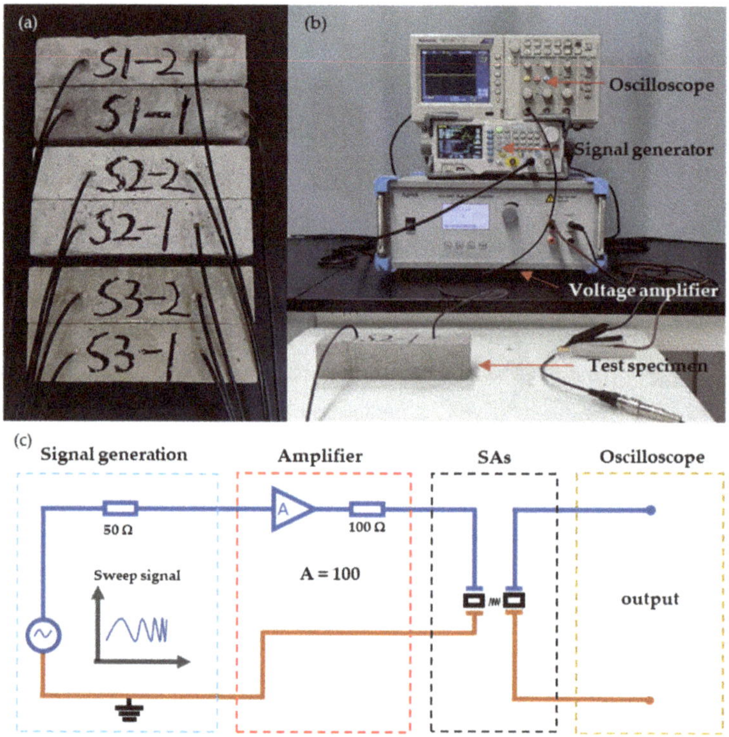

Figure 4. (**a**) Cement mortar specimens with embedded SAs; (**b**) Experimental monitoring system; (**c**) Schematic diagram of the monitoring system with four unit elements: the signal generation module, amplifier module, SAs, and oscilloscope.

Table 2. Ingredient of the cement mortar specimens.

Specimens	Cement Type	Cement Content (g)	Sand (g)	Water (g)
S1	Portland slag cement 32.5			
S2	Portland cement 42.5	300	900	150
S3	Portland cement 52.5			

3.3. Test Procedure

Due to the high stiffness of cement mortar specimens and the high driving voltage requirement of the PZT patch, a 100-volt swept sine signal from 10 Hz to 200 kHz was used to actuate the SA and generate stress waves. The specimens were continuously monitored for 28 days until the cement strength was fully formed and there was no significant change. The experimental monitoring system is shown in Figure 4b. The whole system included four parts: the signal generation module with the signal generator (DG1062Z, RIGOL, Beijing, China); the amplifier module with the voltage amplifier (ATA-2082, Agitek, Xi'an, Shaanxi, China); the SAs; the oscilloscope (TBS1102, Tektronix, Beaverton, OR, USA). The schematic diagram is shown in Figure 4c. The SA as the actuator was connected to the amplifier and the signal generator generated a sweep signal as shown in Figure S1. The other SA, as the sensor, received the stress wave that the actuator transmitted and displayed the waveform on the oscilloscope. Three extra cement mortar specimens for each cement type without embedded SAs were also prepared and tested using a destructive compression test on the 3rd, 5th, 7th, 14th, and 28th days for comparison.

4. Results and Discussion

The characteristics of a resonance signal and the change rule with time can be revealed by analyzing the received signal under a frequency sweep excitation signal. The change in cement strength is most dramatic during the first few days of hydration, and thus the 3rd, 5th, and 7th days were selected as representative time points, in addition to the intermediate and final time points of the strength formation process. The received signal was filtered by FFT to filter out 50 Hz power frequency noise. After the received signal was denoised, the received signals in a sweep cycle on the 3rd, 5th, 7th, 14th, and 28th days are shown in Figures 5, S2 and S3. In the finite element simulation (made by COMSOL Multiphysics, and the detailed setting parameters are given in Table S1), the first eigenfrequencies of P.S 32.5, P.O 42.5 and P.O 52.5 cement mortar specimens were 4738 Hz, 5882 Hz, and 6741 Hz. The second eigenfrequencies were 10,913 Hz, 11,343 Hz, and 12,875 Hz, respectively. In a sweep period signal of experiment, the received signal also has two apparent peak values in the early stage. The first peak value is about 5 kHz, which could be related to the first eigenfrequency of the specimen, and the second peak value is about 10 kHz. The amplitude of the signal from 20 kHz to 100 kHz is relatively stable, which could be caused by the gradual departure from the eigenfrequency of the specimen. The amplitude of a signal with a frequency greater than 100 kHz is minimal, which could be caused by the excitation signal frequency far exceeding the eigenfrequency of the specimen. The signal under sweep excitation also had obvious changes during the strength forming process of the specimen for 28 days. The received signal intensity under the frequency sweep excitation gradually decreased with time after the specimen reached the initial forming. The peak value of the received signal under the sweep excitation also drifted slowly to the higher frequency.

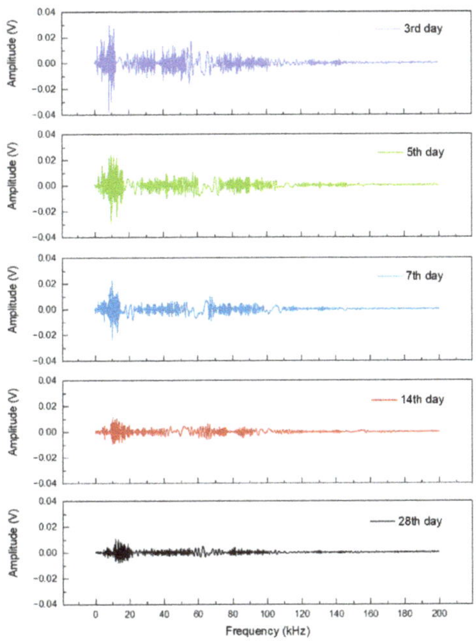

Figure 5. The received signal of P.O 42.5 specimens with swept sine excitation on the 3rd, 5th, 7th, 14th, and 28th days.

In a system of a single degree of freedom without damping, the natural angular frequency is defined as:

$$\omega_0 = 2\pi f_0 = \sqrt{k/m} \quad (1)$$

where ω_0 is the natural angular frequency with unit rad/s, f_0 is the natural frequency, k is the system's stiffness, and m is the mass. The stiffness is proportional to the Young's modulus so that the natural frequency can be defined as:

$$f_0 \propto \sqrt{E/m} \tag{2}$$

where E is the Young's modulus. Furthermore, the Young's modulus E is linearly related to the concrete strength Sm. Therefore, the natural frequency f_0 is correlated with the concrete strength S_m through the status of the Young's modulus [32]. Assuming mass is constant, the relationship between frequency and strength can be defined as:

$$f_0 \propto \sqrt{S_m} \tag{3}$$

The pivotal factors affecting the early strength of concrete or cement mortar are temperature and time, as the raw material and composition ratio are determined. Maturity is a physical quantity combining temperature and time. The relationship is based on the Saul-Bergstrom function (Equation (4)) [45,46].

$$M_s = \sum (T_i + 10)t_i \tag{4}$$

where M_s is the maturity with unit °C·d, T_i is the curing temperature, and t_i is the curing time under T_i temperature. In consideration of constant temperature during the curing process of cement mortar block, the function is simplified as:

$$M_s \propto t \tag{5}$$

where t is the curing time of concrete or cement mortar.

There is a certain functional relationship between strength S_m and maturity M_s [47–49]. The functional relationship between strength S_m and maturity M_s is similar to the power function through experimental verification and regression. Because of the linear relationship between maturity and time, so the strength can be expressed as:

$$S_m \propto t^r \tag{6}$$

where r is exponential coefficient.

With Equations (3) and (6), the function of time and the natural frequency can be defined as:

$$f_0 = a \times t^b \tag{7}$$

where a and b are coefficients.

The resonant frequencies of P.S 32.5, P.O 42.5, and P.O 52.5, obtained by swept signals are shown respectively in Figure 6a–c. After data fitting, the curves of each specimen's first and second resonant frequencies against the curing time illustrate the change of eigenfrequency over curing time. The variation in eigenfrequency with time conforms to the rule of the power function. The square of the correlation coefficients (R^2) of all fitted curves is greater than 0.9 and the absolute value of the correlation coefficients ($|R|$) is greater than 0.95. According to the correlation coefficients, the fitting results are highly correlated. In the comparison of three different cement specimens, the coefficient a of the fitting curve also gradually increases with the increase in cement strength grade. The reason is that the specimens with higher cement grades have higher initial strength in the hydration process.

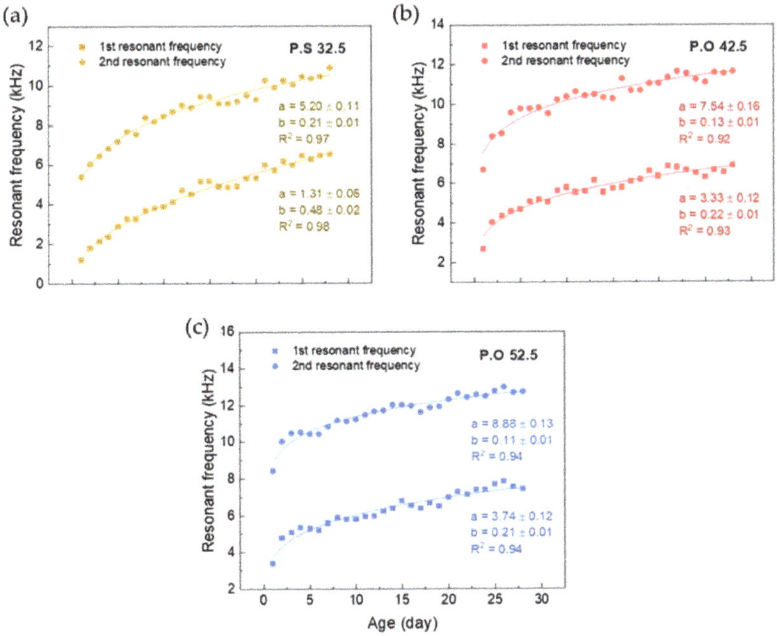

Figure 6. Experimental resonant frequency: (**a**) Resonant frequency of P.S 32.5 variation during time; (**b**) Resonant frequency of P.O 42.5 variation during time; (**c**) Resonant frequency of P.O 52.5 variation during the time.

The finite element model of cement mortar test block embedded with SAs is established by COMSOL Multiphysics. In the simulation, the contact interface between different materials is a fixed constraint. The detailed setting parameters of materials are given in Table S1. The simulation of the cement mortar modal displacement with certain eigenfrequency is normalized and shown in Figure 7. In Figure 7a, the mode at the first eigenfrequency is mainly bending deformation. The position of PZT patches is just at the neutral surface of specimen bending, and the extrusion and tensile deformation are small. In Figure 7b, the mode at the second eigenfrequency is mainly extrusion and tensile deformation. The deformation direction of the second eigenfrequency is consistent with the propagation direction of the stress wave produced by the PZT patch. Therefore, the signal acquisition intensity of the sensor at the working frequency close to the second eigenfrequency is better than that of the first eigenfrequency. This finite element simulation is also in accord with actual experimental results. The amplitude of the receiving signal with a 10 kHz excitation signal (nearly 2nd resonant frequency) is higher than others with different frequency excitation signals.

The cement mortar compressive strength value during curing time is shown in Figure 8a. The compressive strength test of cement mortar specimens was carried out simultaneously with the monitoring by SAs. Each type of cement was tested for compressive strength on days 3, 5, 7, 14, and 28. The curve fitted with the data shows the actual change of cement strength during curing time. The compressive strength test results were compared with the curve of the resonant frequency of the specimens obtained from the SAs with curing time, and it was found that the two are highly similar.

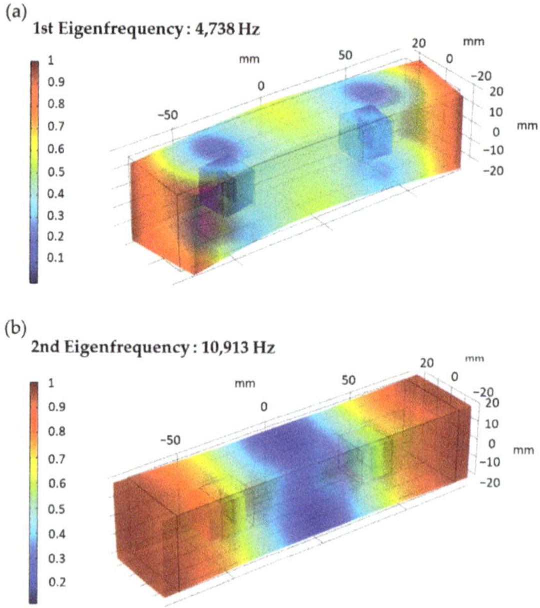

Figure 7. Finite element simulation of specimen eigenfrequency: (**a**) Modal displacement with 1st eigenfrequency of P.S 32.5; (**b**) Modal displacement with 2nd eigenfrequency of P.S 32.5.

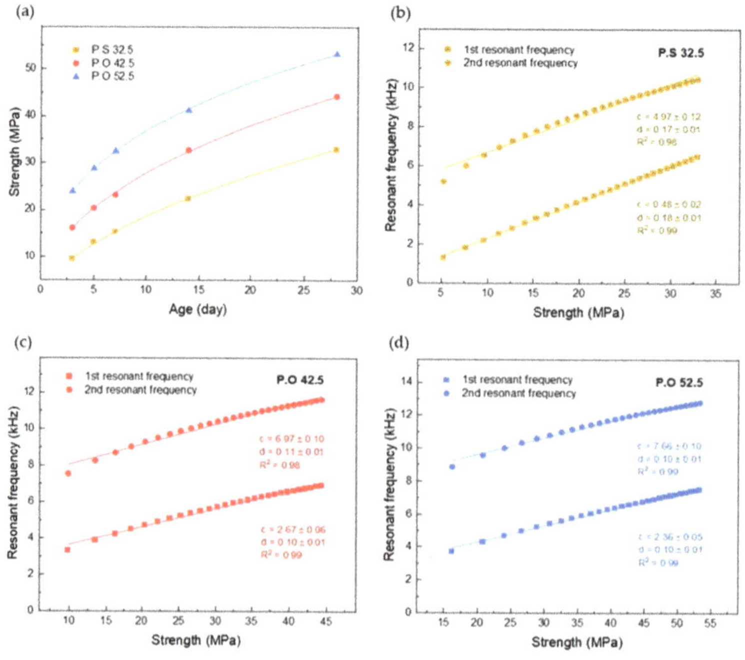

Figure 8. The cement mortar compressive strength and eigenfrequency value during curing time: (**a**) The variation of cement mortar compressive strength over time; (**b**) The variation of P.S 32.5 cement resonant frequency with strength; (**c**) The variation of P.O 42.5 cement resonant frequency with strength; (**d**) The variation of P.O 52.5 cement resonant frequency with strength.

The relationship between daily strength and eigenfrequency of cement mortar in the hydration process was formulated and the linear fittingwas made as shown in Figure 8b–d. The following relationship can be obtained:

$$f_0 = c + d \times S_m \tag{8}$$

where c and d are coefficients. The R^2 of all fitted curves is greater than 0.98. The fitting degree of the first resonance frequency is slightly better than that of the second. With the increase in cement strength grade, the initial strength of the specimen becomes higher, and the initial resonance frequency also increases. The sensor sensitivity, i.e., the slope of the fitting line, is almost the same for the two types of ordinary Portland cement. The sensitivity of Portland slag cement is nearly twice that of ordinary Portland cement, perhaps due to the material characteristics of different types of cement. The analysis method of eigenfrequency is often used in the nondestructive monitoring of structures, which plays a significant role in monitoring the micro change of structures [42–44]. Structural damage and stiffness change both affect the eigenfrequency change. According to the change of characteristic frequency detected by SA, the change in cement mortar strength can be predicted based on experimental analysis.

5. Conclusions

In this paper, piezoelectric SA was used to monitor the early strength formation of cement mortar specimens with different strengths of cement. Excitation signals given to specimens were from 10 Hz to 200 kHz. The emitted stress wave resonated with the cement mortar specimen at the same frequency point as the specimen eigenfrequency. SA also detected the frequency of resonance as a sensor. The relationship between the strength changes and the eigenfrequency was summarized by analyzing the eigenfrequency and comparing the variation of received signals. There is a reasonable linear relationship between strengths and resonance frequencies of cement mortar with R^2 of all fitted curves greater than 0.98. Consequently, the method of describing the early strength development of cement mortar by SA monitoring proves to be feasible. Likewise, the finite element simulation and compressive test were performed for verifying the experimental results. The proposed embedded piezoelectric transducers method with eigenfrequency analysis has the potential to monitor the early strength formation of cement. In the future, some improvements could be made regarding the sensitivity and accuracy of the device.

Supplementary Materials: The following are available online at https://www.mdpi.com/article/10.3390/s22114248/s1, Figure S1: The excitation sweep signal from 10 Hz to 200 kHz generated by the signal generator, Figure S2: The received signal of P.S 32.5 specimens with swept sine excitation at 3rd day, 5th day, 7th day, 14th day, and 28th day, Figure S3: The received signal of P.O 52.5 specimens with swept sine excitation at 3rd day, 5th day, 7th day, 14th day, and 28th day, Table S1: Setting parameters of materials in simulation.

Author Contributions: Conceptualization, M.Z. and G.W.; methodology, G.W. and W.Q.; validation, G.W.; investigation, G.W.; data curation, G.W.; writing—original draft preparation, G.W.; writing—review and editing, M.Z., W.Q., D.W. and H.C.; supervision, M.Z.; project administration, M.Z.; funding acquisition, X.W. and M.Z. All authors have read and agreed to the published version of the manuscript.

Funding: This work is supported by Natural Science Foundation of Guangdong Province (2020A1515010647) and Shenzhen Fundamental Research Funds (WDZC20200817152115001).

Institutional Review Board Statement: The study was conducted according to the guidelines of the Declaration of Helsinki, and approved by the Institutional Ethics Committee of Shenzhen International Graduate School, Tsinghua University.

Informed Consent Statement: Informed consent was obtained from all subjects involved in the study.

Data Availability Statement: The data presented in this study are available on request from the corresponding author.

Conflicts of Interest: The authors declare no conflict of interest.

References

1. Powers, T.C. Structure and physical properties of hardened portland cement paste. *J. Am. Ceram. Soc.* **1958**, *41*, 1–6. [CrossRef]
2. Birgin, H.B.; D'Alessandro, A.; Laflamme, S.; Ubertini, F. Smart Graphite-Cement Composite for Roadway-Integrated Weigh-In-Motion Sensing. *Sensors* **2020**, *20*, 4518. [CrossRef] [PubMed]
3. Liu, X.S.; Gu, Q.H.; Tan, Y.L.; Ning, J.G.; Jia, Z.C. Mechanical Characteristics and Failure Prediction of Cement Mortar with a Sandwich Structure. *Minerals* **2019**, *9*, 143. [CrossRef]
4. Zhang, J.P.; Liu, L.M.; Cao, J.Z.; Yan, X.; Zhang, F.T. Mechanism and application of concrete-filled steel tubular support in deep and high stress roadway. *Constr. Build. Mater.* **2018**, *186*, 233–246. [CrossRef]
5. Zhang, J.P.; Liu, L.M.; Li, Q.H.; Peng, W.; Zhang, F.T.; Cao, J.Z.; Wang, H. Development of cement-based self-stress composite grouting material for reinforcing rock mass and engineering application. *Constr. Build. Mater.* **2019**, *201*, 314–327. [CrossRef]
6. Shi, C.J.; Jimenez, A.F.; Palomo, A. New cements for the 21st century: The pursuit of an alternative to Portland cement. *Cem. Concr. Res.* **2011**, *41*, 750–763. [CrossRef]
7. Arjunan, P.; Silsbee, M.R.; Roy, D.M. Sulfoaluminate-belite cement from low-calcium fly ash and sulfur-rich and other industrial by-products. *Cem. Concr. Res.* **1999**, *29*, 1305–1311. [CrossRef]
8. Marroccoli, M.; Nobili, M.; Telesca, A.; Valenti, G.L. Early hydration of calcium sulfoaluminate-based cements for structural applications. In Proceedings of the International Conference on Sustainable Construction Materials and Technologies, Coventry, UK, 11–13 June 2007; pp. 389–395.
9. Fernandez-Jimenez, A.M.; Palomo, A.; Lopez-Hombrados, C. Engineering properties of alkali-activated fly ash concrete. *ACI Mater. J.* **2006**, *103*, 106–112.
10. Law, D.W.; Adam, A.A.; Molyneaux, T.K.; Patnaikuni, I. Durability assessment of alkali activated slag (AAS) concrete. *Mater. Struct.* **2012**, *45*, 1425–1437. [CrossRef]
11. Paria, S.; Yuet, P.K. Solidification-stabilization of organic and inorganic contaminants using portland cement: A literature review. *Environ. Rev.* **2006**, *14*, 217–255. [CrossRef]
12. Schmidt, T.; Lothenbach, B.; Romer, M.; Neuenschwander, J.; Scrivener, K. Physical and microstructural aspects of sulfate attack on ordinary and limestone blended Portland cements. *Cem. Concr. Res.* **2009**, *39*, 1111–1121. [CrossRef]
13. Guo, S.C.; Hu, J.; Dai, Q.L. A critical review on the performance of portland cement concrete with recycled organic components. *J. Clean. Prod.* **2018**, *188*, 92–112. [CrossRef]
14. Garcia-Lodeiro, I.; Fernandez-Jimenez, A.; Palomo, A. Variation in hybrid cements over time. Alkaline activation of fly ash-portland cement blends. *Cem. Concr. Res.* **2013**, *52*, 112–122. [CrossRef]
15. Ulm, F.J.; Coussy, O. Modeling of thermochemomechanical couplings of concrete at early ages. *J. Eng. Mech.-Asce* **1995**, *121*, 785–794. [CrossRef]
16. Sajedi, F.; Razak, H.A.; Bin Mahmud, H.; Shafigh, P. Relationships between compressive strength of cement-slag mortars under air and water curing regimes. *Constr. Build. Mater.* **2012**, *31*, 188–196. [CrossRef]
17. Eskandari-Naddaf, H.; Kazemi, R. ANN prediction of cement mortar compressive strength, influence of cement strength class. *Constr. Build. Mater.* **2017**, *138*, 1–11. [CrossRef]
18. Bu, J.W.; Tian, Z.H.; Zheng, S.Y.; Tang, Z.L. Effect of Sand Content on Strength and Pore Structure of Cement Mortar. *J. Wuhan Univ. Technol.-Mater. Sci. Ed.* **2017**, *32*, 382–390. [CrossRef]
19. Lu, S.; Ba, H.J. Corrosion Sensor for Monitoring the Service Condition of Chloride-Contaminated Cement Mortar. *Sensors* **2010**, *10*, 4145–4158. [CrossRef]
20. Kim, J.K.; Kee, S.H.; Futalan, C.M.; Yee, J.J. Corrosion Monitoring of Reinforced Steel Embedded in Cement Mortar under Wet-And-Dry Cycles by Electrochemical Impedance Spectroscopy. *Sensors* **2020**, *20*, 199. [CrossRef]
21. Housner, G.W.; Bergman, L.A.; Caughey, T.K.; Chassiakos, A.G.; Claus, R.O.; Masri, S.F.; Skelton, R.E.; Soong, T.T.; Spencer, B.F.; Yao, J.T.P. Structural control: Past, present, and future. *J. Eng. Mech.* **1997**, *123*, 897–971. [CrossRef]
22. Lee, S.J.; You, I.; Zi, G.; Yoo, D.Y. Experimental Investigation of the Piezoresistive Properties of Cement Composites with Hybrid Carbon Fibers and Nanotubes. *Sensors* **2017**, *17*, 2516. [CrossRef] [PubMed]
23. Yoo, D.Y.; You, I.; Lee, S.J. Electrical Properties of Cement-Based Composites with Carbon Nanotubes, Graphene, and Graphite Nanofibers. *Sensors* **2017**, *17*, 1064. [CrossRef] [PubMed]
24. Zhang, T.T.; Zhang, K.P.; Liu, W.D.; Liao, Y.C. Impact Mechanical Response of a 2-2 Cement-Based Piezoelectric Sensor Considering the Electrode Layer Effect. *Sensors* **2017**, *17*, 2035. [CrossRef] [PubMed]
25. Zhang, T.T.; Liao, Y.C.; Zhang, K.P.; Chen, J. Theoretical Analysis of the Dynamic Properties of a 2-2 Cement-Based Piezoelectric Dual-Layer Stacked Sensor under Impact Load. *Sensors* **2017**, *17*, 1019. [CrossRef] [PubMed]
26. Luo, J.L.; Zhang, C.W.; Li, L.; Wang, B.L.; Li, Q.Y.; Chung, K.L.; Liu, C. Intrinsic Sensing Properties of Chrysotile Fiber Reinforced Piezoelectric Cement-Based Composites. *Sensors* **2018**, *18*, 2999. [CrossRef]

27. Li, H.N.; Li, D.S.; Song, G.B. Recent applications of fiber optic sensors to health monitoring in civil engineering. *Eng. Struct.* **2004**, *26*, 1647–1657. [CrossRef]
28. Lopez-Higuera, J.M.; Cobo, L.R.; Incera, A.Q.; Cobo, A. Fiber Optic Sensors in Structural Health Monitoring. *J. Lightwave Technol.* **2011**, *29*, 587–608. [CrossRef]
29. Di Sante, R. Fibre Optic Sensors for Structural Health Monitoring of Aircraft Composite Structures: Recent Advances and Applications. *Sensors* **2015**, *15*, 18666–18713. [CrossRef]
30. Li, S.; Wu, Z. Development of distributed long-gage fiber optic sensing system for structural health monitoring. *Struct. Health Monit.-Int. J.* **2007**, *6*, 133–143. [CrossRef]
31. Chen, Y.Z.; Xue, X.W. Advances in the Structural Health Monitoring of Bridges Using Piezoelectric Transducers. *Sensors* **2018**, *18*, 4312. [CrossRef]
32. Gu, H.; Song, G.; Dhonde, H.; Mo, Y.L.; Yan, S. Concrete early-age strength monitoring using embedded piezoelectric transducers. *Smart Mater. Struct.* **2006**, *15*, 1837–1845. [CrossRef]
33. Hou, S.; Zhang, H.B.; Ou, J.P. A PZT-based smart aggregate for compressive seismic stress monitoring. *Smart Mater. Struct.* **2012**, *21*, 105035. [CrossRef]
34. Meng, Y.Y.; Yan, S.; Sun, W. Experimental research on damage detection of concrete structures based on piezoelectric smart aggregates. In Proceedings of the 2nd International Conference on Civil Engineering, Architecture and Building Materials (CEABM 2012), Yantai, China, 25–27 May 2012; pp. 1145–1151.
35. Divsholi, B.S.; Yang, Y.W. Combined embedded and surface-bonded piezoelectric transducers for monitoring of concrete structures. *Ndt E Int.* **2014**, *65*, 28–34. [CrossRef]
36. Kong, Q.Z.; Fan, S.L.; Bai, X.L.; Mo, Y.L.; Song, G.B. A novel embeddable spherical smart aggregate for structural health monitoring: Part I. Fabrication and electrical characterization. *Smart Mater. Struct.* **2017**, *26*, 095050. [CrossRef]
37. Kong, Q.Z.; Fan, S.L.; Mo, Y.L.; Song, G.B. A novel embeddable spherical smart aggregate for structural health monitoring: Part II. Numerical and experimental verifications. *Smart Mater. Struct.* **2017**, *26*, 095051. [CrossRef]
38. Zou, D.J.; Du, C.C.; Liu, T.J.; Li, W.J. Effects of temperature on the performance of the piezoelectric-based smart aggregates active monitoring method for concrete structures. *Smart Mater. Struct.* **2019**, *28*, 035016. [CrossRef]
39. Zhou, L.Z.; Zheng, Y.; Song, G.B.; Chen, D.D.; Ye, Y.X. Identification of the structural damage mechanism of BFRP bars reinforced concrete beams using smart transducers based on time reversal method. *Constr. Build. Mater.* **2019**, *220*, 615–627. [CrossRef]
40. Gao, W.H.; Li, H.N.; Ho, S.C.M. A Novel Embeddable Tubular Piezoceramics-Based Smart Aggregate for Damage Detection in Two-Dimensional Concrete Structures. *Sensors* **2019**, *19*, 1501. [CrossRef]
41. Gudmundson, P. Eigenfrequency changes of structures due to cracks, notches or other geometrical changes. *J. Mech. Phys. Solids* **1982**, *30*, 339–353. [CrossRef]
42. Lee, Y.S.; Chung, M.J. A study on crack detection using eigenfrequency test data. *Comput. Struct.* **2000**, *77*, 327–342. [CrossRef]
43. Liang, R.Y.; Hu, J.L.; Choy, F. Theoretical-study of crack-induced eigenfrequency changes on beam structures. *J. Eng. Mech. -Asce* **1992**, *118*, 384–396. [CrossRef]
44. Zhang, Y. Eigenfrequency Computation of Beam/Plate Carrying Concentrated Mass/Spring. *J. Vib. Acoust.-Trans. Asme* **2011**, *133*, 021006. [CrossRef]
45. Tomlinson, D.; Moradi, F.; Hajiloo, H.; Ghods, P.; Alizadeh, A.; Green, M. Early age electrical resistivity behaviour of various concrete mixtures subject to low temperature cycling. *Cem. Concr. Compos.* **2017**, *83*, 323–334. [CrossRef]
46. Sidney Mindess, J.F.Y.; David, D. *Concrete*, 2nd ed.; Pearson Education Inc.: Upper Saddle River, NJ, USA, 2003; pp. 404–405.
47. Mariak, A.; Kurpinska, M.; Wilde, K. Maturity curve for estimating the in-place strength of high performance concrete. In Proceedings of the 64th Scientific Conference of the Committee-for-Civil-Engineering-of-the-Polish-Academy-of-Sciences and the Science-Committee-of-the-Polish-Association-of-Civil-Engineers (PZITB) (KRYNICA), Krynica Zdroj, Poland, 16–20 September 2018.
48. Belykh, I.; Sopov, V.; Butska, L.; Pershina, L.; Makarenko, O. Predicting the strength and maturity of hardening concrete. *MATEC Web Conf.* **2018**, *230*, 03001. [CrossRef]
49. Jiang, P.; Jiang, L.H.; Zha, J.; Song, Z.J. Influence of temperature history on chloride diffusion in high volume fly ash concrete. *Constr. Build. Mater.* **2017**, *144*, 677–685. [CrossRef]

sensors

MDPI

Article

Trajectory Control for Vibrating Screen with Magnetorheological Dampers

Szymon Ogonowski * and Piotr Krauze

Department of Measurements and Control Systems, Silesian University of Technology, Akademicka 16, 44-100 Gliwice, Poland; piotr.krauze@polsl.pl
* Correspondence: szymon.ogonowski@polsl.pl

Abstract: The article presents a method of vibrating screen trajectory control based on MR (magnetorheological) dampers applied in a screen suspension. A mathematical description of the dynamic screen model was derived, and parameters of this model were estimated based on experimental data from a semi-industrial vibrating screen. The investigated screen included a single mechanical exciter with unbalanced masses, generating a circular vibration trajectory and operating with over-resonant frequency close to 19 Hz. It was experimentally tested in several phases of operation: start-up, nominal operation at a target vibration frequency and shutdown. The implemented screen model was further extended and included several MR dampers oriented horizontally and vertically in the form of Bouc–Wen models. The Bouc–Wen model was identified based on experiments carried out for an MR damper subjected to harmonic excitations generated by the MTS (material testing system). Dominant frequencies of excitation varied by up to 20 Hz during experiments. The main novelty of the reported solution is that according to the proposed control algorithm, the desired forces generated by MR dampers emulate an additional virtual mechanical exciter of the vibrating screen. In turn, it interacts with the available exciter, resulting in conversion of the trajectory from circular to linear, which was validated in the presented study. For the purpose of simulation accuracy, the desired control force was additionally limited within the simulator by MR damper dissipative domain, which maps the constraints of a semi-active damper. The presented approach allows one to obtain a close to linear trajectory with only one exciter and with semi-active control of suspension stiffness. The results were successfully repeated with different configurations of desired trajectory, indicating that the effectiveness of the desired linear trajectory generation depends on its orientation. The reported findings may lead to the design of new vibrating screen constructions, taking advantage of the semi-active control of a suspension in the attenuation of disturbance resulting from varying processed material parameters.

Keywords: vibrating screen; vibration trajectory control; magnetorheological damper; semiactive suspension; dynamic model; Bouc–Wen model

1. Introduction

Screening and sieving are some of the oldest and extensively used nowadays physical size separation methods for bulk materials [1,2]. The use of such methods is widely present in laboratories for the purpose of particle size distribution analysis, and in the wide range of industries, such as mining, aggregate production, recycling and mineral processing, pharmaceuticals, cosmetics and food, as a unit operation for large scale separation [3,4]. One of the most common devices being applied for the above processes is the vibrating screen. Market analyses indicate that whole vibrating screen market was worth 2.1 billion USD in 2019 and is projected to reach 2.7 billion USD by 2025. The compound annual growth rate (CARG) is projected to go from 3% to 7% during the period of 2019–2025, depending on the source [5,6]. Even the recent COVID-19 pandemic did not manage to dramatically change the predicted growth trend [5]. In the industry, two main classes of screens are used: resonant type and above resonance type. Various constructions of vibrating screens and the tuning of their technological parameters are discussed in [7–11]. The resonance regime is desirable; however, its control is complicated due to vulnerability of the sieving process to changes in bulk material thickness, particle distribution over the deck and their physical properties [12,13]. Moreover, resonance frequencies can be harmful for screen construction. Even in the above resonate solutions, most of the damages occur during startup and stopping procedures, when the device is passing through the resonance frequency. Most vibrating screens use inertial vibrating exciters, which are designed with respect to the specified trajectory (linear, circular, elliptical) and direction of the sieve movement [14]. Such trajectories can be realized with one electric motor and various numbers of unbalances that rotate synchronously using forced kinematic or dynamic synchronization [15]. Different means of kinematic synchronization can be used, e.g., gears [16], belt transmissions or elastic links with nonlinear stiffness [17]. Industrial applications and various research use two-mass systems [18] or systems with different numbers of vibrators and unbalanced masses [19]. Nowadays, most popular solutions are based on drives with two vibrators, providing synchronous rotation of unbalanced masses. Parameters of the trajectory can be controlled by changes in the direction, frequency or phases of the vibrators [20]. Another popular solution uses three or four independently installed vibrators and frequency control during synchronization [21,22]. The research reported in this article focuses on the application of a semi-active element—a magnetorheological damper (MR damper)—in the suspension of a vibrating screen with only one vibrator.

The MR damper consists of a piston and a cylinder which is filled with an MR fluid. The MR fluid is made of magnetizable particles suspended in carrier fluid, e.g., minerals [23]. These particles are subjected to a magnetic field induced by electric coils, which are located in the vicinity of piston gaps, reorganize in a chain-like structure and increase the damping parameter of the MR damper on a macroscopic scale. MR dampers, as an example of a semi-active system, are favored over active solutions for their low energy consumption. They are widely used for vibration control in mountain bikes [24], automotive applications and all-terrain vehicles [25]; for improvement of driving safety [26]; in the construction machinery [27]; and in buildings and bridges [28].

The behavior of MR dampers is commonly analyzed based on characteristics of generated force with respect to axial piston velocity. Here, increased complexity can be justified by the force saturation exhibited for higher piston velocities or hysteresis loops, as presented in [29]. Thus, controlling of the MR damper force is challenging, and it commonly requires preliminary identification of at least a limited MR damper model in order to apply an open-loop of the closed-loop force control approach. Methods of modeling MR dampers are generally divided into phenomenological, input-output and behavioral models. Phenomenological models reflect the internal design of an MR damper and try to describe the occurring phenomena, e.g., by taking into account velocity profiles of MR fluid flowing through a piston gap, as presented in [30]. Additionally, the yield shear stress or magnetic field distribution can be analyzed as reported in [31] for an unbalanced rigid rotor damped by dedicated MR films. More demanding transient analysis of MR damper operation can utilize methods of computational fluid dynamics or distributed modeling of the fluid–structure interaction. Such methods discussed, e.g., in [32,33], can additionally take into account the influences of thermal properties of applied materials or fluid cavitation.

Semi-active vibration control applications which utilize model-based control approaches generally require the MR damper models to exhibit limited computational complexity in order to be easily implemented in a real-time controller. Thus, combinations of input–output and behavioral models are mostly used. The Bingham model presented in [34] consists of two dominant components, i.e., Coulomb friction and viscous damping. Further, the Gamota–Filisko model presented in [35] includes Bingham, Kelvin–Voight and Hooke body models which are connected in series. The well-known Bouc–Wen model [36] of an MR damper and its extension, the Spencer [35] model, are favored for their compact form and ability to map dominant nonlinear features of MR damper behavior. The following should be mentioned: force saturation greenwhich is revealed for higher piston velocities and hysteresis loops greenwhich are noticeable in force–velocity characteristics. Additionally, an input–output modeling approach can be applied based on the hyperbolic tangent function, as presented in [37], which is favored for low computational complexity, and simultaneously, it allows for good adjustment to the shape of force–velocity characteristics.

To make the research concerning vibrating screen modifications more efficient, many researchers use modeling and simulation techniques at the first stage of research. There are reports of this approach being applied for circular vibrating screens [38,39], linear vibrating screens [40–42], an elliptical vibrating screen [43] and banana screens [44,45]. Models are developed for operational diagnosis [46], for diagnosis of spring failures [47] or for vibration exciter operation analysis [48]. The model of a vibrating screen we created was verified with the measurement data from experiments on a semi-industrial vibrating screen (see Figure 1). The model was than expanded with the Bouc–Wen model of the MR damper, identified during previous research based on measurement data obtained from an MTS (material testing system). In the case of identification of both the vibrating screen model and the MR damper model, the methods of analysis of vibration measurements are extremely important [49]. In the case of the screen, additional effects deteriorating the quality of measurements need to be taken into account, e.g., impulsive noise generated by processed material and methods for filtering of such [50].

Figure 1. A semi-industrial vibrating screen with a single mechanical exciter based on rotating unbalanced mass.

The research reported in this article used modeling and simulation techniques to show the possibility of changing and controlling the vibrating screen's trajectory using the semi-active element of its suspension. Various algorithms for trajectory control are well-known in the literature and were applied for various applications. Apart from the classical PID (proportional-integral-derivative) controller, the adaptive backstepping method accompanied with a Lyapunov function was used for trajectory tracking algorithm of an automated guided vehicle in [51]. An MPC (model predictive control) including a novel Hammerstein model was used for controlling the gimbal system mounted on an unmanned aerial vehicle under external disturbances, as presented in [52]. An LTV-MPC (linear time-varying model predictive control) was used for trajectory tracking control for an autonomous vehicle in [53]. However, in order to simplify the trajectory control algorithm which is proposed in this article and developed for future implementation in the real-time controller of the screen suspension, a straight-forward control method was applied. It is intended for emulation of force generated by an additional virtual vibrator. Changing the relationship between the work of the virtual and the real vibrator allows for modification of the vibration trajectory.

The majority of studies presented in the literature dedicated to screen suspension with variable parameters exhibit limited possibilities for adaptation. It is time-consuming to adjust a screen to varying process parameters, making it almost impossible to achieve fast control of vibration trajectory during screen operation. Applications of MR dampers in screen suspension are known in the literature; however, they do not have a control system and they are limited to operate in open-loop configuration or even mostly using a constant control current supplying the MR dampers. A design of MR damper dedicated to a vibrating screen, described in [54], was applied in order to enhance the screening efficiency. Other authors carried out experiments for different screening amplitudes and constant control currents to the MR damper. The key results showed the influences of these parameters on the variance of vibration displacement and screening efficiency. Another study presented in [55] and dedicated to the MR damper installed in a vibrating screen and showed a trispectrum. Correlation dimension analysis was similarly carried out for different constant control currents only. That analysis was extended in [56], where preliminary experiments were carried out for a single MR damper. Further, it was indicated that constant control current can be used for mitigation of extensive vibrations occurring during the start-up and stopping phases of the screen's operation. Relative to the studies currently available in the literature, the main contributions of the approach proposed in this paper are as follows:

- We propose an application with several MR damper models oriented horizontally and vertically and located in the front and rear parts of the vibrating screen model, which allows for efficient vibration control;
- The proposed solution allows for online modification of the vibration trajectory by emulation of force generated by a virtual and additional mechanical exciter;
- Application of MR dampers in the vibrating screen allows faster transitions through the resonant frequencies on startup and stop procedures;
- The Bouc–Wen model was adjusted for a range of vibration frequencies that spans up to 20 Hz, which allows for the application of a single MR damper model during all phases of screen operation;
- The dissipative domain of the MR damper was evaluated based on the Bouc–Wen model, which allowed simplification of the model in implementation and simulation;
- The developed methodology for identifying the screen model is scalable, and it can be applied for future vibrating screens of different physical sizes and for validation of vibration control algorithms.

The article consists of five sections. In Section 2, a dynamic model of the presented vibrating screen is defined. Section 3 presents experimental results obtained for the actual industrial screen and reports results of identification of a dynamic model of the actual screen. Furthermore, a Bouc–Wen model of MR damper is defined and the procedure of Bouc–Wen model identification is presented based on experimental results obtained for harmonic excitation. Section 4 proposes a novel trajectory control approach which is applied to the implemented screen model. The proposed control algorithm is further validated and discussed for different configurations and desired vibration trajectories. Finally, the conclusions are presented in Section 5.

2. Dynamic Model of Vibrating Screen

The considered dynamic model of a screen was implemented and applied in the presented study as a two-dimensional mechanical structure of lumped parameters. A mechanical representation of the discussed model is presented in Figure 2. A mechanical exciter with unbalanced masses driven by a single electric motor is a source of screen vibration. For the purpose of the presented study, the lateral movement of the screen was neglected, since both the construction and unbalanced masses of the exciter are symmetrically located with respect to the longitudinal axis of the screen.

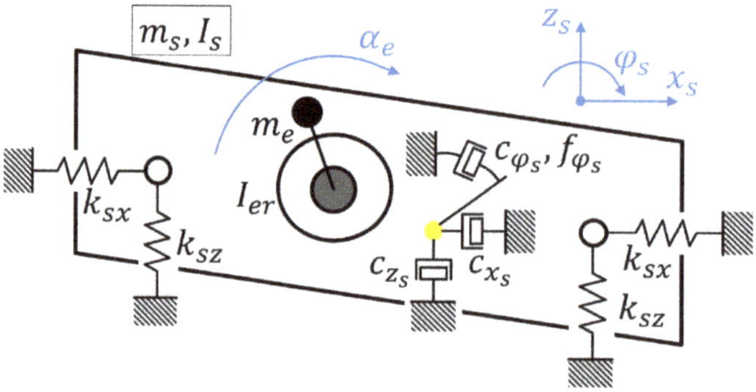

Figure 2. Mechanical representation of the dynamic model of vibrating screen with a single mechanical exciter.

2.1. Structure of the Dynamic Model

The dynamic screen model consists of a rigid body corresponding to the riddle of the screen, including a sieve, which is supported on a viscoelastic suspension. The considered model takes into account the influence of the gravitational field, and it exhibits four DOFs (degrees of freedom) overall. Three DOFs describe the motion of the screen riddle, and they are denoted as x_s, z_s and φ_s. The first two mentioned DOFs correspond to horizontal x_s and vertical z_s displacements of the riddle's center of mass, whose absolute location is defined as (x_s, z_s). Variable φ_s corresponds to the third DOF, and it describes the pitch motion of the riddle. The riddle, including the sieve, which is the key vibrating part of the screen, is characterized by its mass and moment of inertia denoted as m_s and I_s, respectively. For the purpose of further analysis, $\dot{x}$ and $\ddot{x}$ are defined as first and second derivatives of a selected variable x, respectively. Additionally, horizontal and vertical directions are denoted as x and z, respectively.

The assumed model of screen suspension mainly consists of horizontally and vertically oriented linear springs. Here, stiffness parameters are assumed as the same for both front and rear suspension parts depending if the horizontal or vertical direction is considered, and they are denoted as k_{sx} or k_{sz}, respectively. These springs are attached to the selected points of the front and rear parts of the riddle. Thus, coordinates describing locations of these points are denoted as (x_{sf}, z_{sf}) and (x_{sr}, z_{sr}), respectively. These locations are defined based on x_s, z_s and φ_s and with respect to the frame of reference associated with the stationary screen base. They additionally depend on parameters (l_{sf}, h_{sf}) and (l_{sr}, h_{sr}) describing locations of the suspension attachment points with respect to the riddle's center of mass and its frame of reference. Additionally, damping related mainly to the suspension is described by parameters of horizontal and vertical viscous damping acting on the riddle center of mass, denoted as c_{x_s} and c_{z_s}, respectively. The rotation of the riddle is influenced by parameters of rotary viscous damping c_{φ_s} and dry friction f_{φ_s}. As a result, it was assumed that the above-mentioned parameters are sufficient to map the dominant behavior of the actual suspension.

The mechanical exciter with unbalanced masses is an additional component independent of the riddle driven by a electric-motor-related torque denoted as M_m. The electric motor is fixed to the screen base, and the unbalanced masses are mounted on both sides of a shaft which is attached to the screen riddle in lateral direction with the possibility of rotation. The rotating shaft is connected to the electric motor using a rubber belt. The proportion between the diameters of motor shaft and unbalanced-masses-related shaft is denoted as p_e (details about physical parameters of the considered semi-industrial vibrating screen are presented in Section 3). Thus, a torque applied to the unbalanced masses can be defined as $M_e = p_e \cdot M_m$.

Consequently, the fourth DOF, denoted as angle α_e, describes instantaneous position of these unbalanced masses, i.e., their relative inclinations with respect to the longitudinal axis of the riddle. Furthermore, the relationship between the angular velocity of the motor and the unbalanced-masses shaft $\dot{\alpha}_e$ can be defined as $\dot{\alpha}_m = p_e \cdot (\dot{\alpha}_e + \dot{\varphi}_s)$. The unbalanced masses are described with its overall mass m_e and moment of inertia I_{er} evaluated with respect to the axis of its rotation. The exciter is attached to the riddle using a rotating member at location defined by (x_{er}, z_{er}) with respect to the screen base. The imbalance of this structure is described by a distance denoted as r_e and defined as being from the axis of rotation to the center of mass of the exciter m_e. Thus, location of m_e with respect to the stationary screen base is defined as (x_{em}, z_{em}) which depends on all above-mentioned DOFs, r_e and parameters (l_{er}, h_{er}) describing the location of the exciter rotation axis with respect to the screen riddle.

Operation of the electric motor and its torque M_m, which depends on the instantaneous motor slip s, was described by taking torque characteristics generated by a typical actual asynchronous electric motor. The considered motor exhibits a synchronous angular velocity $\dot{\alpha}_{m0}$, nominal torque M_n and nominal slip s_n evaluated for the nominal angular velocity denoted as $\dot{\alpha}_{mn}$. The typical torque characteristics include additional amplification of torque for angular velocities close to zero, i.e., a starting torque denoted as M_a apart from the motor's maximum torque denoted as M_k and corresponding motor slip s_k. Such operation of an electric motor can be mathematically described as a composition of different torque characteristics, as suggested in [57]. Each of these characteristics can be evaluated based on the Kloss equation recalled in [58]. Thus, for the presented study, the composition of switchable Kloss equations evaluated for different values of parameters was implemented as follows:

$$\acute{M}_m(s) = \begin{cases} 2M_a\left(s + \frac{1}{s}\right)^{-1} & \text{for} \quad \dot{\alpha}_m < \dot{\alpha}_{m,th} \\ 2M_k\left(\frac{s}{s_k} + \frac{s_k}{s}\right)^{-1} & \text{for} \quad \dot{\alpha}_m \geq \dot{\alpha}_{m,th} \end{cases}, \tag{1}$$

where a threshold angular velocity denoted as $\dot{\alpha}_{m,th}$ indicates an angular velocity of intersection of constituent characteristics described by the Kloss equation. The torque of the final model of the electric motor $M_m = \acute{M}_m - f_{\alpha_m}$ additionally maps its rotating resistance using a parameter of rotary dry friction f_{α_m}.

2.2. Mathematical Description of the Screen Model

Generalized coordinates describing the screen model were adopted for the purpose of the Euler–Lagrange equations analogously to the definition of the DOFs, i.e., $q_k \in \{x_s, z_s, \varphi_s, \alpha_e\}$. Kinetic T and potential V energies are defined for the considered screen model as follows:

$$\begin{aligned} T =& \frac{1}{2}m_s\dot{x}_s^2 + \frac{1}{2}m_s\dot{z}_s^2 + \frac{1}{2}I_s\dot{\varphi}_s^2 + \frac{1}{2}m_e\dot{x}_{em}^2 + \frac{1}{2}m_e\dot{z}_{em}^2 + \frac{1}{2}I_{em}(\dot{\varphi}_s + \dot{\alpha}_e)^2, \\ V =& \frac{1}{2}k(x_{sf} - x_{sf0})^2 + \frac{1}{2}k(z_{sf} - z_{sf0})^2 + \frac{1}{2}k(x_{sr} - x_{sr0})^2 + \frac{1}{2}k(z_{sr} - z_{sr0})^2 \\ &+ m_sgz_s + m_egz_{em}, \end{aligned} \tag{2}$$

where g denotes gravitational acceleration and I_{em} denotes the moment of inertia defined with respect to the center of exciter mass m_e located at (x_{em}, z_{em}). It can be evaluated using inverted Steiner's theorem about parallel axis as $I_{em} = I_{er} - m_e r_e^2$. Here, initial locations of the front and rear parts of the screen suspension and related to its stiffness are denoted as (x_{sf0}, z_{sf0}) and (x_{sr0}, z_{sr0}), respectively.

A set of non-conservative forces, defined as $F_i \in \{F_{cx_s}, F_{cz_s}, M_{c\varphi_s}, M_e\}$, acts on the corresponding points of the screen along the virtual displacements $r_i \in \{x_s, z_s, \varphi_s, \varphi_e\}$. Symbols F_{cx_s}, F_{cz_s} and $M_{c\varphi_s}$ correspond to viscous damping forces and a torque related mainly to the screen suspension, acting on the riddle's center of mass in horizontal, vertical and angular directions, respectively. The generalized forces $Q_k \in \{Q_{cx}, Q_{cz}, Q_{c\varphi}, Q_{c\alpha}\}$ are evaluated depending on virtual displacements r_i, non-conservative forces F_i and generalized coordinates q_i as follows:

$$\begin{aligned} Q_{cx} &= F_{cx_s}, \\ Q_{cz} &= F_{cz_s}, \\ Q_{c\varphi} &= M_{c\varphi_s} + M_e, \\ Q_{c\alpha} &= M_e. \end{aligned} \tag{3}$$

The mathematical description of the considered screen model was evaluated based on Euler–Lagrange equations and resulted in the following four ordinary differential equations:

$$\begin{aligned} m_s\ddot{x}_s =& - m_e\ddot{x}_{em} + F_{sx}, \\ m_s\ddot{z}_s =& - m_e\ddot{z}_{em} + F_{sz} + F_{gsz}, \\ I_s\ddot{\varphi}_s =& - m_e\ddot{x}_{em}(h_{er}\cos\varphi_s - l_{er}\sin\varphi_s) \\ & + m_e\ddot{z}_{em}(h_{er}\sin\varphi_s + l_{er}\cos\varphi_s) + M_s + M_{gs}, \\ I_{er}\ddot{\alpha}_e =& - m_e\ddot{x}_{er}r_e\sin(\alpha_e + \varphi_s) - m_e\ddot{z}_{er}r_e\cos(\alpha_e + \varphi_s) - I_{er}\ddot{\varphi}_s + M_e + M_{ge}. \end{aligned} \tag{4}$$

Interactions of the screen with its suspension (horizontal F_{sx} and vertical F_{sz} forces, torque M_s) and the influence of the gravitational field (vertical force F_{gsz} and torques related to the riddle M_{gs} and the mechanical exciter M_{ge}) were intentionally generalized in Equations (4). That allowed us to emphasize the internal dynamics of the dynamic screen model and to simplify further model implementations. These variables are derived as follows:

$$\begin{aligned} F_{sx} =& F_{kfx} + F_{krx} + F_{cx_s}, \\ F_{sz} =& F_{kfz} + F_{krz} + F_{cz_s}, \\ F_{gsz} =& - m_s g - m_e g, \\ M_s =& F_{kfx}\cdot(z_{sf} - z_s) + F_{krx}\cdot(z_{sr} - z_s) \\ & - F_{kfz}\cdot(x_{sf} - x_s) - F_{krz}\cdot(x_{sr} - x_s) + M_{c\varphi_s}, \\ M_{gs} =& - m_e g(x_{er} - x_s), \\ M_{ge} =& - m_e g\cdot r_e\cos(\alpha_e + \varphi_s). \end{aligned} \tag{5}$$

Forces generated by the horizontal and vertical stiffness components of the front and rear screen suspension are denoted as F_{kfx}, F_{kfz} and F_{krx}, F_{krz}, respectively. Horizontal F_{cx_s} and vertical F_{cz_s} viscous damping forces depend on parameters c_{x_s} and c_{z_s}, respectively. The damping torque $M_{c\varphi_s}$ depends on parameters of viscous damping c_{φ_s} and dry friction f_{φ_s}.

2.3. Implementation of the Vibrating Screen Simulator

The vibrating screen simulator was implemented based on the ordinary differential equations presented in (4) which were reformulated into the following nonlinear matrix state-space form:

$$\dot{X} = F_A(X, U). \tag{6}$$

The simulated system was described using 8 state variables assumed as a list of generalized coordinates and their derivatives, which are defined in the form of a vector $X = [x_s, z_s, \varphi_s, \alpha_e, \dot{x}_s, \dot{z}_s, \dot{\varphi}_s, \dot{\alpha}_e]^T$. The vector $F_A = [f_{x_s}, f_{z_s}, f_{\varphi_s}, f_{\alpha_e}, \dot{x}_s, \dot{z}_s, \dot{\varphi}_s, \dot{\alpha}_e]^T$ consists of nonlinear functions of X and U, and for some elements, it depends directly on other state variables. The external variables listed in the vector U correspond to the following forces and torques: M_e and those related to interaction with the suspension or gravitational field, F_{sx}, F_{sz}, F_{gsz}, M_s, M_{gs} and M_{ge}, which are defined in Equation (5).

The screen model was implemented using a Matlab environment. The state-space matrix equation, Equation (6), was solved during simulation for a desired simulation time using the ode45 function implemented in Matlab environment, i.e., a fifth-order Runge–Kutta method with a variable integration step. However, it requires reformulation of differential Equations (4), which are nonlinear and complex, to the form of nonlinear functions included in the state-space equation, Equation (6). For this purpose, equations related to $\ddot{x}_s$, $\ddot{z}_s$ and $\ddot{\varphi}_s$ were initially grouped and reformulated into the following form:

$$D(X) \cdot \dot{X}_1 = F_P(X, U), \tag{7}$$

where $X_1 = [\dot{x}_s, \dot{z}_s, \dot{\varphi}_s]^T$ denotes a vector of selected state variables. Symmetrical matrix $D(X)$ depends on nonlinear terms of state variables X; a vector F_P consists of nonlinear functions of X and U.

Equation (7) can be solved analytically mainly by inversion of matrix $D(X)$, which allows for direct calculation of second derivatives of generalized coordinates included in $\dot{X}_1$. However, such an analytical derivation could introduce redundant terms which are difficult to notice and reduce, and it can consequently make further implementation more challenging. Thus, a compromise method was decided on to calculate the inversion of $D(X)$ in every simulation step after the substituting of known instantaneous values of state variables and checking the rank of matrix $D(X)$. Finally, taking advantage of calculated variables $\dot{X}_1$ for a selected simulation step, the $\ddot{\alpha}_e$ can be obtained by reformulation of the corresponding differential Equation (4).

2.4. Procedure of Modeling and Identification of the Vibrating Screen Model

The process of the screen modeling and identification was divided into two main phases, as presented in Figure 3, related to defining the model's structure and final tuning of the model's parameters. First, an initial modeling solution of the electric motor and the vibrating screen was assumed. Further, governing equations were created describing the model's structure. Physical parameters of the model were estimated within the first phase based on observations and physical dimensions of the actual screen design. The updated model was simulated and its response analyzed in order to assess whether the model's structure was satisfying.

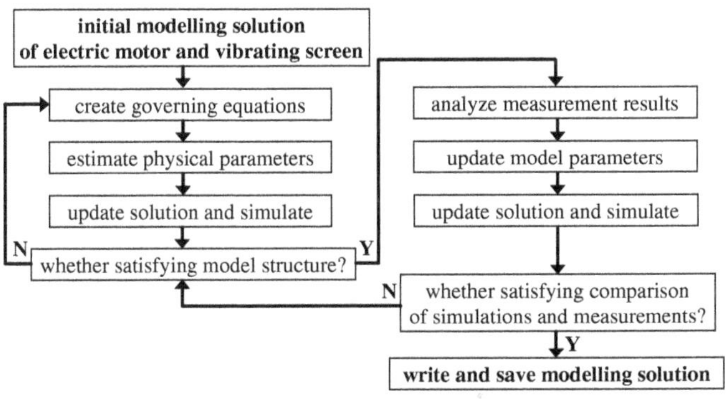

Figure 3. Flowchart for the procedure of modeling and identification of the vibrating screen model.

The second phase was based on observation of measurements, which allowed for further tuning of the model's parameters. Consequently, the screen model was updated and simulated. Validation of the model was based on several quantities describing screen motion, which are presented in the following forms: time diagrams of three degrees of freedom: x_s, z_s and φ_s; and trajectories of vibration displacement and acceleration evaluated for a selected riddle part. Comparison of results obtained for simulations and measurements was used for assessment of whether the model conformed to the measurements to a sufficient degree or whether the model needed further parameter tuning either modification of its structure. Finally, the resultant screen was saved and applied for further analysis of trajectory control algorithm.

3. Identification of Semi-Industrial Vibrating Screen and Magnetorheological Damper

This study is dedicated to the industrial single deck screen presented in Figure 1. The experimental screen was equipped with a single mechanical exciter with unbalanced masses generating a circular trajectory of the riddle vibration. The presented experimental setup is dedicated to the development and validation of a vibration control algorithm desired for improvement of the sieving process. The presented screen was intended to have future modifications to its suspension in order to include semi-active dampers. Such dampers allow for adaptation of the suspension's characteristics during screen operation to production and sieving needs. Future research will take into account applications of MR dampers whose damping parameters can be changed in milliseconds.

The first phase of the study presented in the current manuscript was deriving and validating a screen simulator, including a model of MR damper behavior. The evaluated screen simulator described in the current section needs to map dominant characteristics of the actual screen. It allows one to make the developed control algorithm be representative of an actual screen. Values of screen parameters estimated in the further part of the current section are presented in Table 1, and notation of the model's parameters conforms to what was defined in Section 2. The second phase of the research reported in next sections was dedicated to the development and validation of the proposed semi-active control algorithm of trajectory control using the implemented screen simulator.

Table 1. Estimated parameters of the simulator of the vibrating screen.

	Model of Vibrating Screen		
$l_{er} = 0.011$ m	$h_{er} = 0.048$ m	$m_s = 102.6$ kg	$I_s = 14.78$ kgm^2
$l_{sf} = 0.34$ m	$h_{sf} = 0.14$ m	$l_{sr} = 0.35$ m	$h_{sr} = 0.08$ m
$k_{sx} = 44{,}966$ Nm^{-1}	$k_{sz} = 57{,}094$ Nm^{-1}		
$c_{x_s} = 800$ Nsm^{-1}	$c_{z_s} = 800$ Nsm^{-1}	$c_{\varphi_s} = 31$ Nms rad^{-1}	$f_{\varphi_s} = 1.4$ Nm
	Model of Electric Motor and Mechanical Exciter		
$m_e = 32.4$ kg	$I_{er} = 0.14$ kgm^2	$r_e = 0.007$ m	$P_m = 0.75$ kW
$p_e = 1.25$	$\dot{\alpha}_{m0}/2\pi = 25$ Hz	$\dot{\alpha}_{mn}/2\pi = 23.3$ Hz	$s_n = 0.067$
$M_n = 5.12$ Nm	$M_k = 13.31$ Nm	$M_a = 12.29$ Nm	$s_k = 0.333$
$\dot{\alpha}_{m,th} = 2\pi\, 10.57$ rads^{-1}	$f_{\alpha_m} = 2.24$ Nm		
	Properties of Measurement Signals		
$f_s = 1000$ Hz	$l_a = 0.051$ m	$h_a = 0.117$ m	
	Bouc–Wen Model of the MR Damper		
$l_{mrf} = 0.229$ m	$h_{mrf} = 0.040$ m	$l_{mrr} = -0.111$ m	$h_{mrr} = 0.169$ m
$i_{mr,lb} = 0$ A	$\alpha_{bw,lb} = 211{,}310$	$\beta_{bw,lb} = -167{,}290$	$A_{bw,lb} = 44{,}024$
$\gamma_{bw,lb} = 32.6$	$c_{bw,lb} = 878.0$	$n_{bw} = 2$	
$i_{mr,ub} = 1.0$ A	$\alpha_{bw,ub} = 5840$	$\beta_{bw,ub} = -4620$	$A_{bw,ub} = 1216$
$\gamma_{bw,ub} = 1351.0$	$c_{bw,ub} = 1341.5$		

3.1. Physical Dimensions and Estimation of Selected Parameters

The riddle of the experimental screen was 46 cm wide and 1.26 m long, and it was inclined to the horizontal side by about 18°. Side walls of the riddle were 38 cm high at the front and 40 cm high at the rear. The riddle was mostly made of 6-mm-thick sheet metal. Additionally, three metal pipes were located transversely to the riddle in the vicinity of the mechanical exciter, which were intended to connect the side walls of the riddle with each other. The middle pipe was also used to guide the rotating shaft, which connected the unbalanced masses of the exciter located on both sides of the screen. The rotary axis of the mechanical exciter was located close to the estimated center of mass of the riddle at a distance l_{er} = 0.011 m horizontally and h_{er} = 0.048 m vertically, with respect to the side plane of the screen.

The physical dimensions of the riddle and screen structure were used for estimations of selected parameters required for the screen simulator. The mass of the riddle, which was the key vibrating object of the analyzed system, was obtained as m_s = 102.6 kg assuming the density of steel ρ = 7900 kgm^{-3}, and the riddle's moment of inertia defined for a transverse axis of rotation through the center of mass was estimated as I_s = 14.78 kgm^2. Parameters of the mechanical exciter, which was partly attached to the riddle, were not included in m_s and I_s.

Unbalanced masses of the mechanical exciter were half-circle shaped of radius equal to 10 cm and thickness equal to 22 mm. Apart from unbalanced masses, physical parameters of the pulley and exciter shaft needed to be taken into account. As a result, the overall mass of the mechanical rotating part of the exciter was estimated as m_e = 32.4 kg, and the corresponding moment of inertia defined with respect to the axis of rotation was equal to I_{er} = 0.14 kgm^2. For the given value of m_e, and based on geometrical properties of unbalanced masses, the distance between the center of mass and axis of rotation was calculated as r_e = 7 mm.

The mechanical vibration exciter was driven by a 3-phase electric motor with P_m = 0.75 kW designed for continuous operation. The proportion between diameters of the motor shaft and exciter shaft was p_e = 1.25. The applied motor exhibited 4 poles, and consequently synchronous angular velocity was equal to $\dot{\alpha}_{m0}/2\,\pi$ = 1500 rpm = 25 Hz for frequency of the mains electricity of 50 Hz. Nominal angular velocity of the motor was equal to $\dot{\alpha}_{mn}/2\,\pi$ = 1400 rpm = 23.3 Hz, which gives a nominal motor slip of s_n = 0.067. Consequently, the nominal torque of the motor can be calculated as $M_n = P_m \dot{\alpha}_{mn}$ = 5.12 Nm.

According to the characteristics of electric motors of a similar class, the maximum motor torque was assumed as M_k = 2.6 M_n = 13.31 Nm, and the starting torque was assumed as M_a = 2.4 M_n = 12.29 Nm. Thus, the motor slip related to the maximum torque M_k was evaluated as s_k = 0.333, and consequently, the threshold angular velocity defined in Section 2 was calculated as $\dot{\alpha}_{m,th}$ = 2 π 10.57 rads^{-1}. The characteristics of the applied model of the electric motor can be evaluated using Figure 4.

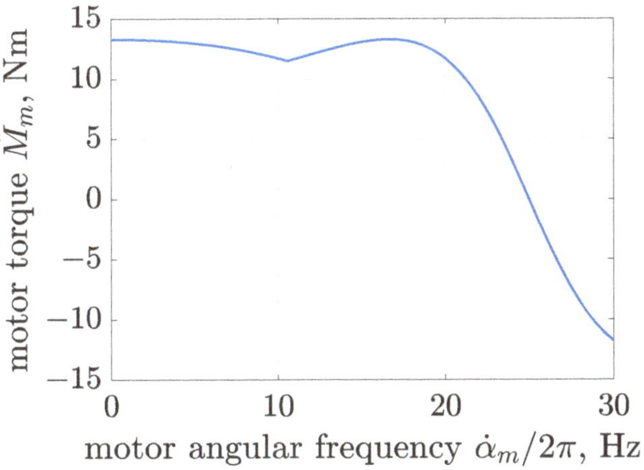

Figure 4. Characteristics of the electric motor model based on switchable Kloss equations.

The modeled electric motor operates in motoring mode for angular frequencies from 0 to 25 Hz ($\dot{\alpha}_{m0}$), and it exhibits positive torque within this frequency range. For angular frequencies greater than 25 Hz, the motor starts to operate as a power generator, in which case the motor torque takes a negative value. The characteristics show two maxima for motor torques corresponding to M_a and M_k. The presented approach of motor modeling allows one to map influence of rotating unbalanced masses on angular velocity of the motor. Instantaneous rotation of the mechanical exciter or varying weight and parameters of the sieved material has an influence on the angular velocity and phase of the electric motor, as in the case of the actual screen.

3.2. Experimental Setup and Identification Experiments

The vibrating screen configured for experiments is presented in Figure 5a, where main components are marked, i.e., the screen suspension and mechanical exciter driven by the electric motor. Vibration of the experimental screen, as presented in Figure 5b, was measured using an MEMS (micro-electro-mechanical system) with 3-axis accelerometers of type ADXL325 and ADXL326 produced by Analog Devices, which exhibit measurement ranges of ±5 g or ±16 g, respectively. Acceleration was assessed by measurement of deflection of an internal moving mass suspended by polysilicon springs within the sensor. The measurement of deflection was carried out using differential capacitors fixed to the moving mass. The resultant measurement signal, which was proportional to acceleration, was conditioned by internal circuits and fed outside the sensor in the form of analog output voltage. Voltage of the output signal for both types of accelerometers could vary within the nominal range from about 0.5 to 2.5 volts; the nominal offset value is equal to 1.5 volts.

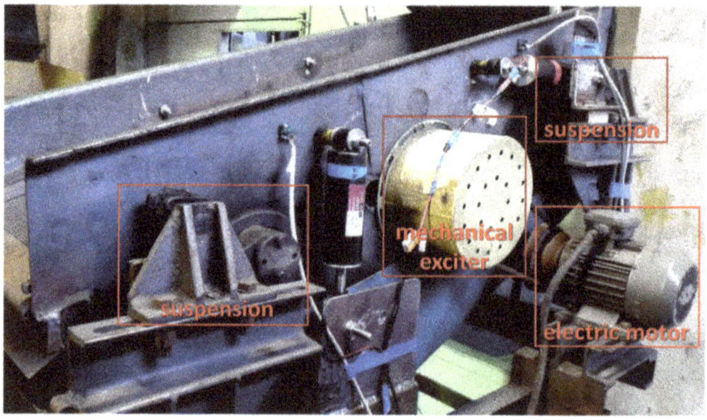

(**a**)

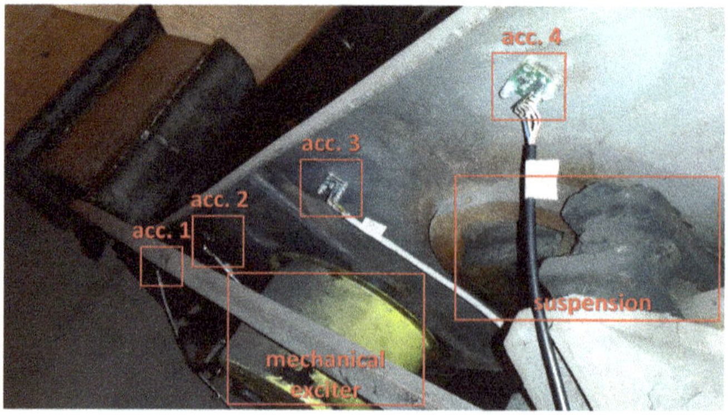

(**b**)

Figure 5. Measurement system applied for identification of screen model: (**a**) mechanical structure of the screen, (**b**) accelerometers attached to the sidewall of the screen riddle at the measurement points.

In the case of accelerometer axes denoted as X and Y, which are parallel to the case of the integrated circuit, the nominal frequency bandwidth of measurements was equal to 1600 Hz, and in the case of the sensor's vertical axis denoted as Z, the frequency bandwidth was equal to 550 Hz. The former X and Y axes of each applied accelerometer were used during the considered measurements of screen vibration. Standard evaluation boards dedicated to applied accelerometers were used during experiments, which allowed for robust assembly of sensors in the screen's structure. Each evaluation board additionally included several capacitors, which in combination with resistors built into the sensor integrated circuit, formed a first-order analog antialiasing filter of cutoff frequency equal to 50 Hz.

Four accelerometers were used for experiments, and they were located in the left sidewalls of the screen riddle, as presented in Figure 5b. The locations of the consecutive sensors were defined by their corresponding horizontal and vertical distances in meters from the center or riddle mass as follows: (−0.501, 0.304), (−0.191, 0.185), (0.339, 0.015) and (0.559, −0.035). These locations allowed for comprehensive analysis of the vibration trajectories of different parts of the riddle and the sieve in a plane defined by the vertical and longitudinal axes of the screen. Measurement signals generated by the sensors and filtered using the antialiasing filter were acquired by a dedicated measurement system with a sampling frequency equal to $f_s = 1/T_s$ = 1000 Hz and preprocessed for further analysis, including conversion of physical units and compensation of signal offset. Velocities and displacements of selected parts of the vibrating screen were estimated based on acceleration measurements by single or double integration with inertia, respectively. An example of a displacement estimation is presented by the following formula:

$$z(i) = H_b(z^{-1}) \cdot \left(\frac{T_s}{1 - z^{-1}} \right)^2 \cdot a(i), \tag{8}$$

where i denotes number of a sample of a selected signal. Displacement, velocity and acceleration digital signals are denoted in the presented study as z, v and a, respectively. Acceleration measurements taken from the experimental setup with vibrating screen were twice numerically integrated in order to estimate displacement signals. These signals were further processed by additional discrete-time Butterworth filter denoted as $H_b(z^{-1})$ defined based on an unit delay operator denoted as z^{-1}. Depending on analysis of processed measurement signals:

- the lowpass filter was used for fine processing of acceleration with cut-off frequency of 50 Hz;
- the highpass filter with cut-off frequency equal to 4 Hz was used for filtering of estimated displacement signals dedicated to time-domain analysis;
- the highpass filter of cut-off frequency 18 Hz was used for signals dedicated to the analysis of vibration trajectory.

3.3. Evaluation of Parameters of the Dynamic Screen Model

Vibration of the screen was tracked by the selected j-th sensor measuring acceleration in horizontal and vertical directions, which can be described as follows:

$$\begin{aligned} \ddot{x}_j &= \ddot{x}_s - h_j(\dot{\varphi}_s^2 \sin\varphi_s - \ddot{\varphi}_s \cos\varphi_s) - l_j(\dot{\varphi}_s^2 \cos\varphi_s + \ddot{\varphi}_s \sin\varphi_s), \\ \ddot{z}_j &= \ddot{z}_s - h_j(\dot{\varphi}_s^2 \cos\varphi_s + \ddot{\varphi}_s \sin\varphi_s) + l_j(\dot{\varphi}_s^2 \sin\varphi_s - \ddot{\varphi}_s \cos\varphi_s). \end{aligned} \tag{9}$$

Evaluation of parameters of the dynamic screen model was performed based on acceleration evaluated independently in horizontal and vertical direction as averages over available accelerometers and denoted as $\ddot{x}_a$ and $\ddot{z}_a$, respectively. The model validation was also carried out based on estimation of the riddle inclination angle φ_s. According to Equation (9), which is linear with respect to l_i and h_i, the above-mentioned averaged acceleration is equivalent to acceleration measured in location as an average of all locations of sensors as follows: $l_a = 1/4 \sum_{j=1}^{4} l_j$ = 0.051 m, $h_a = 1/4 \sum_{j=1}^{4} h_j$ = 0.117 m. The φ_s

angle was evaluated based on displacement signals estimated for accelerometers located in the extreme positions in the front and rear parts of the screen riddle according to the following:

$$\varphi_s = \arcsin \frac{(h_1 - h_4)(x_1 - x_4)}{(l_1 - l_4)^2 + (h_1 - h_4)^2}. \tag{10}$$

Each series of the experiment consisted of three phases of operation of the vibrating screen: start-up, operation with nominal vibration frequency and stopping the mechanical exciter. Time–frequency characteristics evaluated for the averaged horizontal acceleration of the screen riddle are presented in Figure 6. The angular frequency of the mechanical exciter about 1 s after start-up of the electric motor reached the steady-state frequency of about 19.2 Hz. The mechanical exciters rotated with the nominal frequency during 20 s of nominal operation. Finally, the electric motor was turned off, and consequently, the rotating shaft with unbalanced masses began to slow down, making the screen vibrate at successive, decreasing frequencies, which lasted about 5 s.

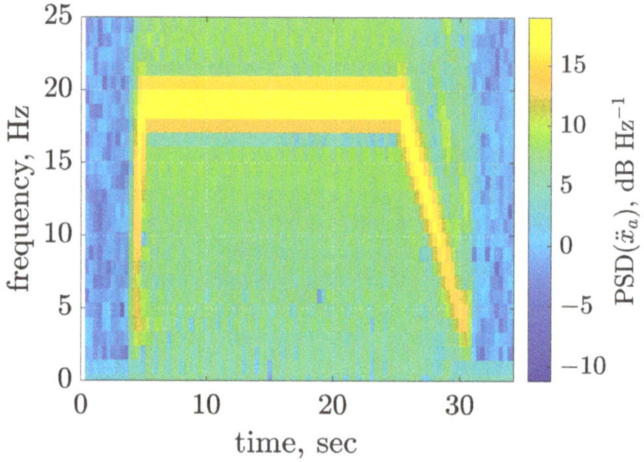

Figure 6. Time–frequency characteristics of the averaged horizontal acceleration of the screen riddle based on measurements.

The considered screen is over-resonant, which means that during start-up and stopping the angular frequency of the mechanical exciter for a moment coincides with the resonance frequency of the screen. Estimated horizontal and vertical displacements of the screen riddle are presented in Figures 7 and 8, where occurrence of resonance is clearly visible for increasing amplitude of vibration at close to 4.5 Hz, slightly dependent on the direction of vibration. Furthermore, it can be noticed that the amplitude of vibration displacement is approximately equal to 1.8 mm during the nominal phase of screen operation. The third degree of freedom φ_s estimated according to Equation (10) is presented in Figure 9. The time diagram of φ_s shows that for the start-up phase, the amplitude reached 0.2°, and for the stopping phase the maximum value of amplitude was close to 0.4° at the moment of resonance. During the nominal operation of the screen, φ_s oscillated within an amplitude close to 0.07°. Despite the fact that maximum amplitudes of φ_s seem to be low, it should be noticed that φ_s = 0.4° resulted in an approximately 2 mm change in vertical displacement of the front suspension z_{sf}, which was about 25% of the overall displacement z_{sf}. Thus, the influence of φ_s on the response of the dynamical screen model was significant in this study.

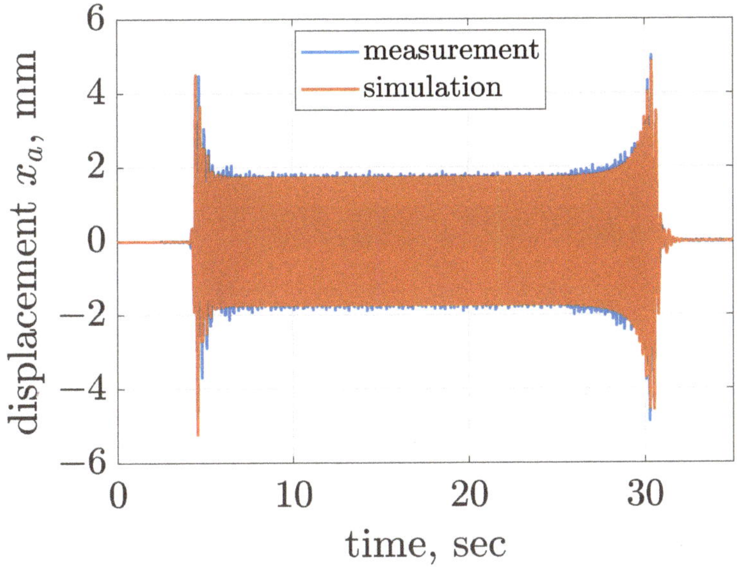

Figure 7. Comparison of time diagrams of horizontal displacement x_a corresponding to the averaged sensor location evaluated for measurement and simulation results.

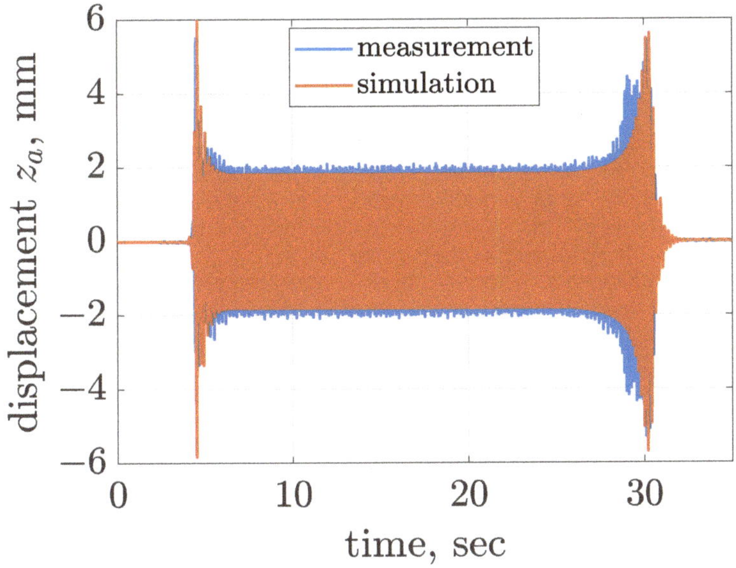

Figure 8. Comparison of time diagrams of vertical displacement z_a corresponding to the averaged sensor location evaluated for measurement and simulation results.

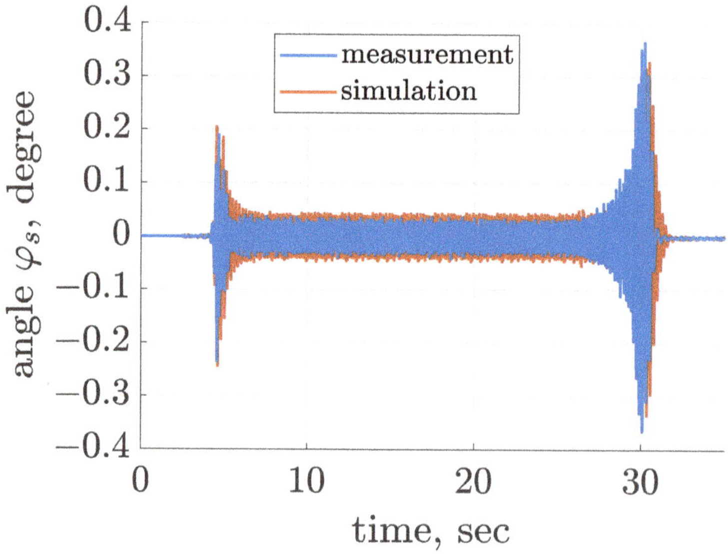

Figure 9. Comparison of time diagrams of riddle angle φ_s evaluated for measurement and simulation results based on acceleration signals for locations at extreme front and rear measurement points of the vibrating riddle.

The remaining parameters of the screen model, i.e., stiffness and damping parameters of the screen suspension, and the rotational friction of the electric motor, were evaluated by comparison of simulation results and measurements, which gave satisfactory results, as presented in Figures 7–9. The suspension of the screen consisted of four parts, each attached to the riddle on one side and to the screen base on the other side. Points of suspension-to-riddle attachment were at distance l_{sf} = 0.34 m horizontally and h_{sf} = 0.14 m vertically at the riddle front and at distance l_{sr} = 0.35 m horizontally and h_{sr} = 0.08 m vertically at the riddle rear. A single part of the screen suspension was designed in the form of a mechanism which consisted of two members and three rotary joints, and the last joint built into the riddle was coupled with a rotary spring.

The equivalent horizontal and vertical stiffness levels of front and rear suspensions were initially assumed based on the mass of the vibrating riddle and measured resonance frequencies, and further tuned which gave the following assumptions: k_{sx} = 44,966 Nm^{-1} and k_{sz} = 57,094 Nm^{-1}. Subsequently, damping parameters related to the screen suspension were tuned as follows: c_{x_s} = 800 Nsm^{-1}, c_{z_s} = 800 Nsm^{-1}, c_{φ_s} = 31 Nms rad^{-1} and f_{φ_s} = 1.4 Nm. Rotational friction of the electric motor was selected as f_{α_m} = 2.24 Nm. The implemented model of the vibration screen was additionally validated by comparing trajectories of vibration acceleration and displacement obtained for experiments and simulations in Figures 10 and 11.

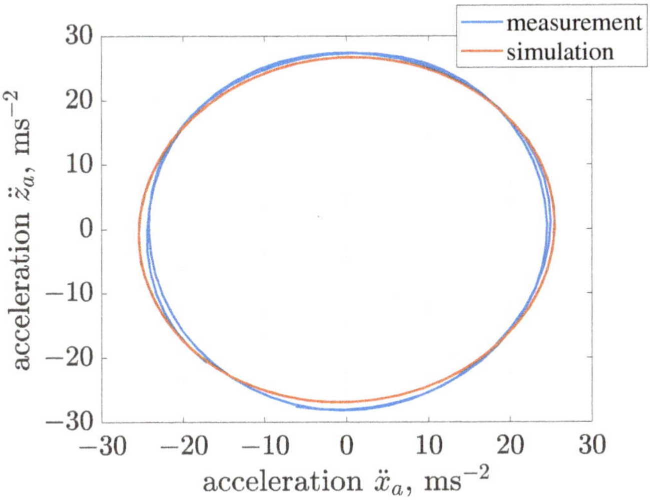

Figure 10. Comparison of trajectories of screen vibration acceleration evaluated for measurement and simulation results.

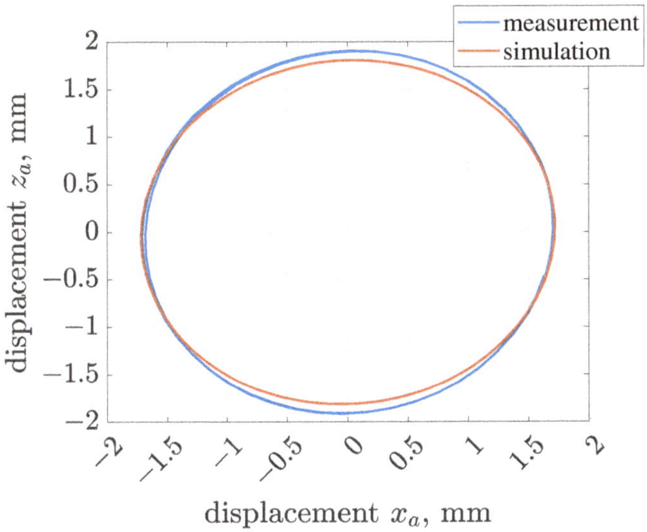

Figure 11. Comparison of trajectories of screen vibration displacement evaluated for measurement and simulation results.

Amplitudes of trajectories of vibration acceleration for both simulations and experiments were close to 25 ms^{-2}. Furthermore, we confirmed that the single mechanical exciter generated riddle vibration trajectory of a circular shape. Commonly, it can be stated that the vibration trajectories of the screen riddle present a phase shift between its horizontal x_a and vertical z_a displacement. Thus, x_a and z_a, which are in-phase results in linear trajectory, and x_a and z_a, shifted in phase by $\pi/2$, correspond to the circular and elliptical trajectories. The presented study relates the synthesis and validation of the trajectory control. Thus, the compatibility of the model with the real object is crucial for the presented application when analyzing the vibration trajectory obtained for nominal screen operation. The significant compatibility of simulation and experimental results justifies the estimates of the parameters of the implemented screen model, which are reported in Table 1, including the parameters of the Bouc–Wen model of the MR damper.

3.4. Non-Dimensional Frequency-Domain Analysis

Additional analysis of a non-dimensional normalized amplitude of vibration obtained for the identified screen model was carried out in the frequency domain for three degrees of freedom, i.e., x_s, z_s and φ_s, as presented in Figure 12. Consecutive points j marked in the characteristics were evaluated in steady-state for the corresponding constant angular frequency of the mechanical exciter $\dot{\alpha}_{e,j}$. Each value of amplitude A was calculated as a root mean squared value of the selected signal within five cycles of its harmonic response. Further, all amplitude values obtained for the selected degree of freedom were normalized by a nominal vibration amplitude denoted as A_n. The value of A_n corresponds to the nominal frequency of screen operation of 19.2 Hz.

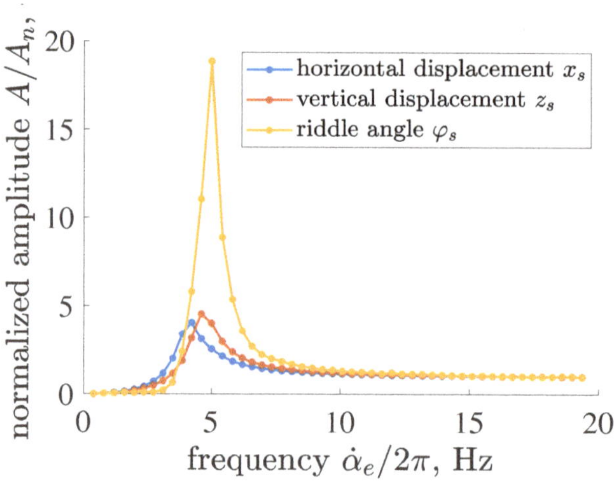

Figure 12. Comparison of non-dimensional normalized amplitudes of horizontal, vertical and angular motion of the vibrating riddle evaluated for the identified screen model in the frequency domain.

The time diagrams can be divided into sub-resonant, resonance and over-resonant ranges. The sub-resonant frequency range is defined up to 4.2–5 Hz, where the normalized amplitude is relatively low. Further, it can be noticed that resonance frequencies are slightly different for corresponding degrees of freedom, i.e., 4.2 Hz for x_s, 4.6 Hz for z_s and 5 Hz for φ_s. The amplitudes of resonance are similar for x_s and z_s—close to 4.0 and 4.5, respectively. The resonance peak of φ_s is significantly greater, which is clearly visible regarding the measurements presented in Figure 9, and approximately equal to 18.9. The vibrating screen is subjected to vibration generated by the mechanical exciter, whose amplitude is proportional to the square of its angular frequency. Thus, despite the increasing damping exhibited by the implemented mechanical model, the amplitudes of all degrees of freedom stabilize at a constant value.

3.5. Bouc–Wen Model of the Magnetorheological Damper

Magnetorheological dampers manufactured by Lord Corporation of type RD-8041-1 presented in Figure 13 were selected for further simulation tests. This type of MR damper exhibits piston stroke equal to 74 mm, and it is appropriate for vibration control in various applications, from industrial machines to road and off-road vehicles.

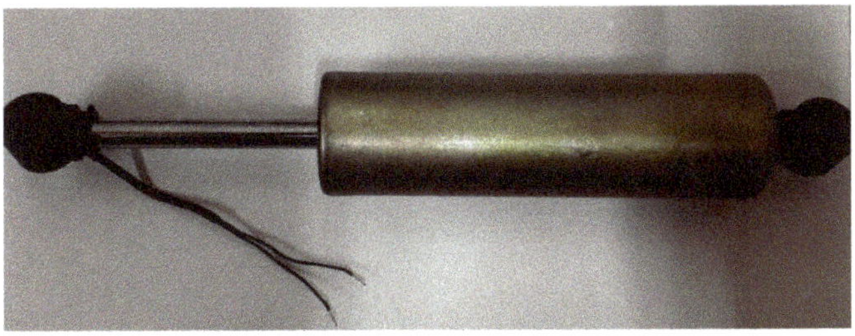

Figure 13. Mechanical structure of a magnetorheological damper, type RD-8041-1, manufactured by Lord Corporation.

According to the corresponding documentation, these MR dampers are recommended to be continuously controlled by electric currents of up to 1 ampere. Thus, further analysis of the MR damper's behavior was focused on the above-mentioned level of control current. The well-known Bouc–Wen model of MR dampers was applied for mapping dominant nonlinear features of MR damper behavior, i.e., force saturation revealed for higher piston velocities and hysteresis loops indicated in force–velocity characteristics. The key part of the Bouc–Wen is the nonlinear differential equation, which can be defined as follows:

$$\begin{aligned} \dot{p}_{bw} &= -\alpha_{bw} \cdot |v_{mr}| \cdot p_{bw} \cdot |p_{bw}|^{n_{bw}-1} - \beta_{bw} \cdot v_{mr} \cdot |p_{bw}|^{n_{bw}} + A_{bw} \cdot v_{mr}, \\ F_{bw} &= -\gamma_{bw} p_{bw} - c_{bw} v_{mr}, \end{aligned} \tag{11}$$

where v_{mr} denotes the axial velocity of the damper's piston; p_{bw} denotes the displacement of the Bouc–Wen model, which is included in the final formula for force F_{bw} generated by the Bouc–Wen model. α_{bw}, β_{bw}, A_{bw}, n_{bw}, γ_{bw} and c_{bw} denote parameters of the model which need to be estimated based on results obtained from an identification experiment.

3.6. Identification of the MR Damper Model

Identification experiments of the MR damper were carried out using a material testing system. The examined MR damper was subjected to axial sinusoidal kinematic excitation at different amplitudes and frequencies. During the experiments, the MR damper was supplied by a control current signal consisting of unit steps of different values generated independently by a dedicated control system. Three configurations were selected for identification experiments, which were characterized by the following amplitudes and frequencies: (1.5 Hz, 15 mm), (6 Hz, 5 mm) and (20 Hz, 2 mm). The range of excitation frequencies was selected intentionally in order to cover the range of frequencies reached during all phases of operation of the screen and its mechanical exciter. Similarly, the selected displacement amplitudes are compatible with those reached by the considered industrial screen.

Estimation of Bouc–Wen parameters for a selected value of control current was carried out based on a cost function defined as the mean squared error calculated between measurements and the outputs of the model simultaneously for all three cases of different excitation frequencies. The next part of the presented analysis focuses on boundary models of the MR damper. Thus, Figures 14–16 present comparisons of measurements and model responses evaluated for control currents equal to 0 and 1 amperes. Estimated parameters of applied Bouc–Wen models are listed in Table 1.

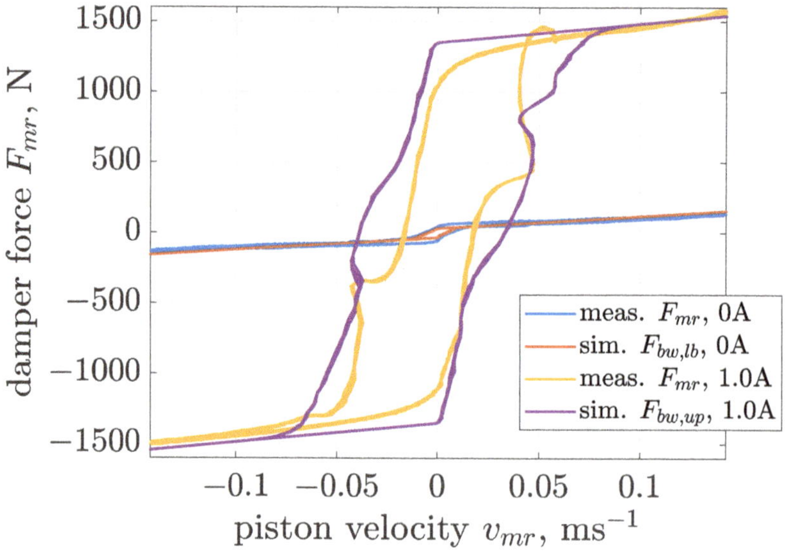

Figure 14. Comparison of force–velocity characteristics obtained for the MR damper and the Bouc–Wen model subjected to the sinusoidal excitation of frequency 1.5 Hz and displacement amplitude 15 mm.

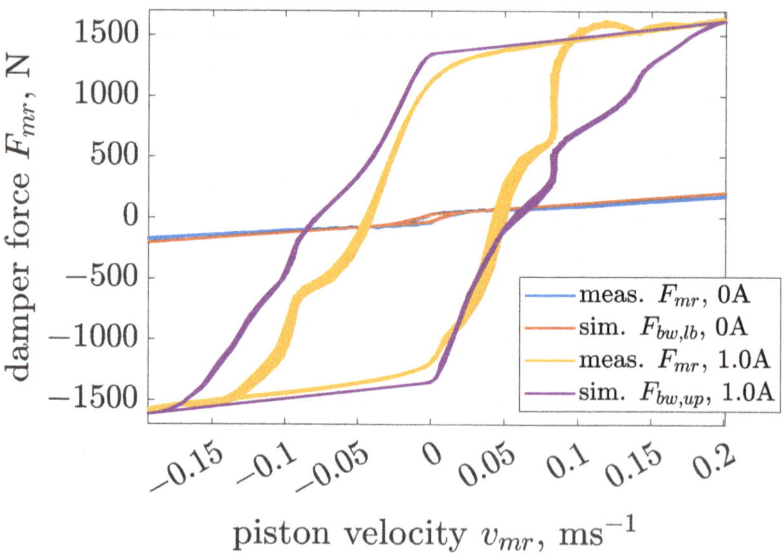

Figure 15. Comparison of force–velocity characteristics obtained for the MR damper and the Bouc–Wen model subjected to the sinusoidal excitation of frequency 6 Hz and displacement amplitude 5 mm.

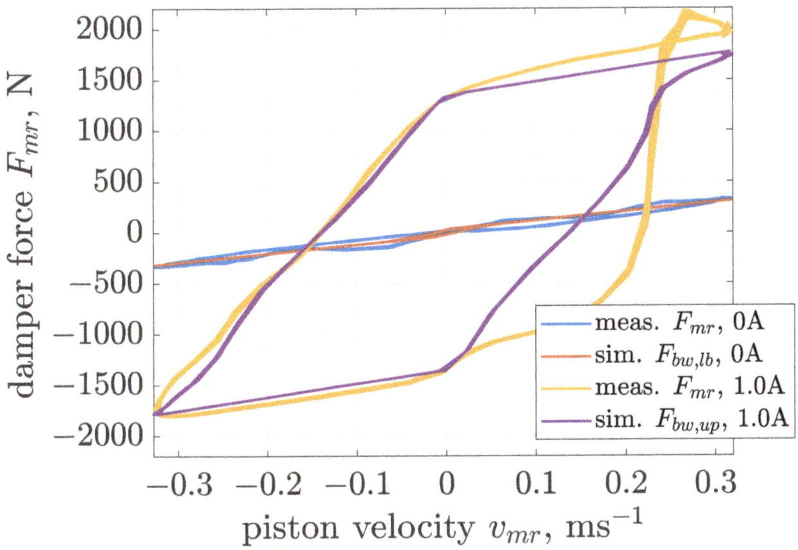

Figure 16. Comparison of force–velocity characteristics obtained for the MR damper and the Bouc–Wen model subjected to the sinusoidal excitation of frequency 20 Hz and displacement amplitude 2 mm.

A characteristic feature of MR damper behavior is significant dependence of the sizes of the hysteresis loop and force amplitudes indicated in the force–velocity curves on sinusoidal excitation frequencies. The applied Bouc–Wen model allows one to map this dependency for the whole range of analyzed excitation frequencies. This advantage is crucial for the considered screen, as comprehensive analysis of its operation requires a single and accurate model of the MR damper which is valid for all operating conditions of the screen.

The hysteresis loop of the MR damper can be described using different approaches, such as a static function of a certain quantity describing the piston motion or a dynamic model of the signal path defined from the piston motion to the output damper force. Some research proposed modeling the velocity–force hysteretic behavior using a first order linear filter is presented in [29]. Thus, similarly to the case of circular trajectory of the riddle vibration discussed previously, the hysteresis loop of the MR damper can be emulated for other studies by the phase shift between signals of the piston velocity v_{mr} and the damper force F_{mr}.

4. Trajectory Control Applied for the Screen Dynamic Model

The features of screen vibration trajectory are crucial for efficiency of sieving process. Moreover, possibility of adaptation of vibration trajectory to varying parameters of the production process and processed material is favored in modern industry. Such adaptability can be introduced into the standard design of the vibrating screen by application of MR dampers in its suspension, as presented in Figure 17. The MR dampers, as semi-active dampers of one type, are favored over active solutions for their low energy-consumption.

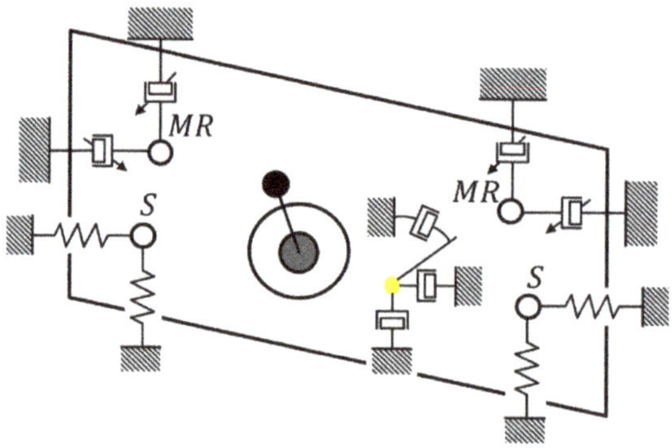

Figure 17. Mechanical representation of the dynamic model of the vibrating screen, including the Bouc–Wen model of the MR damper.

In this study, MR dampers controlled in an appropriate manner were intended to modify the vibration trajectory to a desired shape. Furthermore, in the case of screens which are equipped with only a single mechanical exciter generating a circular vibration trajectory, adaptive control of screen suspension could be the only way to generate a linear vibration trajectory.

4.1. Simulation Results of Passive Suspension

Initial tests of the screen model, including Bouc–Wen models, were carried out assuming constant control current supplying the MR dampers. A similar test procedure consisted of three phases, i.e., start-up and stopping of the mechanical exciter and nominal operation of the vibrating screen. Three configurations of screen suspension were validated: the suspension without an MR damper included, and those with Bouc–Wen models $F_{bw,lb}$ and $F_{bw,ub}$ corresponding to control current equal to 0 and 1 ampere, respectively, as presented in Section 3.6.

The envelopes of horizontal displacement x_s of the center of riddle mass were plotted in Figure 18 for the above-mentioned cases. The first envelope is analogous to time-diagrams presented in Figure 7, where occurrence of resonance is clearly visible for the start-up and stopping phases. However, it can be noticed that including the MR damper model supplied by zero control current significantly mitigated resonant peaks, and at the same time it left the vibration amplitude unchanged for the nominal phase of operation.

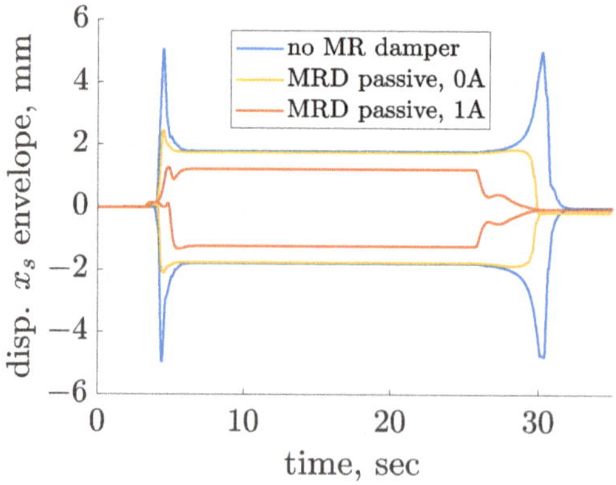

Figure 18. Comparison of time diagrams of envelopes evaluated for the horizontal displacement x_s in the case of the screen model without an MR damper model included and with an MR damper controlled by a constant current equal to 0 or 1 ampere.

The increase in MR damper control current from 0 to 1 ampere significantly influenced vibration amplitude in all phases of screen operation. In the case of the stopping phase, vibration was mitigated clearly faster as soon as the electric motor shut down. The amplitude of vibration for the nominal vibration frequency was significantly decreased from 1.8 to 1.2 and 1.6 mm for the x and z directions, respectively, as presented in Figure 19. The difference between the displacement trajectories obtained for 0 and 1 ampere and the influence of greater control currents are noticeable in Figure 19.

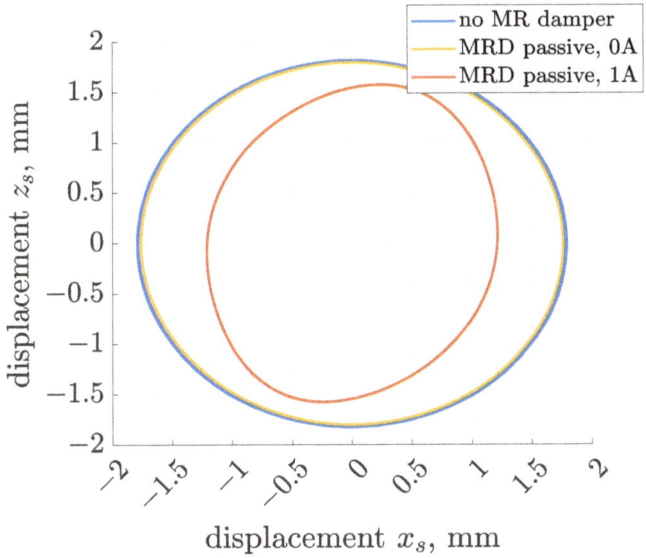

Figure 19. Comparison of trajectories of screen vibration displacement evaluated for the screen model without an MR damper model included and with an MR damper controlled by a constant current equal to 0 or 1 ampere.

This analysis related to passive suspension can be compared to those of other studies available in the literature related to the application of an MR damper for a vibrating screen. Despite the fact that different studies have used slightly different constructions of screens and MR dampers, a valuable qualitative comparison could still be carried out. A study presented by the other authors [56] showed a difference in the damping of the resonance tested for three similar configurations with or without an MR damper in the vibrating screen. The examined vibration screen exhibited an amplitude of vibration for nominal excitation frequency close to 4 mm, which is twice that applied in the current research. However, it was similarly shown in time diagrams of vibration signal that the energized MR damper allowed for significant resonance damping.

Another study presented in [54] discussed the application of a manufactured and tested MR damper whose operation in a vibrating screen was analyzed in frequency and time domains. Different vibration amplitudes were applied, from 3–5 mm, and vibration frequency up to 16 Hz was used. It was also shown that the variance of vibration displacement decreased for increasing MR damper control current. In conclusion, MR dampers included in the screen suspension allows one to modify the amplitude of vibration and consequently the size of the vibration trajectory; however, its circular shape remains the same, independently of the constant control current, when using a single mechanical exciter.

4.2. Description of the Control Algorithm

The proposed control block diagram presented in Figure 20 includes three components, i.e., a dynamic model of the vibrating screen described in Sections 2 and 3, the trajectory control algorithm presented in Section 4.2 and an implementation of semiactive control based on the MR damper dissipative domain defined in Section 4.3.

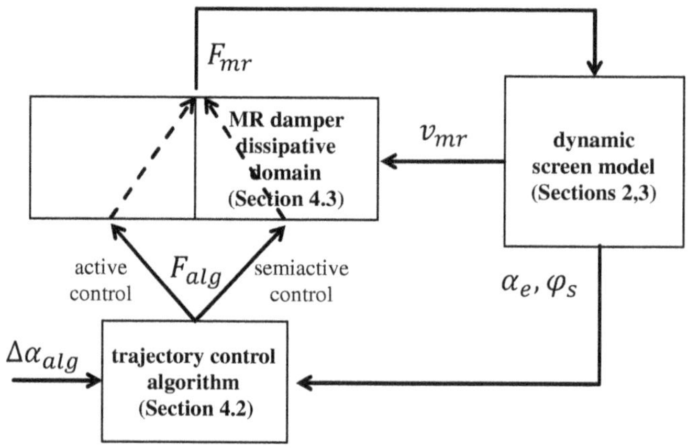

Figure 20. Implemented control block diagram, including the dynamic model of the vibrating screen, the trajectory control algorithm and the MR damper dissipative domain.

In the case of the implemented screen simulator, it was assumed that the MR damper Bouc–Wen model interacts with the screen model in predefined locations which are compatible with those intended for future experiments planned for the considered industrial screen. Thus, it was assumed that single MR dampers were fixed horizontally and vertically to the front and rear parts of the screen riddle in locations defined as follows: l_{mrf} = 0.229, h_{mrf} = 0.040, l_{mrr} = −0.111 and h_{mrr} = 0.169 m, respectively. Consequently, forces denoted as $F_{mr,i}$ generated by four MR dampers included in the implemented model contribute to the resultant balance of forces and moments of forces, which influence the dynamics of the screen.

The goal of the control algorithm is to emulate the force generated by a virtual and additional mechanical exciter which rotates in the opposite direction to the real exciter with an additional phase shift denoted as $\Delta\alpha_{alg}$. The opposite rotation of two mechanical exciters allows for the generation of a linear vibration trajectory. Assuming that a force vector of approximately $F_e = m_e(\dot{\alpha}_e + \dot{\varphi}_s)^2 r_e$ is generated by the real exciter in a direction dependent on angle $\varphi_e = \alpha_e + \varphi_s$, the fictitious exciter rotating opposite would be described horizontally and vertically as follows:

$$\begin{aligned} F_{alg,x} &= -1/N_{mr,x} \cdot F_e \cos(-\alpha_e + \varphi_s + \Delta\alpha_{alg}), \\ F_{alg,z} &= 1/N_{mr,z} \cdot F_e \sin(-\alpha_e + \varphi_s + \Delta\alpha_{alg}). \end{aligned} \tag{12}$$

Forces $F_{alg,x}$ and $F_{alg,z}$ are further distributed into separate MR damper models oriented horizontally or vertically, indicated by $N_{mr,x}$ or $N_{mr,z}$, respectively.

4.3. Dissipative Domain of the MR Damper Model

Industrial or automotive applications with MR dampers require implementation of a damper force control algorithm. Commonly, an open-loop approach to force control is applied based on an inverse MR damper model, since installation of force sensors in such applications often significantly increases costs and weakens the structure of the device.

A typical approach to the simulation research of semi-active control is to apply an inverse model which is fully compatible with the MR damper model. It comes down to a case when desired force generated by the vibration control algorithm can be directly led to the vibrating model after being limited to a region of reachable force generated by the MR damper for instantaneous conditions. The region of reachable forces is called the dissipative domain of an MR damper, and it can be defined as group of damping characteristics obtained for control currents generated within their acceptable ranges.

For the purpose of the presented study, a dissipative domain was generated as presented in Figure 21, based on the response of the identified Bouc–Wen model. The model was subjected to a sinusoidal excitation of frequency 18.85 Hz and an amplitude 1.8 mm, which are related to nominal conditions of screen operation.

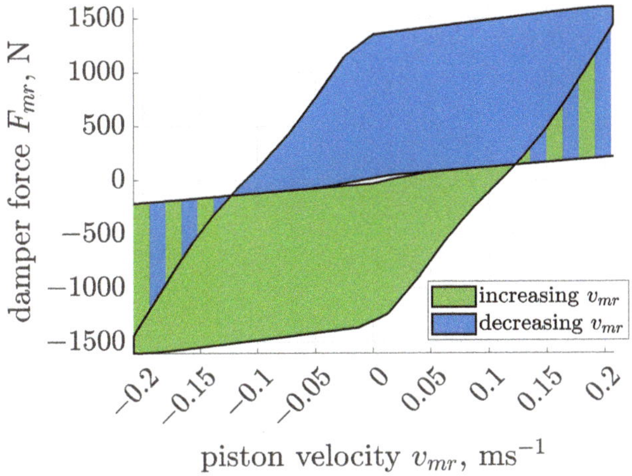

Figure 21. Dissipative domain generated based on the response of the identified Bouc–Wen model, subjected to the sinusoidal displacement excitation of frequency 18.85 Hz and an amplitude of 1.8 mm.

The dissipative domain was defined as a region of reachable forces located between damping characteristics related control currents of 0 and 1.0 ampere. It should be noticed that different force regions are reachable for increasing and decreasing piston velocities. As a result, force generated by the MR damper model is based on limited force F_{alg} generated by the control algorithms, as follows:

$$F_{mr} = \left\{ F_{alg} : \min(F_{bw,lb}, F_{bw,ub}) < F_{alg} < \max(F_{bw,lb}, F_{bw,ub}) \right\}, \tag{13}$$

where $F_{bw,lb}$ and $F_{bw,ub}$ denote force dependent on damper piston velocity v_{mr} generated by the Bouc–Wen model according to Equation (11) and for a set of its parameters related to lower lb or upper bounds up, respectively, listed in Table 1.

4.4. Simulation and Discussion of the Screen Trajectory Control

Analysis of vibration trajectory control was divided into two stages. Firstly, the implemented simulation environment was validated by application of active control without utilizing the dissipative domain, as presented in the control block diagram in Figure 20. Secondly, the dissipative domain was activated and the trajectory control algorithm was tested for the target semi-active configuration related to MR damper behavior.

Results obtained for the active control are presented in the form of trajectories of vibration acceleration and displacement in Figures 22 and 23. The presented simulation cases were obtained for the following values of algorithm phase shifts $\Delta\alpha_{alg}$—0, $1/4\ \pi$, $4/4\ \pi$ and $5/4\ \pi$—and compared with the case of a vibrating screen operating without an MR damper.

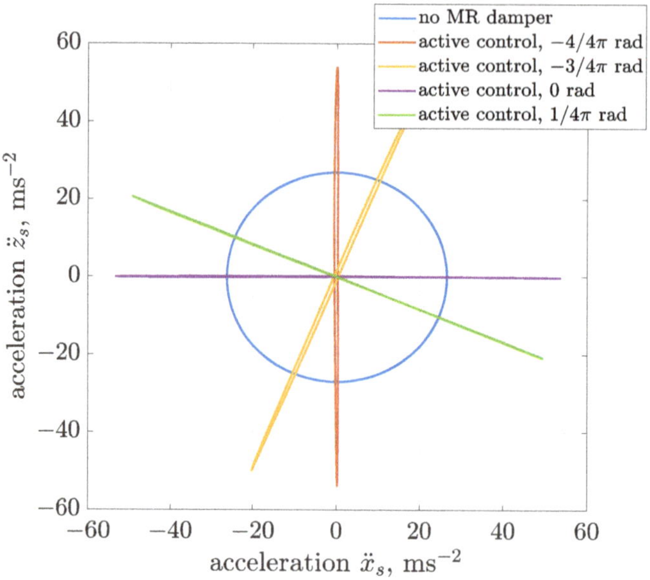

Figure 22. Different configurations of phase shift $\Delta\alpha_{alg}$ dedicated to active generation of linear trajectories of vibration acceleration.

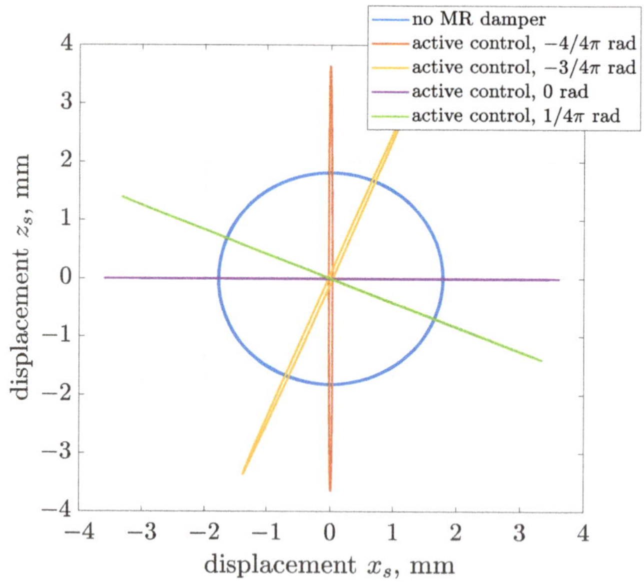

Figure 23. Different configurations of phase shift $\Delta\alpha_{alg}$ dedicated to active generation of linear trajectories of vibration displacement.

It can be seen based on acceleration and displacement trajectories that the proposed control algorithm successfully modifies the shape of the trajectory form circular to linear. The slope of the trajectory can be freely changed from horizontal to vertical or in between. It should be also noticed that the amplitudes of acceleration and displacement doubled, since an active approach of force generation is applied, which adds vibration energy to the considered system.

The trajectory control dedicated to MR damper was successively tested and presented in Figures 24 and 25 for vibration acceleration and displacement, respectively. It can be indicated that the more applicable case of semi-active trajectory control, comparing to previous ideal active control, allows for modification of screen vibration trajectory from circular to linear. The obtained shapes of trajectories include a component of elliptical trajectory. It can also be stated that efficiency in the modification of trajectory depends on the desired algorithm phase shift. For phase shifts equal to $1/4\ \pi$ or $5/4\ \pi$, the shape of the obtained trajectory is closer to linear in comparison to results obtained for phase shifts equal to 0 or π.

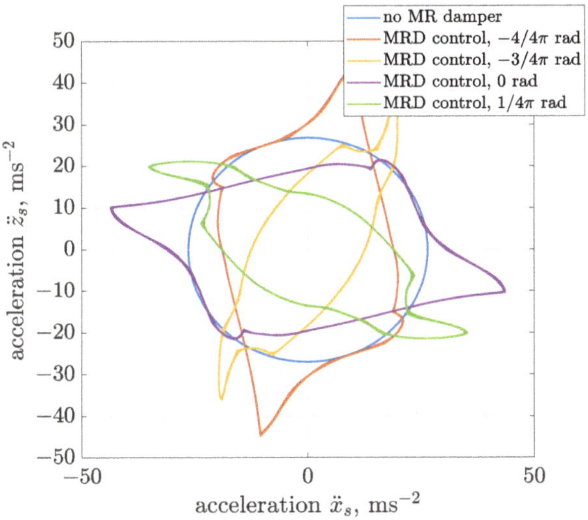

Figure 24. Different configurations of phase shift $\Delta\alpha_{alg}$ dedicated to semi-active generation of linear trajectories of vibration acceleration using the dissipative domain of the Bouc–Wen MR damper model.

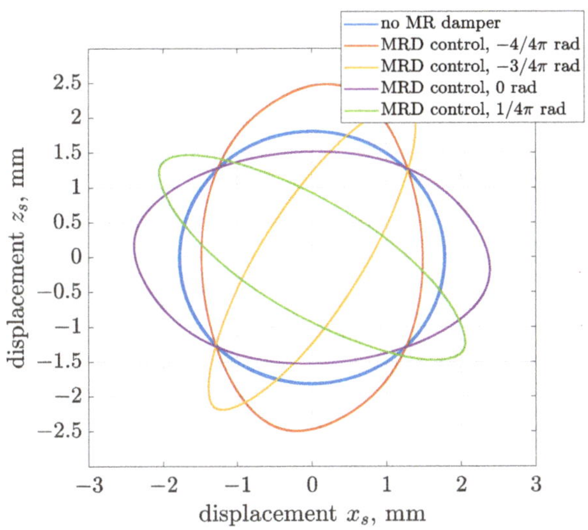

Figure 25. Different configurations of phase shift $\Delta\alpha_{alg}$ dedicated to semi-active generation of linear trajectories of vibration displacement using the dissipative domain of the Bouc–Wen MR damper model.

The quality of the proposed trajectory control algorithm was assessed for different values of control parameter $\Delta\alpha_{alg}$ in more a comprehensive approach based on a quality index of trajectory linearity denoted as J_{TL}, as presented in Figure 26. Its definition is mainly based on phase shift $\Delta\beta_{x,z}$ estimated between horizontal x_s and vertical z_s displacements using a cross-correlation function. Here, we recall for the reader that a zero phase shift results in a linear trajectory and a phase shift equal to $\pi/2$ corresponds to a circular trajectory, as discussed in Section 3. As as result, the quality index J_{TL} was defined by the following:

$$J_{TL} = |\cos(\Delta\beta_{x,z})| \quad \text{where} \quad \Delta\beta_{x,z} \longleftarrow \max(R_{x,z}), \tag{14}$$

where cross-correlation function of signals x_s and z_s is denoted as $R_{x,z}$.

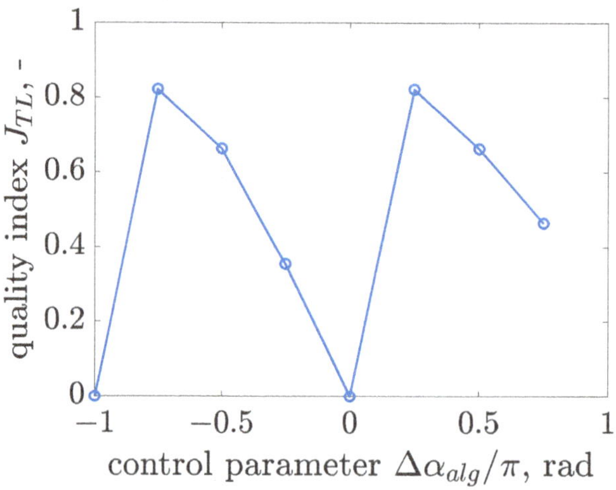

Figure 26. Comparison of values of quality index J_{TL} assessing the linearity the trajectory control algorithm presented for different control parameters $\Delta\alpha_{alg}$ applied to a semi-active screen suspension with the Bouc–Wen model.

The proposed quality index J_{TL} can vary from 0 to 1; the greater the value of J_{TL}, the closer the generated vibration trajectory will be to the linear shape. Evaluated values of J_{TL} indicate two optimized cases of linear trajectory within the analyzed range corresponding to $\Delta\alpha_{alg}$ equal to $-3/4\ \pi$ and $1/4\ \pi$ rad, where J_{TL} is close to 0.8. It can be noticed that the dependence of J_{TL} on $\Delta\alpha_{alg}$ is a smooth function, where the worst case corresponds to $\Delta\alpha_{alg}$ equal to $-\pi$ and 0 rad.

Further analysis gives insight into the results of linear trajectory generation based on a comparison of the desired control force F_{mr} and the generated MR damper force F_{alg}. These signals are presented in Figures 27 and 28 for the case of $\Delta\alpha_{alg} = -3/4\ \pi$ rad with respect to the front suspension part, and they are accompanied with the corresponding limiting forces $F_{bw,lb}$ and $F_{bw,ub}$. The first Figure 27 shows the desired horizontal control force, which was almost ideally represented and generated by the MR damper model. Contrary to that, tracking of the desired vertical force by the actual MR damper force, which is presented in Figure 28, could only be achieved to a limited extent. Despite this fact, the high quality of linear trajectory generation according to the quality index J_{TL} and displacement trajectories presented in Figure 25 was obtained for this case of $\Delta\alpha_{alg}$ and the proposed semi-active control.

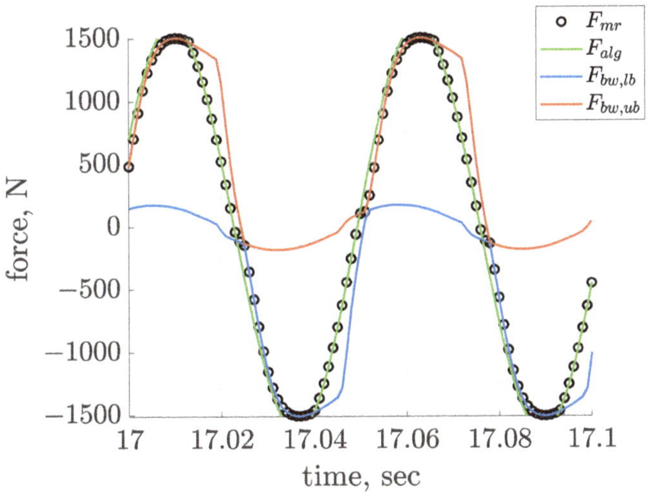

Figure 27. Quality of tracking of the horizontal desired control force $F_{alg,x}$ by the actual MR damper Bouc–Wen $F_{mr,x}$, accompanied by the lower $F_{bw,lb}$ and upper $F_{bw,ub}$ bounds, presented for the front screen suspension part for the control configuration of $\Delta\alpha_{alg} = -3/4\ \pi$ rad.

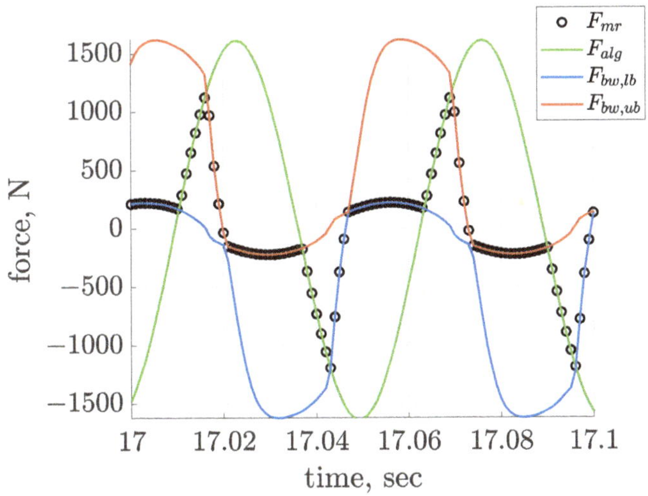

Figure 28. Quality of tracking of the vertical desired control force $F_{alg,z}$ by the actual MR damper Bouc–Wen $F_{mr,z}$, accompanied by the lower $F_{bw,lb}$ and upper $F_{bw,ub}$ bounds, presented for the front screen suspension part for the control configuration of $\Delta\alpha_{alg} = -3/4\ \pi$ rad.

An extended analysis was carried out for different control parameters $\Delta\alpha_{alg}$ and for each horizontally or vertically oriented MR damper applied to the selected front or rear part of the screen model. The quality of tracking the desired control force F_{alg} by the MR damper Bouc–Wen models F_{mr} was assessed based on a normalized quality index of force tracking denoted as J_{FT}, and results are listed in Table 2. The quality index J_{FT} was defined based on a complement to the relative mean squared error calculated between the desired control force F_{alg} and the actual MR damper force F_{mr} as follows:

$$J_{FT} = 1 - \frac{MSE_{FT}}{MSE_{FT,max}} \quad \text{where} \quad MSE_{FT} = \frac{\sum_j \left[F_{mr}(j) - F_{alg}(j)\right]^2}{\sum_j \left[F_{alg}(j)\right]^2}, \tag{15}$$

Table 2. Quality index J_{FT} assessing the tracking of the desired control force F_{alg} by the actual force generated by MR damper Bouc–Wen models F_{mr}. Control configurations in bold correspond to the best quality index of trajectory linearity J_{TL}.

	Location of MR Damper Model: J_{FT}			
$\Delta\alpha_{alg}$	**Front Hor.**	**Front Vert.**	**Rear Hor.**	**Rear Vert.**
$-4/4\ \pi$ rad	0.458	0.476	0.473	0.424
$-3/4\ \pi$ rad	**0.994**	**0.272**	**0.999**	**0.131**
$-2/4\ \pi$ rad	0.953	0.129	0.919	0.052
$-1/4\ \pi$ rad	0.743	0.182	0.697	0.152
0 rad	0.430	0.466	0.390	0.448
$1/4\ \pi$ rad	**0.168**	**0.998**	**0.137**	**0.972**
$2/4\ \pi$ rad	0.079	0.945	0.074	0.968
$3/4\ \pi$ rad	0.165	0.748	0.168	0.760

The symbol $MSE_{FT,max}$ denotes a normalization value which corresponds to the worst-case quality (maximum value) of force tracking evaluated over all considered control configurations and MR damper models. For a single control parameter $\Delta\alpha_{alg}$ and a single MR damper model, the maximum value of MSE_{FT} was found by shifting the desired force F_{alg} in time with respect to lower $F_{bw,lb}$ and upper $F_{bw,ub}$ force bounds, which resulted in a set of estimated $\hat{F}_{mr}$ time diagrams. As a result, the greater value of J_{FT}, the better the quality of force tracking achieved.

Control configurations which are in bold in Table 2 correspond to the best results with respect to trajectory linearity. It is indicated that the quality of force tracking for both the front and rear suspension parts was high for the selected MR damper model, whereas the second MR damper model oriented perpendicularly produced much greater tracking error. Furthermore, when control parameter $\Delta\alpha_{alg}$ was equal to $-3/4\ \pi$ or $1/4\ \pi$ rad, the quality of force tracking for the horizontal or vertical MR damper, respectively, was significantly heightened.

Values of J_{FT} varied from 0.052 for the rear vertical MR damper model and $\Delta\alpha_{alg} = -2/4\ \pi$ rad to 0.999 for the rear horizontal MR damper model, and $\Delta\alpha_{alg} = -3/4\ \pi$ rad. It is worth noting that high overall quality of trajectory linearity depends on dominant force tracking achieved for a single horizontal or vertical MR damper rather than on an average tracking quality: The control configuration corresponding to $\Delta\alpha_{alg} = -\pi$ rad have relatively high J_{FT} values which varied from 0.424 to 0.476. However, it did not provide the best final result according to the displacement trajectories and J_{TL} quality index.

5. Conclusions

The article reported a trajectory control algorithm applied to a vibrating screen equipped with semi-active suspension. The proposed solution allows one to change the shape of the vibrating screen trajectory from circular to linear and control vibration amplitude. The reported results were obtained in simulations performed on the model developed and parameterized based on experimental data from a semi-industrial vibrating screen. The demonstrated solution was applied to the vibrating screen model with only one exciter. The main novelty of the reported solution is that according to the proposed control algorithm, the desired forces generated by MR dampers emulate an additional virtual mechanical exciter of the vibrating screen. In turn, it interacts with the existing exciter, resulting in conversion of the trajectory from circular to linear.

From the future application standpoint, the proposed online trajectory modification may lead to a different vibrating screen design. The desired operating trajectory must be maintained even in the presence of sudden and frequent changes in the processed material throughput or its physical properties (e.g., particles' size distribution, particle shape or moisture). To tackle that problem, the modern vibrating screen's mass is usually much higher then the mass of the processed material on deck to counteract the material flow changes. Lighter constructions with online motion control will require smaller and in turn more economic engines for vibration excitation. Finally, controlled, fast changes in the screen suspension stiffness will allow faster transitions through the resonant frequencies during startup and stop procedures, and may also be used for sieve cleaning. Future research will be focused on the experimental validation of the proposed trajectory control algorithm using different types of vibrating screens. The extension of the presented screen model to its dynamics in space will be also studied.

Author Contributions: Conceptualization, S.O.; methodology, S.O. and P.K.; validation, P.K. and S.O.; formal analysis, P.K.; investigation, P.K.; resources, S.O.; writing—original draft preparation, P.K.; writing—review and editing, S.O.; supervision, S.O.; project administration, S.O.; funding acquisition, S.O. All authors have read and agreed to the published version of the manuscript.

Funding: This research was supported by Silesian University of Technology under Excellence Initiative—Research University programme in 2022.

Institutional Review Board Statement: Not applicable.

Informed Consent Statement: Not applicable.

Acknowledgments: The authors would like to acknowledge the support of Marcin Lutynski from the Faculty of Mining, Safety Engineering and Industrial Automation, Silesian University of Technology, for providing the access to the vibrating screen.

Conflicts of Interest: The authors declare no conflict of interest.

Nomenclature

g	gravitational acceleration
$x, \dot{x}, \ddot{x}$	horizontal displacement, velocity and acceleration with respect to the stationary frame of reference
$z, \dot{z}, \ddot{z}$	vertical displacement, velocity and acceleration with respect to the stationary frame of reference
(x_s, z_s)	horizontal and vertical displacements of riddle center of mass
φ_s	inclination angle of the screen riddle with respect to the stationary frame of reference
m_s, I_s	mass and moment of inertia of the vibrating screen riddle
M_m, α_m	torque and instantaneous angle of the electric motor with respect to the stationary frame of reference
$(x_{er}, z_{er}), (x_{em}, z_{em})$	horizontal and vertical displacements of the rotating shaft of the mechanical exciter and the unbalanced mass
M_e, α_e	torque and instantaneous angle of the mechanical exciter with respect to the riddle frame of reference
m_e, I_{er}, r_e	mass and moments of inertia evaluated with respect to the rotating axis, distance from the axis of rotation to the center of unbalanced mass
(l_{er}, h_{er})	location of the exciter rotation axis with respect to the riddle center of mass and its frame of reference
$(x_{sf}, z_{sf}), (x_{sr}, z_{sr})$	horizontal and vertical displacements of front and rear parts of the screen suspension

k_{sx}, k_{sz}	horizontal and vertical stiffness of the single part of the screen suspension
$c_{x_s}, c_{z_s}, c_{\varphi_s}, f_{\varphi_s}$	horizontal and vertical viscous damping and angular viscous damping and dry friction related to the screen suspension
$(l_{sf}, h_{sf}), (l_{sr}, h_{sr})$	location of the suspension attachment points with respect to the riddle center of mass and its frame of reference
(x_a, z_a)	horizontal and vertical displacement of the averaged sensor
(l_a, h_a)	location of the averaged sensor with respect to the riddle center of mass and its frame of reference
F_{alg}, F_{mr}, v_{mr}	force desired by the control algorithm and actual force generated by the MR damper model, axial piston velocity of the MR damper model
$(l_{mrf}, h_{mrf}), (l_{mrr}, h_{mrr})$	location of the MR damper models attachment points with respect to the riddle center of mass and its frame of reference
$F_{bw,lb}, F_{bw,ub}$	lower and upper bounds generated by the Bouc–Wen model and set on the desired control force
$\Delta\alpha_{alg}$	parameter of trajectory control algorithm corresponding to the phase shift of the emulated mechanical exciter with respect to the real exciter
J_{TL}	quality index assessing linearity of screen vibration trajectory
J_{FT}	quality index assessing tracking of F_{alg} by the F_{mr}

References

1. Standish, N. The kinetics of batch sieving. *Powder Technol.* **1985**, *41*, 57–67. [CrossRef]
2. Jansen, M.L.; Glastonbury, J.R. The size separation of particles by screening. *Powder Technol.* **1968**, *1*, 334–343. [CrossRef]
3. Standish, N.; Bharadwaj, A.K.; Hariri-Akbari, G. A study of the effect of operating variables on the efficiency of a vibrating screen. *Powder Technol.* **1986**, *48*, 161–172. [CrossRef]
4. Coulson, J.M.; Richardson, J.F. *Chemical Engineering V2—Particle Technology and Separation Process*, 4th ed.; Pergamon: Oxford, UK, 1991.
5. Available online: https://www.360researchreports.com (accessed on 20 April 2022).
6. Available online: https://www.futuremarketinsights.com (accessed on 20 April 2022).
7. Wen, B. Recent development of vibration utilization engineering. *Front. Mech. Eng. China* **2008**, *3*, 1–9. [CrossRef]
8. Peng, L.; Jiang, H.; Chen, X.; Liu, D.; Feng, H.; Zhang, L.; Liu, C. A review on the advanced design techniques and methods of vibrating screen for coal preparation. *Powder Technol.* **2019**, *347*, 136–147. [CrossRef]
9. Lyashenko, V.I.; Dyatchin, V.Z.; Franchuk, V.P. Improvement of vibrating feeders-screens for mining and metallurgical industry. *Ferr. Metall.* **2018**, *61*, 470–477. [CrossRef]
10. Peng, L.; Feng, H.; Wang, Z.; Wang, H.; Yang, H.; Huang, H. Screening Mechanism and Properties of a Cantilevered Vibrating Sieve for Particles Processing. *Appl. Sci.* **2019**, *9*, 4911. [CrossRef]
11. Li, Z.; Tong, X.; Zhou, B.; Wang, X. Modeling and parameter optimization for the design of vibrating screens. *Miner. Eng.* **2015**, *83*, 149–155. [CrossRef]
12. Dong, H.; Liu, C.; Zhao, Y.; Zhao, L. Influence of vibration mode on the screening process. *Int. J. Min. Sci. Technol.* **2013**, *23*, 95–98. [CrossRef]
13. Zhang, B.; Gong, J.; Yuan, W.; Fu, J.; Huang, Y. Intelligent prediction of sieving efficiency in vibrating screens. *Shock Vib.* **2016**, *2016*, 7. [CrossRef]
14. Czubak, P. Vibratory conveyor of the controlled transport velocity with the possibility of the reversal operations. *J. Vibroengineering* **2016**, *18*, 3539–3547. [CrossRef]
15. Majewski, T. Vibratory forces and synchronization in physical systems. *Ing. Mecánica Tecnol. Desarro* **2013**, *4*, 119–128.
16. Zhang, X.; Hu, W.; Gao, Z.; Liu, Y.; Wen, B.-C. Composite synchronization on two pairs of vibrators in a far super-resonant vibrating system with the single rigid frame. *J. Low Freq. Noise Vib. Act. Control* **2021**, *40*, 2064–2076. [CrossRef]
17. Hou, Y.; Du, M.; Wang, L. Investigation for synchronization of two co-rotating rotors installed with nonlinear springs in a non-resonance system. *Adv. Mech. Eng.* **2019**, *11*, 1687814019834110. [CrossRef]
18. Liu, Y.; Zhang, X.; Gu, D.; Jia, L.; Wen, B. Synchronization of a dual-mass vibrating system with two exciters. *Shock Vib.* **2020**, *2020*, 9345652. [CrossRef]

19. Zhang, X.; Gu, D.; Yue, H.; Li, M.; Wen, B. Synchronization and stability of a far-resonant vibrating system with three rollers driven by two vibrators. *Appl. Math. Model.* **2021**, *91*, 261–279. [CrossRef]
20. Chen, B.; Yan, J.; Yin, Z.; Tamma, K. A new study on dynamic adjustment of vibration direction angle for dual-motor-driven vibrating screen. *Proc. Inst. Mech. Eng. Part E J. Process Mech. Eng.* **2020**, *235*, 186–196. [CrossRef]
21. Chen, X.; Liu, J.; Li, L. Dynamics of the vibration system driven by three homodromy eccentric rotors using control synchronization. *Appl. Sci.* **2021**, *11*, 7691. [CrossRef]
22. Jia, L.; Kong, X.; Zhang, J.; Liu, Y.; Wen, B. Multiple-frequency controlled synchronization of two homodromy eccentric rotors in a vibratory system. *Shock Vib.* **2018**, *2018*, 4941357. [CrossRef]
23. Ying-Qing, G.; Zhao-Dong, X.; Bing-Bing, C.; Cheng-Song, R.; Wei-Yang, G. Preparation and experimental study of magnetorheological fluids for vibration control. *Int. J. Acoust. Vib.* **2017**, *22*, 194–200.
24. Pierce, R.; Kaul, S.; Friesen, J.; Morgan, T. Mountain bike rear suspension design: Utilizing a magnetorheological damper for active vibration isolation and performance. *Int. J. Acoust. Vib.* **2020**, *25*, 504–512. [CrossRef]
25. Krauze, P.; Kasprzyk, J.; Rzepecki, J. Experimental attenuation and evaluation of whole body vibration for an off-road vehicle with magnetorheological dampers. *J. Low Freq. Noise, Vib. Act. Control.* **2019**, *38*, 852–870. [CrossRef]
26. Krauze, P.; Kasprzyk, J. Driving safety improved with control of magnetorheological dampers in vehicle suspension. *Appl. Sci.* **2020**, *10*, 1–29. [CrossRef]
27. Rafajłowicz, W.; Więckowski, J.; Moczko, P.; Rafajłowicz, E. Iterative learning from suppressing vibrations in construction machinery using magnetorheological dampers. *Autom. Constr.* **2020**, *119*, 1–10. [CrossRef]
28. Chen, Z.H.; Lam, K.H.; Ni, Y.Q. Enhanced damping for bridge cables using a self-sensing MR damper. *Smart Mater. Struct.* **2016**, *25*, 1–13. [CrossRef]
29. Song, X.; Ahmadian, M.; Southward, S.C. Modeling magnetorheological dampers with application of nonparametric approach. *J. Intell. Mater. Syst.* **2005**, *16*, 421–432. [CrossRef]
30. Wereley, N.M.; Pang, L. Nondimentional analysis of semi-active electrorheological and magnetorheological dampers using approximate paralled plate models. *J. Smart Mater. Struct.* **1998**, *7*, 732–743. [CrossRef]
31. Zapoměl, J.; Ferfecki, P.; Forte, P. A computational investigation of the transient response of an unbalanced rigid rotor flexibly supported and damped by short magnetorheological squeeze film dampers. *Smart Mater. Struct.* **2012**, *21*, 1–12. [CrossRef]
32. Lin, Q.; Wei, Z.; Wang, N.; Chen, W. Analysis on the lubrication performances of journal bearing system using computational fluid dynamics and fluid-structure interaction considering thermal influence and cavitation. *Tribol. Int.* **2013**, *64*, 8–15. [CrossRef]
33. Xie, Z.; Wang, X.; Zhu, W. Theoretical and experimental exploration into the fluid structure coupling dynamic behaviors towards water-lubricated bearing with axial asymmetric grooves. *Mech. Syst. Signal Process.* **2022**, *168*, 1–25. [CrossRef]
34. Sapiński, B. Analysis of parametric models of MR linear damper. *J. Theor. Appl. Mech.* **2003**, *41*, 215–240.
35. Spencer, B.F.; Dyke, S.J.; Sain, M.K.; Carlson, J.D. Phenomenological model of a magnetorheological damper. *ASCE J. Eng. Mech.* **1997**, *123*, 230–238. [CrossRef]
36. Elsaady, W.; Oyadiji, S.O.; Nasser, A. A review on multi-physics numerical modelling in different applications of magnetorheological fluids. *J. Intell. Mater. Syst. Struct.* **2020**, *31*, 1855–1897. [CrossRef]
37. Kasprzyk, J.; Wyrwał, J.; Krauze, P. Automotive MR damper modeling for semi-active vibration control. In Proceedings of the IEEE/ASME International Conference on Advanced Intelligent Mechatronics, Besancon, France, 8–11 July 2014; pp. 500–505.
38. Zhao, L.L.; Zhao, Y.M.; Bao, C.Y.; Hou, Q.F.; Yu, A.B. Optimization of a circularly vibrating screen based on DEM simulation and Taguchi orthogonal experimental design. *Powder Technol.* **2017**, *310*, 307–317. [CrossRef]
39. Zhao, L.L.; Zhao, Y.M.; Bao, C.Y.; Hou, Q.F.; Yu, A.B. Laboratory-scale validation of a DEM model of screening processes with circular vibration. *Powder Technol.* **2016**, *303*, 269–277. [CrossRef]
40. Dong, K.; Esfandiary, A.H.; Yu, A. Discrete particle simulation of particle flow and separation on a vibrating screen: Effect of aperture shape. *Powder Technol.* **2017**, *314*, 195–202. [CrossRef]
41. Harzanagh, A.A.; Orhan, E.C.; Ergün, S.L. Discrete element modelling of vibrating screens. *Miner. Eng.* **2018**, *121*, 107–121. [CrossRef]
42. Davoodi, A.; Bengtsson, M.; Hulthén, E.; Evertsson, C. Effects of screen decks' aperture shapes and materials on screening efficiency. *Miner. Eng.* **2019**, *139*, 105699. [CrossRef]
43. Yin, Z.; Zhang, H.; Han, T. Simulation of particle flow on an elliptical vibrating screen using the discrete element method. *Powder Technol.* **2016**, *302*, 443–454. [CrossRef]
44. Jahani, M.; Farzanegan, A.; Noaparast, M. Investigation of screening performance of banana screens using LIGGGHTS DEM solver. *Powder Technol.* **2015**, *283*, 32–47. [CrossRef]
45. Li, Z.; Tong, X. Applications of the discrete element method and Fibonacci sequence on a banana screen. *J. Eng. Des. Technol.* **2017**, *15*, 2–12. [CrossRef]
46. Krot, P.; Zimroz, R.; Michalak, A.; Wodecki, J.; Ogonowski, S.; Drozda, M.; Jach, M. Development and Verification of the Diagnostic Model of the Sieving Screen. *Shock Vib.* **2020**, *2020*, 8015465. [CrossRef]
47. Krot, P.; Zimroz, R. Methods of springs failures diagnostics in ore processing vibrating screens. In *IOP Conference Series: Earth and Environmental Science*; IOP Publishing: Bristol, UK, 2019; Volume 362, p. 012147. [CrossRef]
48. Gursky, V.; Krot, P.; Korendiy, V.; Zimroz, R. Dynamic Analysis of an Enhanced Multi-Frequency Inertial Exciter for Industrial Vibrating Machines. *Machines* **2022**, *10*, 130. [CrossRef]

49. Więckowski, J.; Rafajłowicz, W.; Moczko, P.; Rafajłowicz, E. Data from vibration measurement in a bucket wheel excavator operator's cabin with the aim of vibrations damping. *Data Brief* **2021**, *35*, 1–13. [CrossRef]
50. Wiora, J. Least lp-norm low-pass filter. *Automatica* **2021**, *133*, 1–10. [CrossRef]
51. Pratama, P.S.; Luan, B.T.; Tran, T.P.; Kim, H.K.; Kim, S.B. Trajectory tracking algorithm for automatic guided vehicle based on adaptive backstepping control method. In *AETA 2013: Recent Advances in Electrical Engineering and Related Sciences*; Lecture Notes in Electrical Engineering; Springer: Berlin/Heidelberg, Germany, 2014; pp. 535–544.
52. Altan, A.; Hacıoğlu, R. Model predictive control of three-axis gimbal system mounted on UAV for real-time target tracking under external disturbances. *Mech. Syst. Signal Process.* **2020**, *138*, 1–23. [CrossRef]
53. Cao, J.; Song, C.; Peng, S.; Song, S.; Zhang, X.; Xiao, F. Trajectory tracking control algorithm for autonomous vehicle considering cornering characteristics. *IEEE Access* **2020**, *8*, 59470–59484. [CrossRef]
54. Zheng, G.; Huang, Y.; Gan, B. The application of MR fluid damper to the vibrating screen as the enhance of screening efficiency. *Adv. Mater. Res.* **2010**, *97–101*, 2628–2633. [CrossRef]
55. Wu, F.-S.; Huang, Y.-J.; Huang, K.; Xu, S. Trispectrum and correlation dimension analysis of magnetorheological damper in vibration screen. *J. Cent. South Univ.* **2012**, *19*, 1832–1838. [CrossRef]
56. Wu, M.; Chen, F.; Li, A.; Chen, Z.; Sun, N. A novel vibration isolator for vibrating screen based on magnetorheological damper. *J. Mech. Sci. Technol.* **2021**, *35*, 4343–4352. [CrossRef]
57. Mujal-Rosas, R.; Orrit-Prat, J. General analysis of the three-phase asynchronous motor with spiral sheet rotor. Operation, parameters and characteristics values. *IEEE Trans. Ind. Electron.* **2011**, *58*, 1799–1811. [CrossRef]
58. Ćalasan, M.; Alqarni, M.; Rosić, M.; Koljčević, N.; Alamri, B.; Abdel Aleem, S.H.E. A novel exact analytical solution based on Kloss equation towards accurate speed-time characteristics modeling of induction machines during no-load direct startups. *Appl. Sci.* **2021**, *11*, 1–15. [CrossRef]

sensors

MDPI

Article

Multiferroic Cantilevers Containing a Magnetoactive Elastomer: Magnetoelectric Response to Low-Frequency Magnetic Fields of Triangular and Sinusoidal Waveform

Gašper Glavan *, Inna A. Belyaeva and Mikhail Shamonin *

East Bavarian Centre for Intelligent Materials (EBACIM), Ostbayerische Technische Hochschule (OTH) Regensburg, Seybothstr. 2, D-93053 Regensburg, Germany; inna.belyaeva@oth-regensburg.de
* Correspondence: gasper.glavan@oth-regensburg.de (G.G.); mikhail.chamonine@oth-regensburg.de (M.S.)

Abstract: In this work, multiferroic cantilevers comprise a layer of a magnetoactive elastomer (MAE) and a commercially available piezoelectric polymer-based vibration sensor. The structures are fixed at one end in the horizontal plane and the magnetic field is applied vertically. First, the magnetoelectric (ME) response to uniform, triangle-wave magnetic fields with five different slew rates is investigated experimentally. Time and field dependences of the generated voltage, electric charge, and observed mechanical deflection are obtained and compared for four different thicknesses of the MAE layer. The ME responses to triangular and sinusoidal wave excitations are examined in contrast. Second, the ME response at low frequencies ($\leq$3 Hz) is studied by the standard method of harmonic magnetic field modulation. The highest ME coupling coefficient is observed in the bias magnetic field strength of $\approx$73 kA/m and it is estimated to be about 3.3 ns/m (ME voltage coefficient $\approx$ 25 V/A) at theoretically vanishing modulation frequency ($f \to 0$ Hz). Presented results demonstrate that the investigated heterostructures are promising for applications as magnetic-field sensors and energy harvesting devices.

Keywords: magnetoactive elastomer; piezoelectric polymer; multilayer cantilever; direct magnetoelectric effect; magnetic field sensing

1. Introduction

The direct magnetoelectric (ME) effect is the induced change in the electric polarization P due to a change in the applied magnetic field strength H [1,2]. Natural, single-phase materials demonstrating the direct ME effect are rather rare and the ME coupling is very low (ME voltage coefficient $\alpha_V^H \approx 0.1 - 10$ mV/A) [3]. These materials are usually designated as multiferroic, which means that at least two types of spontaneous ferroic orders, such as ferroelectricity and ferromagnetism, are present [3,4]. The ME effect can be significantly increased in the composite (two-phase) materials, which exhibit much higher ME coupling [1,5]. Most conventional multiferroic composite materials are layered heterostructures made from rigid materials (Young's modulus $Y \sim 10^{11}$ Pa), such as piezoelectric (PE) ceramics and ferromagnetic (FM) alloys [6–8]. In applications, ME functional elements usually operate at electromechanical-resonance frequencies, where the ME coupling is enhanced due to the increased deformation amplitude [6,9–12]. Because of the potential applications of ME heterostructures as vibration energy harvesters and biomedical sensors, it would be desirable to reduce the resonance frequency (typically, 1–100 kHz in conventional composites) by three orders of magnitude. This can be achieved if the constitutive materials are soft ($Y \leq 10^9$ Pa). The lack of soft ME materials in nature stimulates the search for new concepts. Soft PE materials can be obtained from polymers, e.g., polyvinylidene fluoride (PVDF) [13]. A soft replacement for the rigid FM material in ME heterostructures is somewhat of a challenge [14]. Very recently, attention has been drawn to the soft ($Y \sim$ 30–300 kPa), magnetoactive elastomers (MAEs) as promising candidates for the usage as magnetostrictive layers in layered multiferroic heterostructures [15,16]. MAEs consist of micrometer-sized ferromagnetic particles embedded in a soft polymer matrix. They exhibit exceptional magneto-mechanical effects [17–20]. In [21], helically coiled MAE stripes were used to realize untethered soft robots.

In an external magnetic field, a compliant MAE object may change its dimensions or shape (morph) [22]. If an applied magnetic field is homogeneous, the corresponding strain during the process of magnetization is often designated as magnetostrictive strain. A giant extensional strain up to ≈20% was recently observed in soft MAE cylinders [23], which makes the MAEs promising candidates for ME composites. MAEs can be combined with PE polymers in layered composite materials, where a significant direct ME effect is observable [24,25]. If a magnetic field is applied perpendicularly to the plane of the layered cantilever, the cantilever is bent due to the torque acting on the MAE layer [26,27], and the resulting deformation of the PE layer leads to the generation of an electric voltage. The proof of concept has been performed in inhomogeneous magnetic fields [16], which did not allow one to measure the ME voltage coefficient. The quasi-static ME voltage coefficient for the discussed type of ME heterostructures was later estimated to be about 50 V/A in a magnetic field of ≈100 kA/m [24]. The experiments in [24] were performed in uniform pulsed magnetic fields. In [16,24], the magnetic field was applied perpendicularly to the plane of the MAE samples. Such an arrangement is commonly designated as the T-T (transverse magnetic field and transverse electric field) configuration [1]. Very recently, a conventional L-T (longitudinal magnetic field and transverse electric field) configuration was also investigated, where a superposition of constant and small alternating magnetic fields (standard method of sinusoidal field excitation [28]) was applied in the plane of the composite heterostructure [25]. The highest ME voltage coefficient of about 7.85 V/A was measured in a sample where the thickness of the MAE layer was ≈2 mm. The corresponding resonance frequency in the absence of a magnetic field was low ($f_r \approx 29$ Hz). The first results [16,24,25], obtained on multiferroic layered structures containing an MAE layer and a PE polymer, confirmed that these heterostructures are promising for the development of magnetic field sensors and energy harvesting devices. It seems that contrary to the conventional paradigm, the T-T configuration leads to a higher ME voltage coefficient than the L-T configuration. Mechanically soft, multiferroic cantilevers, such as those presented in this paper, can easily morph (bend to a great extent) towards the direction of an applied magnetic field so that their demagnetizing factor drastically decreases and the ME coupling becomes highly efficient [25]. Hitherto, multiferroic layered composites containing an MAE layer are not well investigated and not all the specific features of ME coupling in them are well understood, for example, the unexceptionally large increase in the resonance frequency in magnetic fields [25] or the retarded deflection and voltage response with respect to the magnetic field [24].

One specific feature which distinguishes the heterostructures in the present paper from conventional multiferroic laminates is the viscoelastic behavior of the magnetostrictive layer (MAE slab), which is field-dependent. As far as MAEs are concerned, it has been recently pointed out that special attention to deformation and magnetic rates is required [29]. However, even in the rapidly evolving field of MAEs, this aspect seems to be largely neglected.

The main goal of this paper is to investigate in detail the ME response of multiferroic heterostructures containing a PE polymer layer and an MAE slab in a triangle-wave magnetic field, in particular, the dependence on the magnetic field rate. Furthermore, we use the conventional method of harmonic (sinusoidal) magnetic field excitation at very low frequencies (≲3 Hz) and determine the optimum magnetic field for the effective ME coupling. In particular, a high ME voltage coefficient of ≈25 V/A is demonstrated at the excitation frequency of 0.1 Hz in the constant magnetic field of about 73 kA/m.

2. Materials and Methods

2.1. Multiferroic Cantilevers

Cantilevers comprised an MAE layer of varying thickness x and a commercially available PVDF-based vibration sensor (LDT0-028K, Measurement Specialties, Hampton, VA,

USA) [30] (Figure 1a). For the sake of simplicity, the vibration sensor will be denoted in the following as a PE polymer (PEP). PEP was a flexible component comprising a thin (thickness t_{PE} = 28 µm) PE PVDF polymer film with screen-printed silver ink electrodes, laminated to a 0.125 mm polyester substrate, and fitted with two crimped contacts. Because the PE film was displaced from the mechanical neutral axis, bending created a very high strain within the PVDF layer and high voltages were generated [31,32]. The fabrication procedure of heterostructures was described in detail in [24]. A thin, silicone-based adhesive layer, which is not sensitive to the magnetic field, was employed between the MAE layer and the vibration sensor. This layer highly increased the durability of bonding between different layers and enhanced the magnitude of ME coupling [24]. The MAE material contained 80 wt% (≈33 vol%) of carbonyl iron powder (CIP, type SQ, BASF SE, Ludwigshafen, Germany). The low-frequency shear storage modulus of the MAE material, obtained using a commercial rheometer (model Physica MCR 301, Anton Paar, Graz, Austria), was 50.4 kPa. This particular MAE material for composite cantilevers was selected because, in the previous work [24], it resulted in the highest voltage response in comparison to alternative compositions.

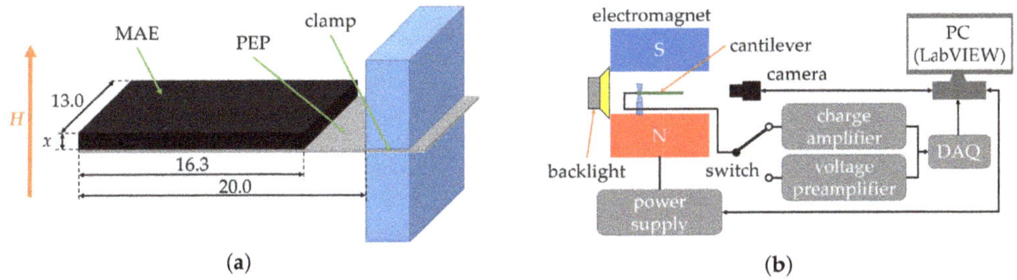

Figure 1. Sketch of a clamped cantilever [24] (**a**) and schematic of the experimental setup (**b**). Dimensions are given in mm.

2.2. Experimental Setup

A cantilever was clamped on the side of electrical contacts in a 3D printed holder from a polylactic acid (PLA) material (Figure 1a). It was placed horizontally between the electromagnet's poles. A uniform magnetic field was applied vertically. The cantilever's deflection was recorded using a camera (Alvium 1800 U-319m, Allied Vision Technologies GmbH, Stadtroda, Germany) with a suitable lens (Edmund Optic Double Gauss Focusable, 25 mm C-mount F4.0 1.3″, Barrington, NJ, USA). The backlight illumination employed light-emitting diodes (LED, Illuminant G4 Pen, Conrad Electronics, Hirschau, Germany) with a diffuser (Perspex diffuse, 2.5 mm, 3A Composites GmbH, Sins, Switzerland) (Figure 1b). A graph paper screen was placed between the LED and a cantilever. The graph paper is visible as the background in Figure 2a. The LED illumination had no significant effect on the temperature of a cantilever because of the absence of direct illumination and short measurement times less than ≈100 s. All experiments were performed under normal laboratory conditions at 22 ± 1 °C. The output electrical signal from the PE material was directed either to a voltage preamplifier (KISTLER 5165A, Winterthur, Switzerland) with an amplification coefficient k = 1 and an input impedance of 10 MΩ or to a charge amplifier (KISTLER 5018, Winterthur, Switzerland), whose outputs were connected to a data acquisition (DAQ) board (NI USB-6212, National Instruments, Austin, TX, USA) linked to a personal computer (PC). The experiment was automated with LabVIEW software (National Instruments, Austin, TX, USA). By these means, either the generated voltage U or charge Q were measured as functions of time t.

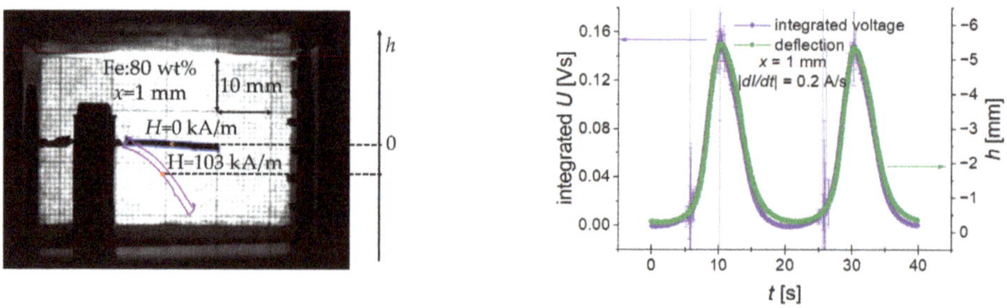

(**a**) (**b**)

Figure 2. (**a**) Example of determining the deflection h from measurements of the cantilever's "mass centre" [24]. (**b**) Comparison of the numerically integrated output voltage U (purple line) and the deflection h (green line) for the case of a slew rate of 12.2 kA/sm and $x = 1$ mm.

2.3. Image Processing

The deflection of a cantilever was determined by image analysis using the OpenCV library in Python programming language. The deflection h was defined as the difference between the positions of the so-called "mass centre" which is the average of the vertical coordinate of the cantilever's contour (Figure 2a). It is a measure of the cantilever's deformation, which causes strain in the PE layer. We have found that the reference to the "mass centre" is a more robust procedure than the strategy to follow the tip of the cantilever. Furthermore, such a definition through the "mass centre" is meaningful because the generated voltage is practically proportional to the time derivative of the cantilever's deflection [24]. An example is demonstrated in Figure 2b for the triangle-wave excitation and a magnetic slew rate of 12.2 kA/sm.

2.4. Measurement Protocol

The experimental setup was calibrated in such a way that the external magnetic field strength H was calculated from the electric current flowing through the electromagnetic coils. The resulting magnetic field strength was almost directly proportional to the excitation current. An alternating current had a symmetrical triangular waveform. However, there was a slight offset of not more than $\approx$2.4 kA/m, observed in the dependence of the magnetic field on the electric current due to some remanent magnetization of electromagnet's pole shoes. The compensation of such a small offset was not possible due to the lack of necessary precision in setting the excitation current. The bi-polar power supply (FAST-PS 1k5, CAENels s.r.l., Basovizza, Italy) allowed one to adjust a constant rate for the rising or falling electrical current.

Five different slew rates for the electric current were used, as shown in Figure 3a and Table 1. The frequency f of the resulting magnetic field depended on the maximum electric current and the slew rate. The frequency in the last column of Table 1 was calculated for the maximum current $I_{max} = 2.0$ A, which corresponds to the peak value of the magnetic field strength $H_{max} \approx 103$ kA/m.

Table 1. Slew rates and frequencies of the magnetic field for cases shown in Figure 3a, where the maximal field amplitude was $H_{max} \approx 103$ kA/m.

$\left\|\frac{dI}{dt}\right\|$ [A/s]	$\left\|\frac{dH}{dt}\right\|$ [kA/sm]	$f(H_{max} = 103$ kA/m) [Hz]
0.2	12.2	0.025
1.0	52.5	0.125
5.0	253.7	0.625
10.0	505.3	1.250
15.0	756.9	1.875

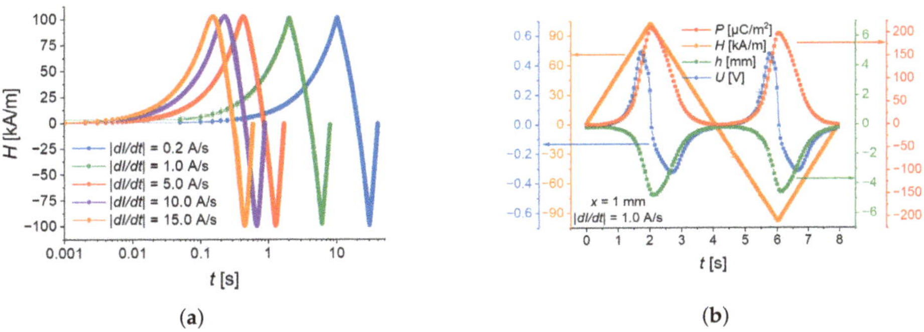

Figure 3. (**a**) One cycle of magnetic field oscillations in the logarithmic time scale for the case $H_{max} \approx 103$ kA/m. (**b**) Typical responses of polarization (red, P), deflection (green, h) and voltage (blue, U) to triangular waveform with a slew rate of 53 kA/sm (orange, H).

Two sampling rates were used for measurements of physical quantities. The electric polarization or voltage were obtained at 1000 Hz, while the magnetic field was measured at 20 Hz. For the lower two slew rates in Table 1, the corresponding electrical quantities (charge, voltage) were taken at the same time points as the magnetic field and the deflection was measured. However, for the higher three rates, a cubic spline interpolation was used to interpolate from 20 Hz onto 1000 Hz.

Voltage and charge were monitored with a DAQ board. To remove some noise at mains frequency, a notch filter, centered at 50 Hz with a 3dB bandwidth of 10 Hz, was applied to measured data. The electrical charge Q was re-calculated into the polarization as $P = \sigma\left(1 - \varepsilon_r^{-1}\right)^{-1}$, where $\sigma = Q/A$, A is the effective area of the PE layer, ε_r is the relative permittivity of the PE material, and a typical value $\varepsilon_r \approx 15$ was assumed [33].

A train of bi-polar current pulses comprising five periods of a symmetric triangular waveform was used at a fixed slew rate and an amplitude. The output signal (generated voltage or charge) and the applied magnetic field strength were recorded. Since MAEs exhibit hysteretic behavior of measured quantities with respect to the applied magnetic field [24,25,34,35], the response to each period of an applied magnetic field is somewhat different. A hysteresis in MAEs is commonly attributed to the magnetic-field-induced restructuring of filler particles, which is also the origin of magnetic hysteresis in such a material. The hysteresis of magnetic properties in MAEs, filled with soft-magnetic particles, was explained in [36] by the hysteresis of the consolidation of filler particles into elongated aggregates. However, it has been previously noticed that only an initial (magnetic field increase from zero to the peak value) curve significantly differs from the subsequent cycles [24,35]. A possible explanation is that during the first cycle major restructuring of the filler takes place, whereas further changes are minor [35]. To take into account the hysteretic behaviour of MAEs, the following approach was used for the evaluation of data. The results were presented over one period of the excitation current. The values at a particular time point within a period were averaged over the last four periods of the excitation current. The differences between the measured values at a particular time point in four subsequent periods are reflected by the error bars, with a confidence level of 95%.

The ME coupling coefficient $\alpha^H = \partial P/\partial H$ [28] at time point $i \cdot \Delta t$ was numerically calculated as $(P_i - P_{i-1})/(H_i - H_{i-1})$, where Δt is the time interval between measurement points and i is the consecutive number of a measurement point.

As an example, Figure 4 presents the result of measurements of the ME coupling coefficient α^H for the triangular waveform with a slew rate of 53 kA/sm and a peak magnetic field strength of 103 kA/m. Here and in the following Figures, the field dependences are constructed by using the values at the same time moment. It is the only example of α^H, where error bars will be shown. On all further graphs presenting the ME coupling coefficient α^H, error bars are not shown. These error bars may become too large due to the use of a numerical derivative of raw data, which is sensitive to the presence of some noise. This is typically the case in the vicinity of the local maxima and minima of the magnetic field, where its time derivative changes its sign. In the following, an additional running average was run over ME coupling coefficient data to smoothen the first derivative. The moving average was run over five points for the lower two rates in Table 1 and over fifteen points for the higher three rates in Table 1.

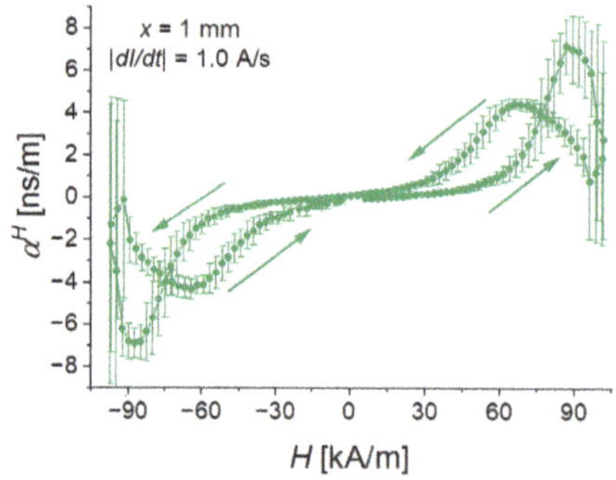

Figure 4. Dependence of the magnetoelectric coupling coefficient for $x = 1$ mm and a slew rate of 53 kA/sm on the applied magnetic field strength. Here and in the following Figures, the arrows designate the direction of magnetic-field change.

3. Results

3.1. Effect of Magnetic-Field Slew Rate

Five different slew rates were investigated for a sample with the MAE thickness $x = 1$ mm. In Figure 5, you can see the field dependences of measured physical quantities. The generated voltage is shown in Figure 5a, where one can notice come difference between peak values in positive and negative magnetic fields. This difference is caused by a slight remanent magnetization of the electromagnet's shoe poles and it is clearly visible only in the voltage measurement. The highest ME voltage was achieved at an intermediate slew rate of 254 kA/sm.

As all measured physical quantities are almost perfectly symmetric either with respect to the y axis (voltage, deflection, polarization) or with respect to the origin of the graph (ME coupling coefficient), Figures 5c,d and the following Figures show only the results for the positive values of the applied magnetic field.

In Figures 5b and 6a, it is seen that the deflection amplitude decreases with the increasing rate of the magnetic field. This can be explained by the delayed magneto-mechanical response of the viscoelastic material. With increasing slew rate, it becomes increasingly difficult for the cantilever to follow the magnetic field. Because the time derivative of the cantilever's deflection h can be written as $dh/dt = (dh/dH) \cdot (dH/dt)$, the interplay between the delayed magneto-mechanical response (at fixed external magnetic field the derivative dh/dH decreases with increasing magnetic rate) and the increase in the magnetic rate (dH/dt) leads to the appearance of the highest peak value $|U_{max}|$ of the generated voltage $U(t)$ at an intermediate value of $|dH/dt|$ (Figure 6b). Note that the deflection has negative values due to the selected direction of the coordinate axis (Figure 2). How the direction of deflection can be influenced was discussed in [24].

Figure 5c presents the field dependence of the (averaged) polarization. It can be observed that its field behavior is similar to the field behavior of the cantilever's deflection. Figure 5d compares the field dependences of the ME coupling coefficient at different magnetic-field rates. The magnitude of a ME coupling coefficient has two local maxima—one ("advancing") maximum for the ascending branch of the magnetic field magnitude and another ("receding") maximum for the descending branch of the magnetic field magnitude.

Advancing maximum is higher and narrower than receding maximum. Both maxima of $\alpha^H(H)$ "shift" to the higher values of magnetic field with increasing slew rate. For the two fastest rates (10 A/s, 15 A/s), the local maxima in the field dependences of α^H are not observable in the measurement range of the magnetic field. Remarkably, the local maxima of α^H correspond to particular values of polarization P: the maximum value of α^H for the ascending magnetic field occurs at $P = (110 \pm 13)$ μC/m^2, while the maximum value of α^H for the descending magnetic field occurs at $P = (97 \pm 10)$ μC/m^2, independently of the magnetic rate (Figure 7). We believe that this observation should be related to the ferroelectric hysteresis loop of the PE PVDF film and this phenomenon should be investigated in detail in future works.

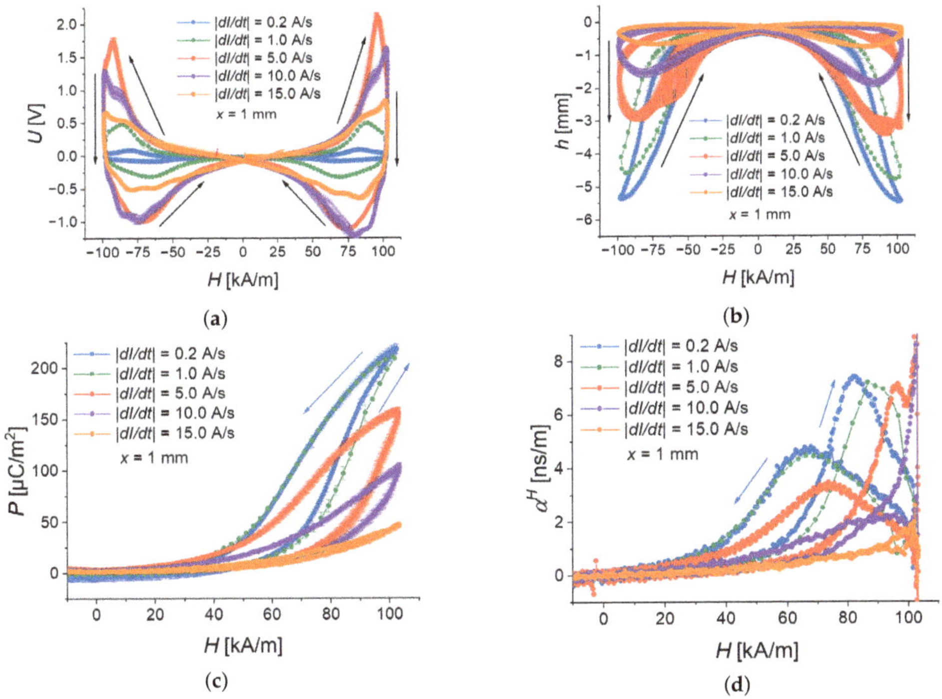

Figure 5. Average field dependences of generated voltage (**a**), deflection (**b**), polarization (**c**), and ME coupling coefficient (**d**) for different slew rates.

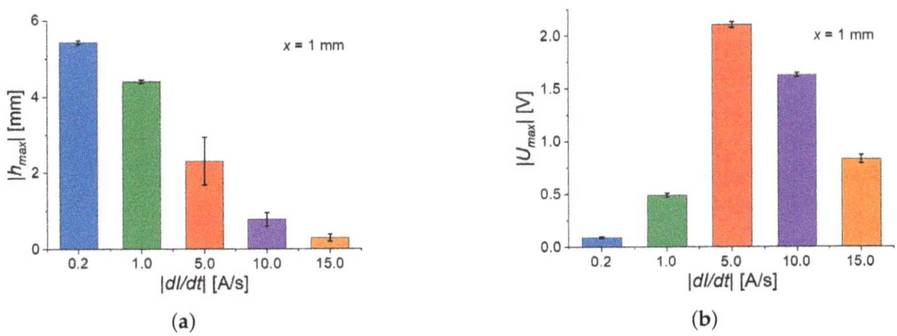

Figure 6. Peak deflection (**a**) and peak induced voltage (**b**) for different slew rates.

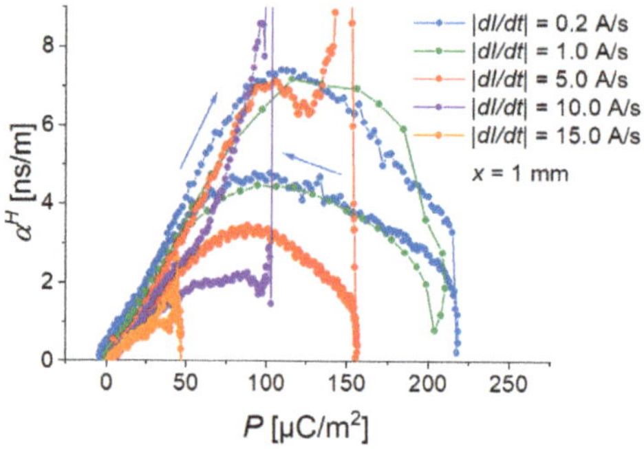

Figure 7. Dependences of the ME coupling coefficient on the electric polarization for different slew rates.

3.2. Effect of MAE Thickness

Next, we investigated the effect of MAE thickness on the ME response. Four cantilevers with different MAE thicknesses x were compared at a fixed slew rate of 12.2 kA/sm. As reported in [24], in a fixed magnetic field, the deflection increases with the increasing thickness of the MAE layer. The reason is that the torque in a magnetic field increases with increasing amount of magnetic material (total mass of iron particles). Since the maximum possible deflection of a cantilever ($h \leq 6$ mm) is limited by the geometry of the experimental setup, the amplitude of the magnetic field was adjusted according to the thickness of the MAE layer (Table 2). Obviously, this results in different frequencies of the applied magnetic field. Figure 8 depicts the results of measurement of the field dependences of the electrical polarization and the ME coupling coefficient.

Table 2. Amplitudes of triangle-wave magnetic fields and their corresponding frequencies for different thicknesses of MAE layer at the magnetic slew rate of 12.2 kA/sm.

x [mm]	1	2	3	4
H_{max} [kA/m]	103	83	73	63
f [mHz]	25.0	31.3	35.7	41.7

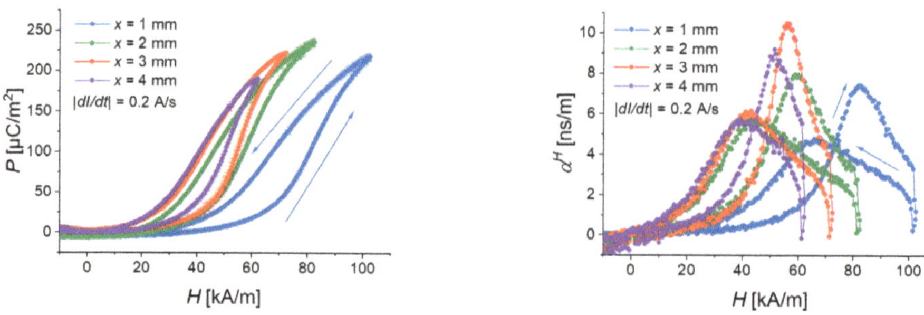

(**a**) (**b**)

Figure 8. Field dependences of the electric polarization (**a**) and the ME coupling coefficient (**b**) for different thicknesses of MAE layer. The magnetic field slew rate is 12.2 kA/sm.

The highest "advancing" maximum of the ME coupling coefficient was observed for a cantilever with an MAE thickness of 3 mm. At a fixed magnetic field rate and a fixed deflection amplitude, there seems to be an optimum MAE thickness for the maximal ME coupling. Similar to what is observed in Section 3.1, the local maxima of α^H correspond to particular values of polarization P: the maximum value of α^H for the ascending magnetic field occurs at

$P = (114 \pm 10)$ μC/m^2, while the maximum value of α^H for the descending magnetic field occurs at $P = (99 \pm 10)$ μC/m^2, independently of the MAE thickness (Figure 9).

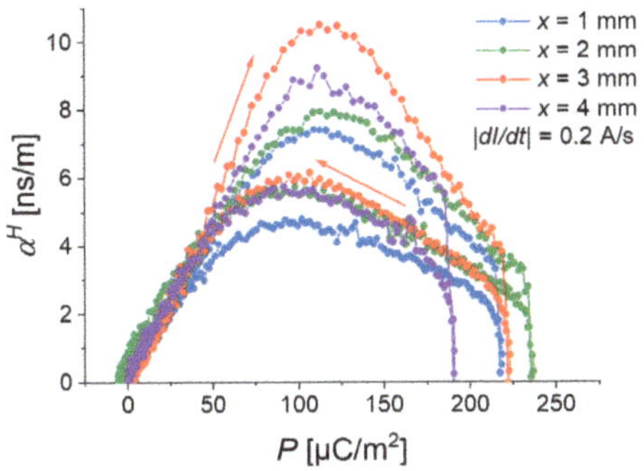

Figure 9. Dependences of ME coupling coefficient on electric polarization for different thicknesses of MAE layer and the magnetic slew rate of 12.2 kA/sm.

3.3. Effect of Magnetic-Field Amplitude

How does the amplitude of magnetic field affect the polarization and the ME coupling coefficient? The measurements were performed for a cantilever with an MAE thickness of 1 mm at a fixed magnetic field rate of 12.2 kA/sm. Figure 10 presents the measurement results. From Figure 10a, it is seen that the field dependence of the polarization for the ascending magnetic field remains the same (the curves practically overlap), while the polarization for the descending magnetic field is significantly different due to the effect of hysteresis. A similar effect is observed in Figure 10b for the ME coupling coefficient. Figure 11 shows the dependences of the ME coupling coefficient on the electric polarization. For the two smallest amplitudes of the triangle-wave magnetic field (73 kA/m, 83 kA/m), the local maximum of $\alpha^H(H)$ is not observed for the ascending magnetic field, because the optimum value of polarization of ≈110 μC/m^2 is not reached. For these two magnetic-field amplitudes, the local maximum of $\alpha^H(H)$ for the descending magnetic field is an artifact, related to the numerical differentiation needed to calculate $\alpha^H(H)$. The optimum polarization value of ≈100 μC/m^2 for the descending magnetic field is not reached. The highest ME coupling coefficient of ≈7.5 ns/m is obtained in the ascending magnetic field for the two highest peak values of magnetic field (93 kA/m, 103 kA/m) at $P \approx 110$ μC/m^2.

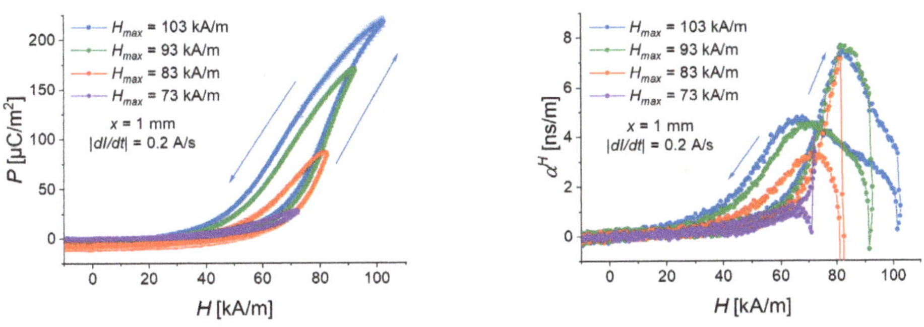

(a) (b)

Figure 10. Field dependences of polarization (**a**) and ME coupling coefficient (**b**) for four peak values of magnetic field (H_{max} = 73, 83, 93, and 103 kA/m) for a sample with 1 mm thick MAE layer.

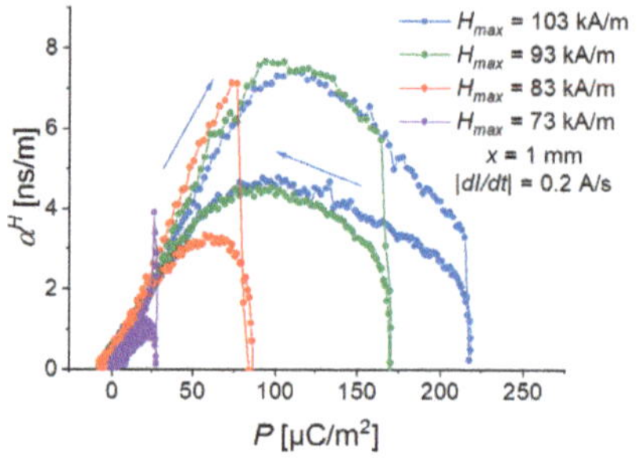

Figure 11. Dependence of the ME coupling coefficient on the electric polarization for four peak values of magnetic field ($H_{max} = 73, 83, 93$, and 103 kA/m) and a sample with an MAE layer thickness of 1 mm at the magnetic slew rate of 12.2 kA/sm.

3.4. Comparison of Triangular and Sinusoidal Waveforms

To investigate the effects of magnetic-field slew rate, the triangular waveform is an obvious choice. However, conventional experimental characterizations involve sinusoidal wavefunction. To investigate the difference between triangular and sinusoidal magnetic fields, we used a feature of the available power source. To obtain comparable data, a point-by-point protocol was created with a sampling rate of 20 Hz. The slew rate of the power supply was set to the highest possible value of 15 A/s (757 kA/sm). The current was measured at 15 ms after it was set. The period duration of the sinusoidal voltage was 8 s ($f = 0.125$ Hz), this corresponded to 160 measurement points per cycle. For the fast Fourier transform (FFT) analysis, 12 cycles were used. The data in Figures 12 and 13 are the averaged values over the last 11 cycles, from which also the uncertainties of measurements have been obtained.

Figure 12 presents the time behavior of the averaged polarization for one cycle of applied magnetic fields. As was expected, this behavior was different for triangular and sinusoidal waveforms. The polarization had a double repetition frequency with respect to the applied magnetic field (Figure 12a,c) because the cantilever always deflects in the same direction, independently of the polarity of an applied magnetic field. The corresponding spectra (Figures 12b,d) demonstrated the presence of the double frequencies of the excitation field. Figure 12d clearly shows that our synthesis of a sinusoidal field was successful because there was one dominating frequency in the FFT spectrum. As expected, the Fourier spectrum of the symmetric triangle-wave function also had higher harmonics [37,38].

However, Figure 13 demonstrates that the field dependences of the measured polarization and calculated ME coupling are very similar. Therefore, the results presented in this paper can be reasonably applied to sinusoidal magnetic fields with the same frequency and amplitude if field dependences are considered.

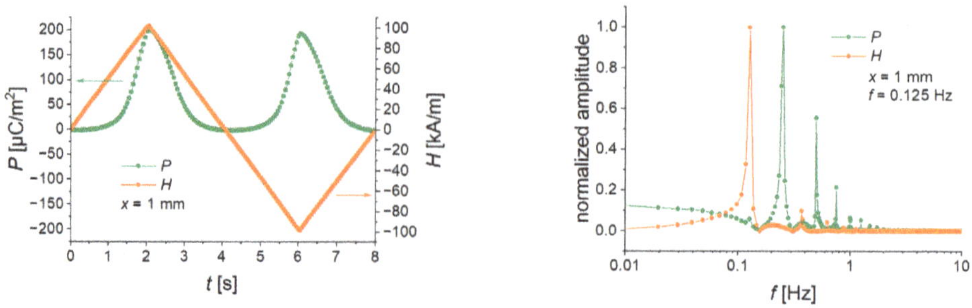

(a) (b)

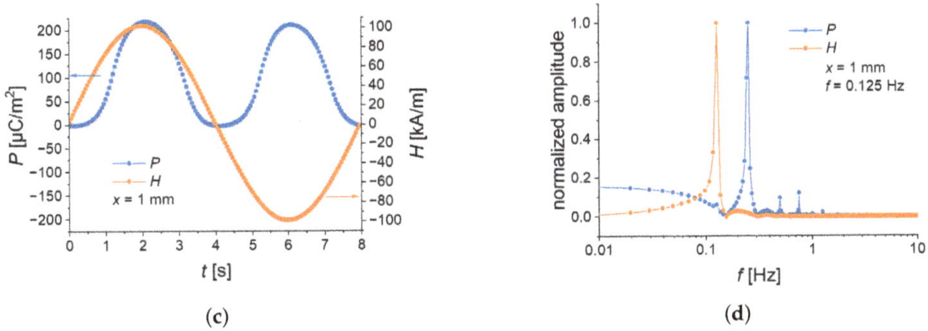

Figure 12. One average pulse of magnetic field and time dependence of the polarization response for symmetric triangle-wave (**a**) and sine-wave (**c**) magnetic field. FFT of magnetic field and electric polarization for symmetric triangular (**b**) and sinusoidal (**d**) waveforms of magnetic field excitation. The amplitudes of the FFT spectra are normalized to their maximum values.

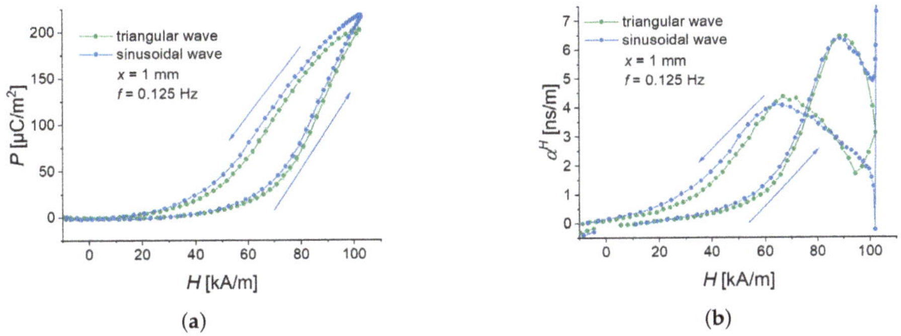

Figure 13. Comparison of field dependences of the polarization (**a**) and the ME coupling coefficient (**b**) for two different magnetic-field excitations. Symmetric triangular and sinusoidal waves have a frequency of 0.125 Hz.

4. Discussion

The above results clearly demonstrate that the ME response of multiferroic cantilevers containing an MAE layer depends on the slew rate and the amplitude of a magnetic field strength. We attribute this fact to the viscoelastic nature of MAE materials as it was pointed out in [29].

Furthermore, as already reported in [24], measured physical quantities (voltage, deflection, polarization) and calculated ME coupling coefficient have hysteresis behavior with respect to the applied magnetic field.

At a particular momentary value of the calculated external magnetic field, most of the measured physical quantities (deflection, polarization) and ME coupling coefficient did not reach their steady-state (maximum) value, which would be possible for the step magnetic field excitation because the magneto-mechanical response is retarded with respect to the magnetic field. Obviously, this is also the origin of observed hysteresis behavior. As far as the measured voltage is concerned, under a step field excitation, it vanishes in the steady-state due to the finite electrical impedance of the used pre-amplifier.

In the above considerations, alternating magnetic fields were employed. How are the determined values of the ME coupling coefficient compared with those obtained by a conventional method of harmonic field modulation (HFM) [28]? To answer this question, an AC magnetic field with an amplitude H_{AC} of 5 kA/m, which was in the linear region of voltage response $U(H_{AC})$ (Figure 14), was superimposed on a constant (bias) magnetic field H_{DC}. Note that $H_{AC} \ll H_{DC}$.

$$H(t) = H_{DC} + H_{AC}\cos(2\pi f t + \varphi_H), \tag{1}$$

$$P(t) = P_{slow}(t) + P_{AC}\cos(2\pi f t + \varphi_P). \tag{2}$$

$P(t)$ is the time dependence of the polarization P, P_{AC} is the amplitude of the sinusoidally varying component of the polarization, $P_{slow}(t)$ is the transient component of polarization, which will be explained below. φ_H and φ_P are the initial phases of magnetic field and polarization, respectively.

The essential difference between the above experiments is that the constant field H_{DC} was not zero. Therefore, the generated electric charge oscillated at the excitation frequency and not at the double excitation frequency as in the experiments above. H_{DC} varied between 53 and 108 kA/m with a step of 5 kA/m. The value of H_{AC} was equal to the step size. The frequency value f were 0.1, 0.5, 1.0, 2.0, and 3.0 Hz. The frequency $f = 3.0$ Hz was maximal possible due to the time discretization of the excitation current and the time resolution of our measurement setup (20 Hz). At this frequency, the applied AC magnetic field still had a reasonable sinusoidal waveform (7 points per cycle).

As an example, Figure 14 demonstrates that the amplitude H_{AC} of the sinusoidal magnetic field was small enough, and that the linear ME effect was investigated.

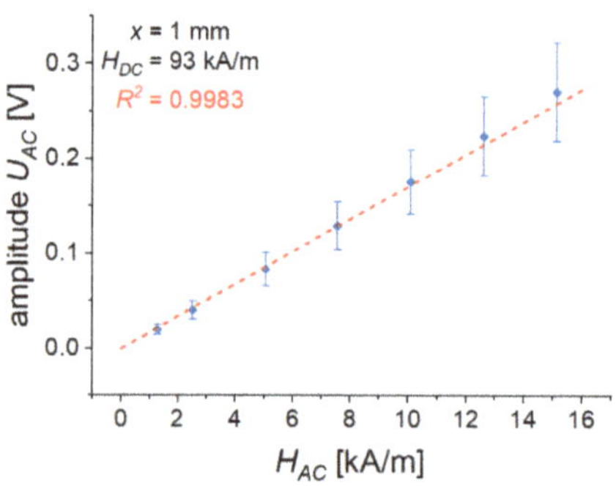

Figure 14. Dependence of the AC voltage amplitude U_{AC} on the amplitude H_{AC} of AC magnetic field. R^2 denotes the determination coefficient of the linear regression.

Similar to the measurements above, charge, magnetic field, and deflection were acquired. The average deflection for different frequencies and magnetic fields is shown in Figure 15a. The electric charge was re-calculated into the polarization. The transient component $P_{slow}(t)$ of $P(t)$ was subtracted from the signal, so that only the sinusoidal function remained. The examples of the transient behaviors are shown in Figure 15c,d. Finally, the sinusoidal functions were fitted to the experimentally obtained magnetic field and polarization according to Equations (1) and (2). From the obtained amplitudes of fits, the ME coupling coefficient was calculated as:

$$\alpha^H = \frac{P_{AC}}{H_{AC}}. \tag{3}$$

The obtained ME coupling coefficient is shown in Figure 15b. At the lowest frequency, the ME coupling coefficient reaches 3.2 ns/m (the ME voltage coefficient $\approx$ 25 V/A). Similarly to the electric polarization, the cantilever's deflection oscillates at a given frequency. Therefore we introduced the deflection $\bar{h}$, which is the mean value of deflection over several oscillation cycles. Figure 15a also shows numerically calculated derivatives of the averaged deflection $\bar{h}$ with respect to the applied DC magnetic field $d\bar{h}/dH_{DC}$. The maximum of this derivative is always observed at $H_{DC} \approx 73$ kA/m. The derivative $d\bar{h}/dH_{DC}$ can be seen as an analog of the piezomagnetic coefficient in the conventional theory of the ME effect in composite multiferroic materials. It is visible from Figure 15b, that the maximum of the ME coupling coefficient is observed at approximately the same value of the constant magnetic field as for the derivative $d\bar{h}/dH_{DC}$. The mechanism of the optimum low-frequency ME coupling seems to be similar to conventional ME multiferroic layered composites. We also recalculated the ME coupling coefficient α^H into the ME voltage coefficient $\alpha_V^H = \alpha^H(\varepsilon_0\varepsilon_r)^{-1}$ [28,39], where ε_r is the relative permittivity of the PE material, and a typical value $\varepsilon_r = 15$ was assumed [33]. The ME voltage coefficient can be easily expressed in CGS units, taking into account that 1 Oe corresponds to 79.577 A/m.

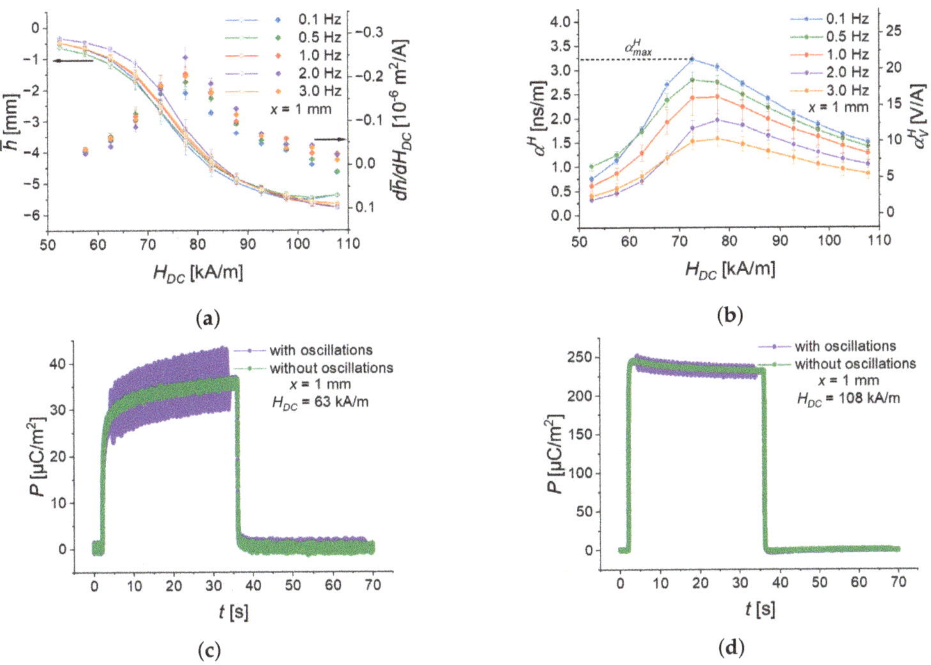

Figure 15. (**a**) Field dependences of the deflection and its derivative with respect to the constant magnetic field. (**b**) Field dependences of the ME coupling coefficient for five different frequencies. Examples of the transient behaviour of polarization for the excitation current of 1.2 A (**c**) and 2.1 A (**d**). As an example, α^H_{max} shows the maximum value of α^H for $f = 0.1$ Hz.

We also observed the characteristic behaviour of the transient component $P_{slow}(t)$ changes around the magnetic field value of $H_{DC} = 88$ kA/m. Below this value, the polarization $P_{slow}(t)$ reaches the steady-state exponentially as $P_{slow}(t) = P_{max}\left(1 - e^{-kt}\right)$ (Figure 15c). Above $H_{DC} = 88$ kA/m, first an overshoot was observed and then $P_{slow}(t)$ relaxed exponentially as $P_{slow}(t) = P_0 e^{-kt}$ (Figure 15d). Qualitatively similar changes in the transient behavior of the magnetodielectric effect (magnetic-field-dependent permittivity) [35] and the engineering stress [29] were previously observed. In [29], four different magnetic rates were compared at constant step magnitude. At lower magnetic rates (2 and 20 mT/s), the behavior of engineering stress was qualitatively similar to the transient behavior in Figure15c. For the higher two magnetic rates (200 and 1000 mT/s), the transient behavior of stress was similar to that in Figure 15d. When applying a magnetic field with a slow slew rate, the particular transient behavior was attributed to the balance between magnetic forces between magnetized particles and the viscoelasticity of the polymer matrix. For the higher two slew rates, an alternative explanation was suggested. In this case, the behavior was assigned to the microstructural stiffening of polymeric matrix, which results in a "collapse" of magnetic particles being unable to properly rearrange along magnetic field lines, leading to a microstructural blockade [29].

Figure 16 presents the frequency dependencies of the maximum ME coupling coefficient α^H_{max}, extracted from Figure 15a and the phase difference $\Delta\varphi = \varphi_H - \varphi_P$ between the alternating magnetic field and the alternating component of polarization. The maximum ME coupling coefficient increases with the decreasing excitation frequency. The reason for that can be the decreasing phase delay between the magnetic field and the polarization so that the ME coupling becomes more efficient. The data shown in Figure 16 have been fitted to the function $y(f) = a + be^{-kf}$. The results of fitting are summarized in Table 3.

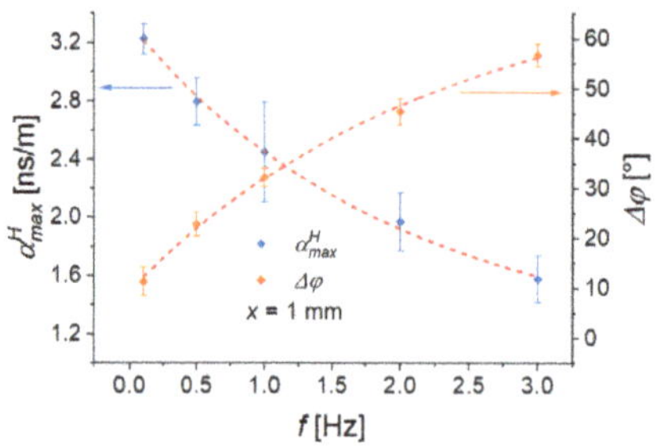

Figure 16. Frequency dependences of the peak value of the ME coupling coefficient α^H_{max} and the phase shift $\Delta\varphi$. Dashed lines indicate fitted exponential functions ($y(f) = a + be^{-kf}$), where y is either α^H_{max} or $\Delta\varphi$.

Table 3. Parameters of fitted functions in Figure 16. R^2 denotes the determination coefficient.

Parameter	α^H_{max}	$\Delta\varphi$
a	1.10 ± 0.22	74.24 ± 7.22
b	2.23 ± 0.20	-64.69 ± 6.33
k	0.50 ± 0.10	0.43 ± 0.09
R^2	0.998	0.997

From the exponential fit, the quasi-static ($f \rightarrow 0$ Hz) phase difference $\Delta\varphi$ of $9.6° \pm 13.5°$ can be estimated. Therefore, it can be reasonably assumed that in a quasi-static state, the ME response can follow the applied magnetic field ($\Delta\varphi \rightarrow 0°$). The phase delay of the polarization increases with increasing frequency and tends asymptotically to roughly 74.24°, if higher resonance frequencies are neglected. Similarly, the quasi-static ME coupling coefficient can be estimated to be $\approx$3.3 ns/m (ME voltage coefficient $\approx$25.1 V/A). One can conclude that the low-frequency ME voltage coefficient, obtained using the conventional method of sinusoidal magnetic-field excitation [28], has the same order of magnitude as previously reported value from magnetic-field excitations [24]. The maximum obtained value of the low-frequency ME voltage coefficient is more than three-fold higher than the largest ME voltage coefficient obtained at a resonance frequency on similar structures in the L-T geometry [25]. Note that the results presented at very low excitation frequencies ($\leq$3 Hz); the multiferroic heterostructures were not operating at resonance. It can be expected that the ME voltage coefficient can be further enhanced at higher frequencies, where electromechanical resonances can be observed [25].

We are aware of only a few papers, where ME coefficients at such low frequencies have been reported. The measured maximum value of α^H_V is two orders of magnitude higher than that obtained [40] at $\approx$1 Hz utilizing the concept of the so-called magnetoelectric electret. In [41], the reported ME coupling coefficient of a heterostructure composed of six layers of a magnetostrictive amorphous alloy (Metglas®, $Y \sim 10^{11}$ Pa) and a piezoelectric composite core of five PMN-PT fibers, interrogated by a pair of Kapton inter-digited (ID) electrodes, was $\alpha_Q^H \approx 34$ ps·m at $f = 1$ Hz, which is about two orders of magnitude higher than measured here ($\alpha_Q^H \approx 0.35$ ps·m). Several interesting experiments were performed [42] at low frequencies, but the ME coefficient was not reported.

Finally, Figure 17 compares the highest values of ME voltage coefficient obtained with MAE/PEP heterostructures for different magnetic field excitations. It is seen that investigated structures compare well with an alternative technology [40].

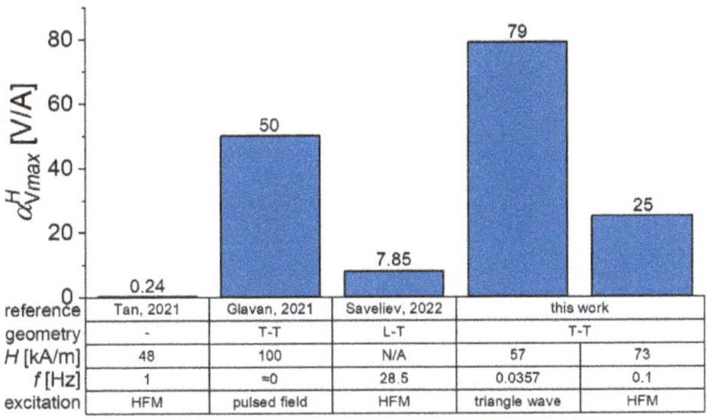

Figure 17. Comparison of different reported values of the ME coupling coefficient at low frequencies. (N/A stands for not available; Tan, 2021 stands for [40]; Glavan, 2021 stands for [24] and Saveliev, 2022 stands for [25]).

5. Conclusions

We have investigated the direct ME effect in multiferroic cantilevers containing an MAE layer in the alternating triangle-wave magnetic field. Furthermore, we measured the ME voltage coupling coefficient in the T-T geometry using sinusoidal magnetic-field excitation at low frequencies ($f \leq 3$ Hz). The following main conclusions can be made:

- The interplay between the delayed magneto-mechanical response and the increase in the magnetic slew rate led to the highest generated voltage at an intermediate value of the magnetic slew rate of ≈254 kA/sm;
- Two local maxima in the field dependence of the ME coupling coefficient were observed. One maximum appeared at ascending (increasing) magnetic field, the other maximum appeared at descending (decreasing) magnetic field. The maxima corresponded to particular polarization values: $P \approx 110$ μC/m^2 for the ascending magnetic field and $P \approx 100$ μC/m^2 for the descending magnetic field. Polarization values, where maxima of α^H were located, varied neither with the magnetic field slew rate nor with the thickness of the MAE layer;
- Field dependences of ME responses to triangular and sinusoidal waveforms of applied magnetic field were very similar. The electric polarization oscillated at double excitation frequency because the cantilever's magnetomechanical response did not distinguish between different polarities of the magnetic field;
- Systematic measurements of the ME coupling coefficient at sinusoidal magnetic-field modulation were performed at very low frequencies (up to 3 Hz). The highest ME voltage coefficient was observed in the constant magnetic field $H_{DC} \approx 73$ kA/m. From these results, it was estimated that the highest ME coupling coefficient α^H at $f \to 0$ Hz is ≈3.3 ns/m (ME voltage coefficient ≈25 V/A or ≈20 Vcm^{-1}Oe^{-1} in CGS units);
- Due to the viscoelastic properties of the MAE material, a phase delay between the applied magnetic field and electric polarisation was observed for sinusoidal magnetic-

field excitations in the presence of a constant magnetic field. The phase delay increased with the increasing frequency f.

Further research is required to better understand and further enhance the ME coupling in the presented heterostructures. In particular, the increase in the excitation frequency of sinusoidal magnetic-field modulation may lead to a significant increase in the ME coupling in the T-T geometry due to the resonance enhancement.

Author Contributions: Conceptualization, M.S.; methodology, M.S., G.G. and I.A.B.; software, G.G.; validation, G.G., I.A.B. and M.S.; investigation, G.G. and I.A.B.; resources, I.A.B. and M.S.; data curation, G.G.; writing—original draft preparation, M.S. and G.G.; writing—review and editing, M.S., G.G. and I.A.B.; visualization, G.G.; supervision, M.S.; project administration, M.S.; funding acquisition, M.S. All authors have read and agreed to the published version of the manuscript.

Funding: The research of G.G. and M.S. was funded by the Deutsche Forschungsgemeinschaft (DFG, German Research Foundation)—Project No. 466920132.

Institutional Review Board Statement: Not applicable.

Informed Consent Statement: Not applicable.

Data Availability Statement: The data that support the findings of this study are available from the corresponding author, G.G., upon reasonable request.

Acknowledgments: We thank V. Krause and T. Probst for technical assistance.

Conflicts of Interest: The authors declare no conflict of interest.

References

1. Srinivasan, G. Magnetoelectric composites. *Annu. Rev. Mater. Res.* **2010**, *40*, 153–178. [CrossRef]
2. Fiebig, M. Revival of the magnetoelectric effect. *J. Phys. D Appl. Phys.* **2005**, *38*, R123. [CrossRef]
3. Liang, X.; Matyushov, A.; Hayes, P.; Schell, V.; Dong, C.; Chen, H.; He, Y.; Will-Cole, A.; Quandt, E.; Martins, P.; et al. Roadmap on magnetoelectric materials and devices. *IEEE Trans. Magn.* **2021**, *57*, 1–57. [CrossRef]
4. Wang, Y.; Hu, J.; Lin, Y.; Nan, C.W. Multiferroic magnetoelectric composite nanostructures. *NPG Asia Mater.* **2010**, *2*, 61–68. [CrossRef]
5. Nan, C.W.; Bichurin, M.; Dong, S.; Viehland, D.; Srinivasan, G. Multiferroic magnetoelectric composites: Historical perspective, status, and future directions. *J. Appl. Phys.* **2008**, *103*, 1. [CrossRef]
6. Rajaram Patil, D.; Chai, Y.; Kambale, R.C.; Jeon, B.G.; Yoo, K.; Ryu, J.; Yoon, W.H.; Park, D.S.; Jeong, D.Y.; Lee, S.G.; et al. Enhancement of resonant and non-resonant magnetoelectric coupling in multiferroic laminates with anisotropic piezoelectric properties. *Appl. Phys. Lett.* **2013**, *102*, 062909. [CrossRef]
7. Ryu, J.; Priya, S.; Uchino, K.; Kim, H.E. Magnetoelectric effect in composites of magnetostrictive and piezoelectric materials. *J. Electroceramics* **2002**, *8*, 107–119. [CrossRef]
8. Chu, Z.; PourhosseiniAsl, M.; Dong, S. Review of multi-layered magnetoelectric composite materials and devices applications. *J. Phys. D Appl. Phys.* **2018**, *51*, 243001. [CrossRef]
9. Zeng, M.; Wan, J.; Wang, Y.; Yu, H.; Liu, J.M.; Jiang, X.; Nan, C. Resonance magnetoelectric effect in bulk composites of lead zirconate titanate and nickel ferrite. *J. Appl. Phys.* **2004**, *95*, 8069–8073. [CrossRef]
10. Wan, J.; Li, Z.; Wang, Y.; Zeng, M.; Wang, G.; Liu, J.M. Strong flexural resonant magnetoelectric effect in Terfenol-D/epoxy-Pb (Zr, Ti) O_3 bilayer. *Appl. Phys. Lett.* **2005**, *86*, 202504. [CrossRef]
11. Bichurin, M.; Petrov, V.; Petrov, R. Direct and inverse magnetoelectric effect in layered composites in electromechanical resonance range: A review. *J. Magn. Magn. Mater.* **2012**, *324*, 3548–3550. [CrossRef]
12. Fetisov, L.; Chashin, D.; Saveliev, D.; Afanas'ev, M.; Simonov-Emel'yanov, I.; Vopson, M.; Fetisov, Y. Magnetoelectric direct and converse resonance effects in a flexible ferromagnetic-piezoelectric polymer structure. *J. Magn. Magn. Mater.* **2019**, *485*, 251–256. [CrossRef]
13. Ramadan, K.S.; Sameoto, D.; Evoy, S. A review of piezoelectric polymers as functional materials for electromechanical transducers. *Smart Mater. Struct.* **2014**, *23*, 033001. [CrossRef]
14. Elhajjar, R.; Law, C.T.; Pegoretti, A. Magnetostrictive polymer composites: Recent advances in materials, structures and properties. *Prog. Mater. Sci.* **2018**, *97*, 204–229. [CrossRef]
15. Feng, J.; Xuan, S.; Ding, L.; Gong, X. Magnetoactive elastomer/PVDF composite film based magnetically controllable actuator with real-time deformation feedback property. *Compos. Part A Appl. Sci. Manuf.* **2017**, *103*, 25–34. [CrossRef]
16. Makarova, L.; Alekhina, Y.A.; Isaev, D.; Khairullin, M.; Perov, N. Tunable layered composites based on magnetoactive elastomers and piezopolymer for sensors and energy harvesting devices. *J. Phys. D Appl. Phys.* **2020**, *54*, 015003. [CrossRef]

27. Metsch, P.; Romeis, D.; Kalina, K.A.; Raßloff, A.; Saphiannikova, M.; Kästner, M. Magneto-Mechanical Coupling in Magneto-Active Elastomers. *Materials* **2021**, *14*, 434. [CrossRef] [PubMed]
28. Li, Y.; Li, J.; Li, W.; Du, H. A state-of-the-art review on magnetorheological elastomer devices. *Smart Mater. Struct.* **2014**, *23*, 123001. [CrossRef]
29. Chertovich, A.; Stepanov, G.; Kramarenko, E.Y.; Khokhlov, A. New composite elastomers with giant magnetic response. *Macromol. Mater. Eng.* **2010**, *295*, 336–341. [CrossRef]
30. Odenbach, S. (Ed.) *Magnetic Hybrid-Materials: Multi-Scale Modelling, Synthesis, and Applications*; Walter de Gruyter GmbH & Co KG: Berlin, Germany, 2021.
31. Park, J.E.; Jeon, J.; Cho, J.H.; Won, S.; Jin, H.J.; Lee, K.H.; Wie, J.J. Magnetomotility of Untethered Helical Soft Robots. *RSC Adv.* **2019**, *9*, 11272–11280. [CrossRef]
32. Bastola, A.K.; Hossain, M. The shape–morphing performance of magnetoactive soft materials. *Mater. Des.* **2021**, *211*, 110172. [CrossRef]
33. Saveliev, D.V.; Belyaeva, I.A.; Chashin, D.V.; Fetisov, L.Y.; Romeis, D.; Kettl, W.; Kramarenko, E.Y.; Saphiannikova, M.; Stepanov, G.V.; Shamonin, M. Giant extensional strain of magnetoactive elastomeric cylinders in uniform magnetic fields. *Materials* **2020**, *13*, 3297. [CrossRef] [PubMed]
34. Glavan, G.; Belyaeva, I.A.; Ruwisch, K.; Wollschläger, J.; Shamonin, M. Magnetoelectric Response of Laminated Cantilevers Comprising a Magnetoactive Elastomer and a Piezoelectric Polymer, in Pulsed Uniform Magnetic Fields. *Sensors* **2021**, *21*, 6390. [CrossRef] [PubMed]
35. Saveliev, D.V.; Glavan, G.; Belan, V.O.; Belyaeva, I.A.; Fetisov, L.Y.; Shamonin, M. Resonant Magnetoelectric Effect at Low Frequencies in Layered Polymeric Cantilevers Containing a Magnetoactive Elastomer. *Appl. Sci.* **2022**, *12*, 2102. [CrossRef]
36. Kalita, V.; Dzhezherya, Y.I.; Cherepov, S.; Skirta, Y.B.; Bodnaruk, A.; Levchenko, G. Critical bending and shape memory effect in magnetoactive elastomers. *Smart Mater. Struct.* **2021**, *30*, 025020. [CrossRef]
37. Sheridan, R.; Roche, J.; Lofland, S.E.; von Lockette, P.R. Numerical simulation and experimental validation of the large deformation bending and folding behavior of magneto-active elastomer composites. *Smart Mater. Struct.* **2014**, *23*, 094004. [CrossRef]
38. Vopson, M.; Fetisov, Y.; Caruntu, G.; Srinivasan, G. Measurement techniques of the magneto-electric coupling in multiferroics. *Materials* **2017**, *10*, 963. [CrossRef]
39. Moreno, M.; Gonzalez-Rico, J.; Lopez-Donaire, M.; Arias, A.; Garcia-Gonzalez, D. New experimental insights into magneto-mechanical rate dependences of magnetorheological elastomers. *Compos. Part B Eng.* **2021**, *224*, 109148. [CrossRef]
40. LDT with Crimps Vibration Sensor/Switch. Available online: https://eu.mouser.com/datasheet/2/418/5/NG_DS_LDT_with_Crimps_A1-1130083.pdf (accessed on 24 March 2022).
41. Uddin, M.; Alford, S.; Aziz, S.M. Evaluating Energy Generation Capacity of PVDF Sensors: Effects of Sensor Geometry and Loading. *Materials* **2021**, *14*, 1895. [CrossRef]
42. Gusarov, B.; Gusarova, E.; Viala, B.; Gimeno, L.; Cugat, O. PVDF piezoelectric voltage coefficient in situ measurements as a function of applied stress. *J. Appl. Polym. Sci.* **2016**, *133*. doi: 10.1002/app.43248 [CrossRef]
43. Arshad, A.; Wahid, M.; Rusop, M.; Majid, W.; Subban, R.; Rozana, M. Dielectric and structural properties of poly (vinylidene fluoride)(PVDF) and poly (vinylidene fluoride-trifluoroethylene)(PVDF-TrFE) filled with magnesium oxide nanofillers. *J. Nanomater.* **2019**, *2019*, 5961563. [CrossRef]
44. Belyaeva, I.A.; Kramarenko, E.Y.; Stepanov, G.V.; Sorokin, V.V.; Stadler, D.; Shamonin, M. Transient magnetorheological response of magnetoactive elastomers to step and pyramid excitations. *Soft Matter* **2016**, *12*, 2901–2913. [CrossRef] [PubMed]
45. Belyaeva, I.A.; Kramarenko, E.Y.; Shamonin, M. Magnetodielectric effect in magnetoactive elastomers: Transient response and hysteresis. *Polymer* **2017**, *127*, 119–128. [CrossRef]
46. Zubarev, A.Y.; Chirikov, D.; Borin, D.Y.; Stepanov, G. Hysteresis of the magnetic properties of soft magnetic gels. *Soft Matter* **2016**, *12*, 6473–6480. [CrossRef] [PubMed]
47. James, J.F. *A Student's Guide to Fourier Transforms: With Applications in Physics and Engineering*, 3rd ed.; Cambridge University Press: Cambridge, UK, 2011.
48. Serov, V. *Fourier Series, Fourier Transform and Their Applications to Mathematical Physics*; Springer: Berlin/Heidelberg, Germany, 2017; Volume 197.
49. Lawes, G.; Srinivasan, G. Introduction to magnetoelectric coupling and multiferroic films. *J. Phys. D Appl. Phys.* **2011**, *44*, 243001. [CrossRef]
50. Tan, K.; Wen, X.; Deng, Q.; Shen, S.; Liu, L.; Sharma, P. Soft rubber as a magnetoelectric material—Generating electricity from the remote action of a magnetic field. *Mater. Today* **2021**, *43*, 8–16. [CrossRef]
51. Wang, Y.; Gray, D.; Berry, D.; Gao, J.; Li, M.; Li, J.; Viehland, D. An extremely low equivalent magnetic noise magnetoelectric sensor. *Adv. Mater.* **2011**, *23*, 4111–4114. [CrossRef]
52. Reermann, J.; Durdaut, P.; Salzer, S.; Demming, T.; Piorra, A.; Quandt, E.; Frey, N.; Höft, M.; Schmidt, G. Evaluation of magnetoelectric sensor systems for cardiological applications. *Measurement* **2018**, *116*, 230–238. [CrossRef]

sensors

MDPI

Article

Design and Analysis of a Hybrid Annular Radial Magnetorheological Damper for Semi-Active In-Wheel Motor Suspension

Olivier Munyaneza [1], Pacifique Turabimana [1], Jong-Seok Oh [2], Seung-Bok Choi [3,4,*] and Jung Woo Sohn [5,*]

[1] Department of Aeronautics, Mechanical and Electronic Convergence Engineering, Graduate School, Kumoh National Institute of Technology, Gumi 39177, Gyeongbuk, Korea; olivismnza@gmail.com (O.M.); turabimanapacifique@gmail.com (P.T.)
[2] Department of Future Automotive Engineering, Kongju National University, Cheonan 31080, Chungnam, Korea; jongseok@kongju.ac.kr
[3] Department of Mechanical Engineering, The State University of New York, Korea (SUNY Korea), Incheon 21985, Korea
[4] Department of Mechanical Engineering, Industrial University of Ho Chi Minh City (IUH), Ho Chi Minh City 70000, Vietnam
[5] Department of Mechanical Design Engineering, Kumoh National Institute of Technology, Gumi 39177, Gyeongbuk, Korea
* Correspondence: seungbok.choi@sunykorea.ac.kr (S.-B.C.); jwsohn@kumoh.ac.kr (J.W.S.)

Abstract: In this study, a novel hybrid annular radial magnetorheological damper (HARMRD) is proposed to improve the ride comfort of an electric vehicle (EV) powered by an in-wheel motor (IWM). The model primarily comprises annular-radial ducts in series with permanent magnets. Mathematical models representing the governing motions are formulated, followed by finite element analysis of the HARMRD to investigate the distribution of the magnetic field density and intensity of the magnetorheological (MR) fluid in both the annular and radial ducts. The optimized model generates a damping force of 87.3–445.7 N at the off-state (zero input current) with the excitation velocity ranging between 0 and 0.25 m/s. By contrast, the generated damping force varies from 3386.4 N to 3753.9 N at an input current of 1.5 A with the same velocity range as the off state. The damping forces obtained using the proposed model are 31.4% and 19.2% higher for the off-field and on-field states, respectively, compared with those of the conventional annular radial MR damper. The efficiency of the proposed model is evaluated by adopting two different vehicles: a conventional vehicle powered by an engine and an EV powered by an IWM. The simulation results demonstrate that the proposed HARMRD along with the skyhook controller significantly improves both the ride comfort and road-holding capability for both types of vehicles.

Keywords: magnetorheological fluid; annular radial ducts; electric vehicle; in-wheel motor; permanent magnet

1. Introduction

Over the past two decades, vehicle comfort and road holding have become important considerations for car manufacturers owing to their ability to mitigate the effects caused by road imperfections. Semi-active suspensions have received considerable attention in this context. Semi-active devices use smart materials, such as magnetorheological (MR) fluids, which can shift from Newtonian to Bingham behavior within a microsecond under the influence of an external magnetic field. MR fluids offer a high yield strength and a wide range of operating temperatures [1,2]. MR dampers with MR fluids have advantages such as rapid response, low-voltage required to control the damping force, and structural simplicity [3–5]. Different structures of MR dampers have been studied and applied in control systems. For instance, Solomon et al. [6] designed a mono-tube MR damper by focusing on the performance of indices such as damping force, dynamic range, and valve ratio. They concluded that the design of an MR damper is highly dependent on the performance indices. The mono-tube-type MR damper is more commonly used than other models. However, it requires a larger volume to generate a high damping force, which is inconvenient for many applications [7,8]. Studies have attempted to enhance the damping force using the twin-tube model with low volume; however, this has proved difficulties owing to its complexity, hydraulic imbalance, and limited dynamic range [3,9]. Generating a highly controllable damping force with a compact model remains challenging. Jiang et al. [10] presented a novel design of a multichannel bypass MR damper operating under three working modes. Compared with conventional MR models, their model offers a larger adjustment range and improves the vibration response. However, this model is complex, requires a larger volume, and exhibits a high friction force, which can decay its damping force. By contrast, hybrid models generate a high damping force and address the problem of failure safety for field-off [11–13]. To achieve model simplicity and compactness while improving its damping characteristics, Bai et al. [14] proposed an annular-radial-duct MR damper to improve the efficiency of the MR damper. Their experimental results were compared with those obtained using an MR damper manufactured by LORD Corp. (Cary, NC, USA) (commercially available on the market), and the proposed model provided both a larger damping force and damping range compared with the commercial damper. Several studies have reported that the annular-radial model improves the controllable damping force and has a wide adjustable range [15–17]. However, research on the application of MR dampers based on the annular-radial model to conventional engine-powered or electric vehicles remains limited.

The main technical contribution of this study is the introduction of a novel type of MR damper to improve ride comfort designed based on the annular-radial model and its application to electric vehicles (EVs) powered by in-wheel motors (IWMs). Currently, several manufacturers are attempting to adopt this promising future technology in EVs. The vehicles powered by IWMs offer many benefits compared with internal combustion engine vehicles, such as high torque and vehicle efficiency up to 98% [18]. However, this technology has some serious limitations that can affect its reliability, such as increased sprung mass, which can deteriorate the dynamic performance of the vehicle, braking problems, and deterioration of motor bearing [19–21]. To address the challenges faced by annular-radial models that lack failure safety, we proposed a hybrid annular radial-type MR damper that achieves failure safety using a permanent magnet in addition to the annular and radial ducts that are connected in series. The model was designed and mathematically formulated to represent the governing motion. The performance of the new model was then analytically validated, and the results were compared with those reported in [14] to demonstrate the advantages of the proposed hybrid annular radial magnetorheological damper (HARMRD).

2. Configuration of HARMRD

The structural principle of the proposed HARMRD model is depicted in Figure 1. This figure presents a 2D schematic drawing of the HARMRD. The proposed model comprises a piston rod, permanent magnets (PMs), bobbin core, electromagnetic coil, two magnetic poles around the PM, magnetic core cylinder (flux return), damper cylinder, and other non-magnetic components such as washers and pins. Pins are used to connect the circular disc or poles to the bobbin core and ensure that a constant radial gap is formed. The annular gap is located between the outer core and the PM circumference and the inner circumference of the flux return. A permanent ring magnet is installed between the two poles at the top and bottom of the valve model. The material properties and dimensions are listed in Table 1.

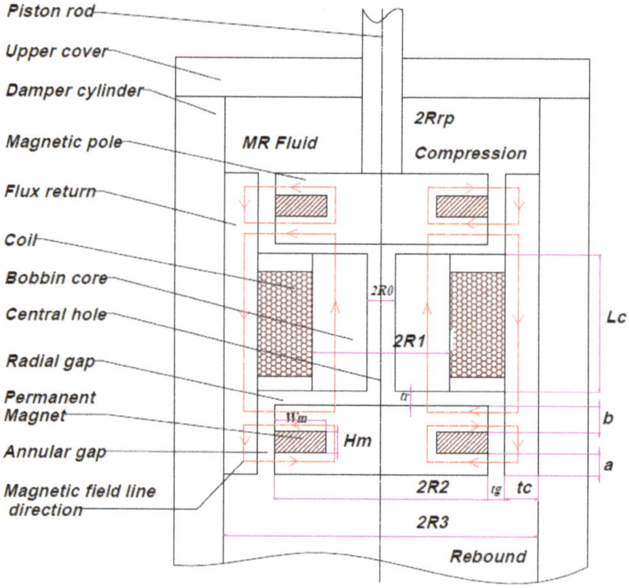

Figure 1. Schematic diagram of the proposed HARMRD.

Table 1. Structural dimensions of the proposed HARMRD.

Parameter	Symbol	Value
Piston rod radius	R_{rp}	5 [mm]
Piston radius	R_3	20 [mm]
Annular duct	t_g	1 [mm]
Radial duct	t_r	1 [mm]
Flux return	t_c	5 [mm]
Central hole radius	R_o	3 [mm]
Magnetic pole radius	R_2	14 [mm]
Height of magnet	H_m	3 [mm]
Height of core	L_c	26 [mm]
Core radius	R_1	12 [mm]
Minor pole	a	2 [mm]
Big pole	b	5 [mm]
Piston valve length	L	48 [mm]
Coil turns	N	304 turns

The initial magnetic field is supplied by the PMs, wherein the field generated from the bobbin core passes through the annular gap (lower or upper), continues to the magnetic core cylinder (flux return), and returns to the core to complete a magnetic circuit cycle. The field generated by the PM has a greater influence on the annular channel than on the radial channel. To enhance the magnetic field generated by the magnets, a positive/negative current is applied to the coil to generate the primary magnetic field. Unlike the field generated by the magnets, that from the coil can be gradually increased by varying the current input. Depending on the position of the PM poles, the magnetic field of the coil can cancel or strengthen the field induced by PM. The PM field from the magnets creates yield stress in the MR fluid around the annular channel, which prevents the model from acting as a passive damper in the case of electric failure. As is evident from Figure 1, as the piston is compressed, the MR fluid flows from the upper annular gap into the upper radial gap. The fluid from the upper radial gap passes through the central hole of the core to the lower radial and annular gaps. This process is repeated during the rebound process. In addition, according to [22] the MR fluid fills the space between the channels because of the pressure drop, which is controlled by the magnetic field of the PM and electromagnetic coil.

3. Modeling of Hybrid Annular Radial MR Damper

The proposed model operates based on the principles of valve mode [8,23,24]. As shown in Figure 1, the model comprises two annular gaps, two radial gaps, and one central hole to facilitate the flow of the MR fluid from the low-pressure to the high-pressure side, and vice versa. The total pressure drop inside the valve is expressed as follows:

$$\Delta_p = 2(\Delta_{pa} + \Delta_{pr}) + \Delta_{p0}, \tag{1}$$

where Δ_{pa}, Δ_{pr}, Δ_{p0} are the pressure drops in the annular gap, radial gap, and central hole, respectively. The pressure in the annular flow gap is expressed as [17]

$$\Delta_{pa} = \Delta p_{\eta a} + \Delta p_{\tau a}, \tag{2}$$

where $\Delta p_{\eta a}$ and $\Delta p_{\tau a}$ are the viscous pressure drop and field-dependent pressure drop, respectively, and are expressed as

$$\Delta p_{\eta a} = \frac{6\eta Q_a t_f}{\pi t_g{}^3 (R_2 + 0.5 t_g)}, \tag{3}$$

$$\Delta p_{\tau a} = \frac{C_a t_e \tau_{\gamma a}}{t_g}. \tag{4}$$

The expressions for Q_a, t_f, and t_e are deduced from Figure 1.

$$Q_a = V_p A_p, \tag{5}$$

$$t_f = a + b + H_m, \tag{6}$$

$$t_e = a + b, \tag{7}$$

where η is the viscosity of the MR fluid; Q_a is the volumetric flow rate of the MR fluid through the annular gap; V_p is the piston velocity; A_p is the effective area of the piston valve; t_f is the length of the annular gap; t_e is the effective length of the annular gap; a and b are the effective pole lengths below and above the magnet, respectively; and H_m is the height of the magnet. The pressure drop in the radial gap is determined as

$$\Delta_{pr} = \Delta p_{\eta r} + \Delta p_{\tau r}, \tag{8}$$

where $\Delta p_{\eta r}$ and $\Delta p_{\tau r}$ are the pressure drops due to the MR fluid flowing in the radial gap and the yield stress, respectively.

$$\Delta p_{\eta r} = \frac{6\eta Q_r}{\pi t_g{}^3} \ln\left(\frac{R_2}{R_0}\right) \tag{9}$$

$$\Delta p_{\tau r} = \frac{C_r (R_2 - R_0) \tau_{\gamma r}}{t_g} \tag{10}$$

The pressure drop, Δ_{p0}, in the central hole is defined as

$$\Delta_{p0} = \frac{8\eta Q L_c}{\pi R_0{}^4}, \tag{11}$$

where L_c denotes the length of the central hole. By substituting Equations (3), (4), (9), (10) and (11) into Equation (1), the pressure drop becomes

$$\Delta_p = 2\left[\frac{6\eta Q_a t_f}{\pi t_g{}^3 (R_2 + 0.5t_g)} + \frac{6\eta Q_r}{\pi t_g{}^3} \ln\left(\frac{R_2}{R_0}\right) + \frac{C_a t_e \tau_{\gamma a}}{t_g} + \frac{C_r (R_2 - R_0)\tau_{\gamma r}}{t_g}\right] + \frac{8\eta Q L_c}{\pi R_0{}^4}. \quad (12)$$

The viscous damper coefficient can be calculated from Equation (12) as follow

$$C_{vic} = \left(2\left[\frac{6\eta t_f}{\pi t_g{}^3 (R_2 + 0.5t_g)} + \frac{6\eta}{\pi t_g{}^3} \ln\left(\frac{R_2}{R_0}\right)\right] + \frac{8\eta Q L_c}{\pi R_0{}^4}\right) \times A_p{}^2 \quad (13)$$

The total damping force can be calculated using Equation (12) as

$$F_D = 2A_p\left\{\frac{6\eta Q_a t_f}{\pi t_g{}^3 (R_2 + 0.5t_g)} + \frac{6\eta Q_r}{\pi t_g{}^3} \ln\left(\frac{R_2}{R_0}\right) + \frac{C_a t_e \tau_{\gamma a}}{t_g} + \frac{C_r (R_2 - R_0)\tau_{\gamma r}}{t_g}\right\} + \frac{8\eta Q L_c}{\pi R_0{}^4} A_p. \quad (14)$$

To evaluate the performance of the HARMRD, the dynamic range (K) was calculated as the ratio of the field-dependent force to the viscous damping force.

$$K = \frac{F_\tau}{F_\eta} + 1 \quad (15)$$

4. Finite Element Analysis of HARMRD

Mathematically, calculating the damping force and dynamic range of the HARMRD is challenging owing to the nonlinearity of the materials that make up the model. For convenience, finite element analysis (FEA) was performed using the ANSYS/Maxwell software. Figure 2 shows the axisymmetric model of the HARMR valve used in the simulation. The results of the two cases considered in the simulation are presented in Figure 3a,b. The distribution of the magnetic flux lines and magnetic flux density for the HARMR valve in the absence of excitation current to the coil is shown in Figure 3a. Evidently, the magnetic flux lines and density are concentrated around the PMs, and in this case, the proposed model continuously experiences a magnetic field around the annular channel. In addition, the magnetic fields generated in the annular gap and radial gap are 0.06 T and 0.01 T, respectively. Figure 3b shows the generated magnetic field lines and flux density when both the magnets and the electrical coil are operating. In this case, the initial magnetic field from the magnets is strengthened by the field generated by the coil after excitation. A current variation of 0–1.5 A with an interval of 0.25 A is supplied to the coil.

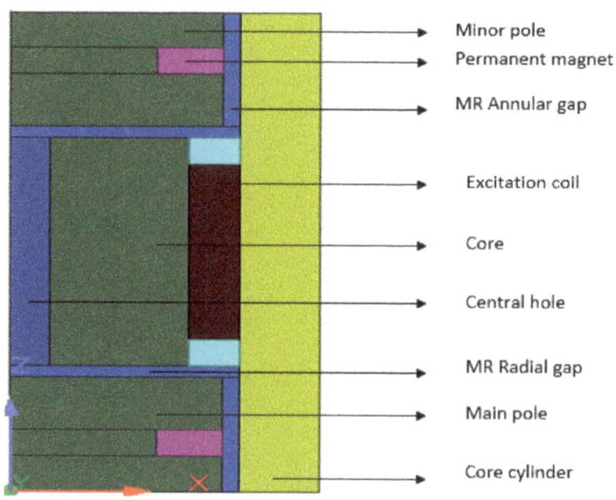

Figure 2. Axisymmetric model of the HARMRD.

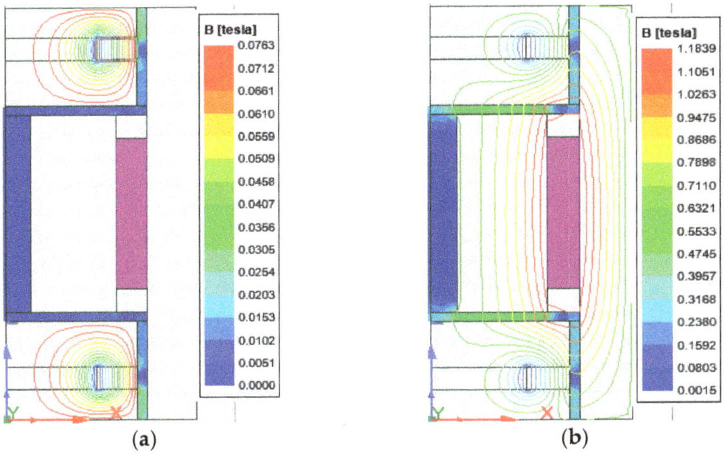

Figure 3. Distribution of magnetic field lines and flux density when: (**a**) only the PM is operating; (**b**) both the PM and coil are operating.

Unlike the magnetic field from the magnets, that from the coil, passes through both radial and annular directions and extends to the core cylinder to form a complete cycle, as shown in Figure 3a. After supplying a current of 1.5 A to the coil, the magnetic field densities generated for both radial and annular gaps were 0.56 T and 0.44 T, respectively. A typical magnetic field problem is described by defining its geometry, material properties, magnetic input, and boundary conditions. Based on the properties of MR fluid (MRF-132DG) [25] and FEM analysis results for electromagnetic characteristics shown in Figure 3, relationship between yield stress and magnetic field intensity can be obtained as presented in Figure 4. Figure 4a,b depict the relationship between magnetic flux density (B) and yield stress (τ_γ) in the annular and radial channels, respectively. The data points in Figure 4 were fitted into a polynomial curve using the least square method in MATLAB simulation software. Evidently, the radial channel generates a higher yield stress and magnetic field than the annular duct. The yield stresses in the two channels were deduced from Figure 4a,b as follows.

$$\tau_{ya} = \left(-1.2324 \times B_a^3 + 1.4074 \times B_a^2 + 0.1604 \times Ba + 0.0005\right) \times 10^5, \quad (16)$$

$$\tau_{yr} = \left(-1.1602 \times B_r^3 + 1.3716 \times B_r^2 + 0.1619 \times Br + 0.0010\right) \times 10^5 \quad (17)$$

where τ_{ya} and τ_{yr} are the yield stresses in the annular and radial channels, respectively. To evaluate the performance of the proposed model, the damping force was calculated based on the equations in Section 3. Figure 5 shows the minimum damping force when no current is supplied to the coil (field-off condition) and the maximum damping force generated when the coil is excited (field-on condition). When the field is off (I = 0 A), a damping force of 70.507–365.98 N is generated for piston velocities of 0 and 0.25 m/s, respectively. By contrast, when the field is on (I = 1.5 A), the damping force increases from 2696.3 N to 2991.8 N for the same piston velocities. In addition, the highest dynamic range of 3.19 and 47.63 was obtained at a piston velocity of 0.05 m/s for field-off and field-on states, respectively. However, as the piston velocity increases, the dynamic range decreases significantly. For a constant speed of 0.25 m/s, dynamic ranges of 2.24 and 11.13 were obtained for the field-off and field-on states, respectively.

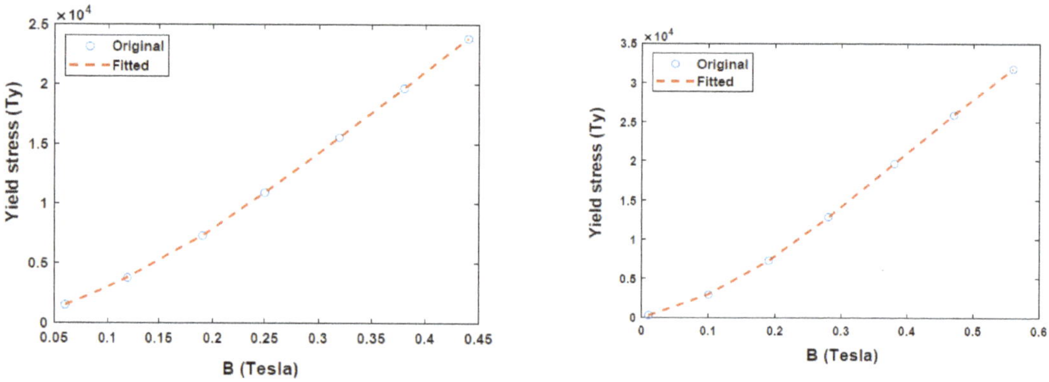

(a) (b)

Figure 4. Polynomial curve fitting for variation of yield stress with magnetic flux density of MRF-132DG: (**a**) Annular channel; (**b**) Radial channel.

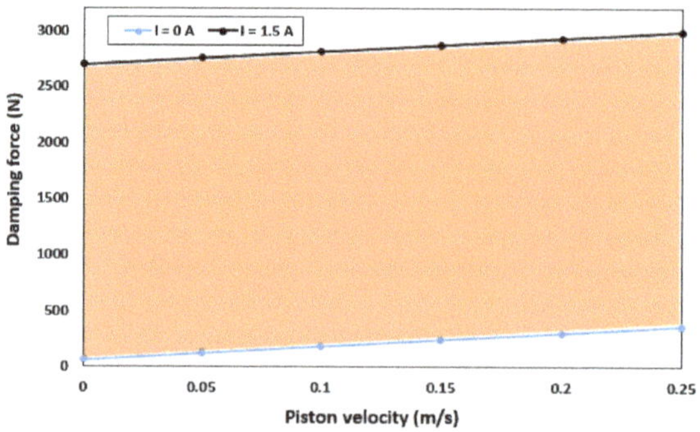

Figure 5. Variation of damping force of the HARMRD with piston velocity for initial design values.

5. Optimization of the HARMR Valve

After obtaining the performance indices from the initial design parameters, a multi-objective genetic algorithm (MGA) is used to optimize the HARMRD. MGA is based on concepts from natural selection and genetics. It is commonly used to obtain high-quality solutions to search and optimization problems. The optimization process used in this study is depicted in Figure 6. During optimization, seven design variables (DVs) were selected based on their effects on the performance indices, including the damping force and dynamic range. The selected DVs are the piston radius (R_3), radius of circular disc (R_2), radius of the core from the central hole (R_1), coil width (W), piston radius (R_{pr}), annular duct (t_g), and main pole (b). The upper and lower bounds of the design variables are listed in Table 2. In this study, the damping force and dynamic range were selected as the fitness functions. Ride comfort and suspension travel are two important factors in vehicle suspension design. To maintain the suspension travel short, a larger damping force is required, whereas to achieve better ride comfort, a low damping force and larger dynamic range are required. The fitness function is expressed as follows [26]:

$$OBJ = \alpha \frac{F_D{}^*}{F_D} + \beta \frac{D_r{}^*}{D_r}, \tag{18}$$

where $F_D{}^*$, $D_r{}^*$ are the reference damping force and reference dynamic range, respectively, and α, and β are the weighting factors. In addition, the summation of both weighting factors should be equal to 1. To balance both damping force and dynamic range, the weighting factors were chosen as 0.5 in this work.

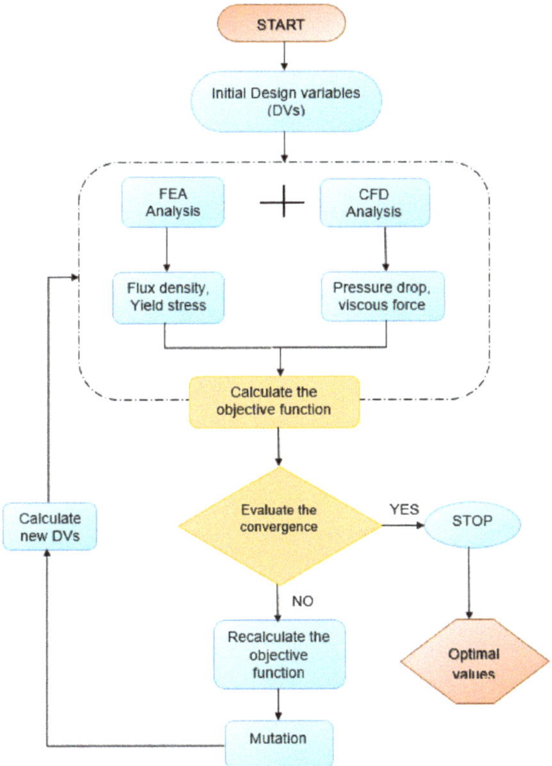

Figure 6. Flowchart of genetic algorithm used for optimization.

Table 2. Lower and upper bounds of design variables.

Parameter	Lower Bound	Upper Bound
Piston rod radius (R_{rp})	5 mm	7 mm
Piston radius (R_3)	20 mm	23 mm
Annular duct (t_g)	0.8 mm	1 mm
Magnetic pole radius (R_2)	14 mm	16 mm
Core radius (R_1)	12 mm	14 mm
Big pole (b)	5 mm	5 mm

Figure 7 shows the Pareto front generated between the fitness functions. Evidently, the dynamic range decreases with increasing damping force. The optimal design parameters are listed in Table 3. Figure 8 shows the dynamic ranges of the optimal HARMRD under various velocity excitations. As it is apparent from this figure, the dynamic range decreases as the piston velocity increases. The wide range of controllability is confirmed by the highest dynamic range experienced at the excitation current of I = 1.5 A. Moreover, the dynamic range obtained during the off-field state indicates that the proposed model addresses the fail-safe problem. The relationship between the dynamic range and piston velocity for the optimal model was determined using MATLAB/Simulink under current excitations ranging from 0 to 1.5 A. During the simulation, a sinusoidal displacement with an amplitude (A) of 0.075 m and frequency of 2 Hz was used.

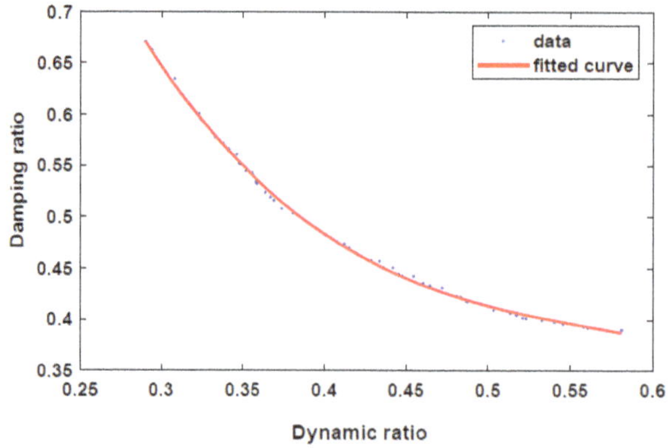

Figure 7. Pareto front between objective functions.

Table 3. Optimal geometric parameters for the HARMRD.

SN	Parameters	Dimensions
1	Annular duct (t_g)	0.97 mm
2	Piston radius (R_3)	20.7 mm
3	Piston road radius (R_{rp})	6 mm
4	Magnetic pole radius (R_2)	16 mm
5	Big pole (b)	6 mm
6	Core radius (R_1)	13 mm
7	Damping force	3753.9 N
8	Total length (L)	48 mm

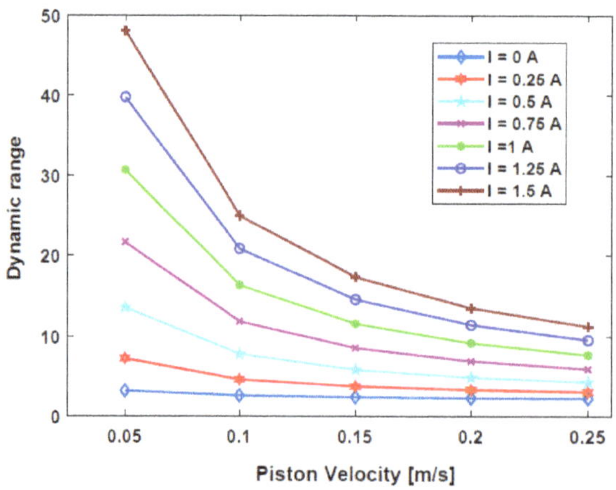

Figure 8. Dynamic ranges of the HARMRD under various excitation velocities.

The force-piston stroke and force-piston velocity are shown in Figure 9a,b, respectively. Figure 9a depicts the relationship between the damping force and piston stroke (FD-S) for various current excitations. Evidently, FD-S curves gradually rise with the current supplied. Moreover, the damping forces for the compression and extension strokes are symmetric. The small deformation around the top-left corner and bottom-right corner is due to insufficient gas pressure in the accumulator. As shown in Figure 9a, a minimum damping force of 445.7 N was generated in the off-field state (I = 0 A). By contrast, a maximum damping force of 3753.9 N was obtained when the maximum current (I = 1.5 A) was applied. This indicates that the optimal HARMR model is capable of generating a wide range of damping force (445.7–3753.9 N), which is necessary for achieving better driving comfort. Figure 9b shows the variation of damping force with piston velocity under various current inputs. It is evident that the proposed HARMR model can mimic the behavior of an MR damper. Generally, it is difficult to replicate the MR property in an off-field state. In addition, the damping force increased as the excitation and piston velocities increased. The proposed model generates a damping force range of 445.7–3753.9 N under the piston velocity of 0.25 m/s. Figure 10 illustrates the damping force range under various velocity inputs. The wide damping force range enables the use of short stroke which is essential for improving the ride comfort and vehicle handling. When the field is off (I = 0 A), a damping force of 87.25–445.7 N is generated for piston velocities of 0 and 0.25 m/s, respectively. By contrast, when the field is on (I = 1.5 A), the damping force increases from 3386.4 N to 3753.9 N for the same piston velocities. Compared with the results in Figure 5, it can be clearly seen that the damping force range increased after optimization. After optimization, the damping force is 3753.9 N, which is 20% higher than that of 2991.8 N before optimization, and the suspension travel can be restricted with sufficient damping force. The dynamic range under the condition of piston velocity of 0.05 m/s is evaluated by 3.19 and 47.63 at field-off and field-on, respectively, before optimization. However, it increased to values of 3.19 and 48.08 under the same conditions after optimization. These results mean that the damping force can be greatly improved without reducing the dynamic range through optimization. A typical passive damper has a damping force of about 130 to 400 N at a piston velocity of 0.025 to 0.2 m/s [27]. In the case of the model proposed in this paper, it was confirmed to have a damping force of 147 to 437 N for the same piston speed in the field-off state. From these results, it is confirmed that the proposed model has adequate performance for fail-safe issues. When the damping force per unit volume is calculated based on the maximum damping force, it has values of 0.0535 N/mm^3, 0.0388 N/mm^3, and 0.0597 N/mm^3 for Bai's model, RD-1005-3 model of LORD Corp. (Cary, NC, USA), and the model proposed in this paper, respectively. The maximum damping force under the same current input of 1.5 A and the piston speed of 0.19 m/s is considered. The volume is calculated using the piston radius, piston rod radius and stroke. From this, it is confirmed that improved damping force can be obtained by using the proposed model compared with both model in previous research work and commercial model.

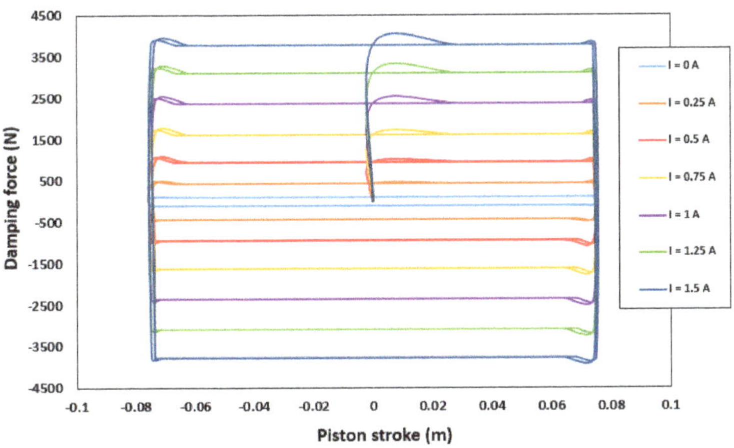

(a)

Figure 9. *Cont.*

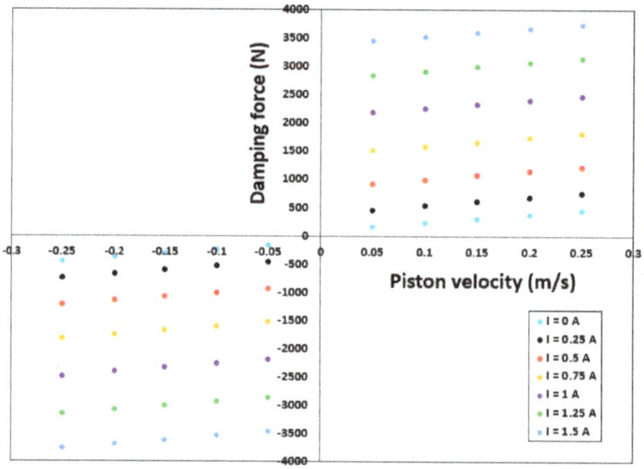

(**b**)

Figure 9. Damping force characteristics: (**a**) damping force versus piston stroke; (**b**) damping force versus piston velocity.

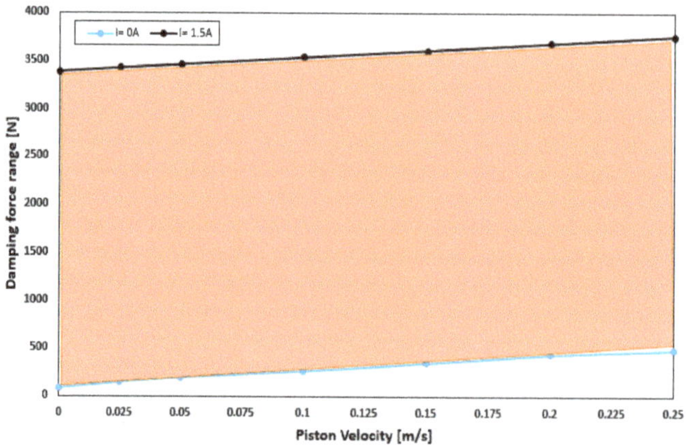

Figure 10. Variation of damping force of the HARMRD with piston velocity with optimized parameter.

6. Performance Evaluation of the HARMRD

EVs are typically powered by either a centralized motor or an IWM, depending on the vehicle design. Motors arrangement is the primary difference between these two EV designs. In the IWM design, a motor is installed on each wheel to directly supply torque to the wheels. By contrast, in a conventional layout, the motor is centralized, and the torque is transferred to the wheels via the drive shaft. The IWM configuration offers advantages over conventional models, such as improved efficiency, easy generation of forward and reverse torques, and improved performance of systems such as the anti-lock brake system and traction control system [28]. However, driving comfort and handling stability are primary considerations in the design of the suspension system. In previous studies, the IWMs integrated into the wheels increased the unsprung mass, which resulted in the deterioration of the tire dynamic road and body vibration acceleration. Several methods have been proposed to tackle the weight issue, including the in-wheel switched reluctant motor (IWSRM) and permanent magnetic synchronous motor (PMSM) [29]. However, both the IWSRM and PMSM models have been found to experience radial force imbalance due to rotor eccentricity, which affects the ride comfort. However, an IWSRM is affordable and simple. In this study, an SRM coupled inside a wheel was modeled, and a dynamic vibration absorbing structure was connected to the unsprung mass to suppress the vibration of the unsprung mass. Figure 11 shows the proposed EV suspension powered by the IWM. The proposed IWM comprises 13 components, which are grouped into three parts (wheel and tire, SRM, and suspension dampers). To suppress motor vibration, a small passive damper was installed inside the stator cover, which was connected to the hub shaft. In addition to the small damper in the DVA structure, HARMRD was installed between the lower and upper arms of the suspension to control road-induced vibration.

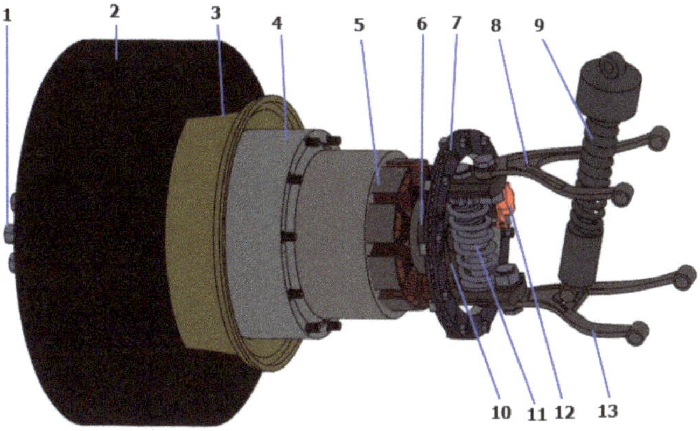

Figure 11. Physical structure of the HARMRD for electrical IWM suspension. 1. Nuts, 2. Tire, 3. Wheel, 4. Motor rotor, 5. Motor stator, 6. Hub shaft, 7. Brake disk, 8. Upper control arm, 9. HARMRD, 10. Hub, 11. DAV damper, 12. Brake caliper, 13. Lower control arm.

A co-simulation software (CarSim, V 9.0, Mechanical Simululation Corporation, Ann Arbor, MI, USA) was used to evaluate the effectiveness of the proposed HARMR model for an EV powered by an IWM. The CarSim interface is divided into three main parts: vehicle parameters and drive cycle, MATLAB/Simulink interface, and simulated results. CarSim is widely used to simulate the vehicle performance in terms of ride comfort, handling stability, and power. A D-class sedan with independent suspensions was used in the simulation. Figure 12 illustrates the simulation in MATLAB/Simulink in conjunction with CarSim. The damping forces (F_D) supplied on each tire were considered as an external input to CarSim, and they were controlled by the Skyhook controller. In CarSim, the full vehicle parameters are specified based on a real vehicle. The considered parameters include the sprung mass, unsprung mass, wheelbase, vehicle width, and vehicle height. The parameters for the full vehicle are listed in Table 4. Five outputs were obtained from the sensor, namely force rate, body velocity, body deflection, tire deformation, and vertical acceleration. During the simulation, bump and random input roads were adopted from the CarSim software. The moving vehicle velocity for the bump and random road was fixed to 60 km/h. Here, a normal vehicle powered by an engine and an EV powered by an IWM were compared under the same road conditions. For simplicity, only the front-left suspension responses were considered for the full vehicle. For the riding test, a bump road input with amplitudes of 35 mm and 400 mm was used. A vehicle velocity of 60 km/h was used.

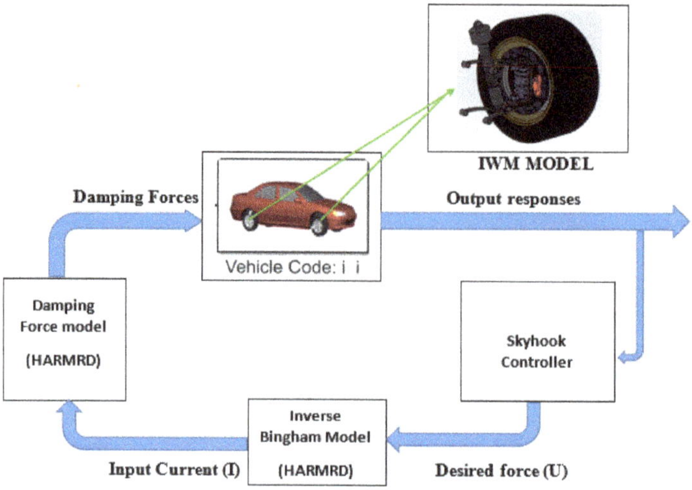

Figure 12. Block diagram of co-simulation of EV suspension.

Table 4. Passenger vehicle parameters [30].

Parameter	Value	Parameter	Value
Sprung mass	1370 kg	Unprung mass (normal vehicle)	80 kg
Longitudinal inertia moment	356.3 kg·m^2	Unsprung mass (IWM)	140 kg·m^2
Front suspension stiffness	153 kN/m	rear suspension stiffness	82 kN/m
Tire stiffness constant	230 kN/m	Damper coefficient	1534 N·s/m
Distance from C.G to the front suspension	1.11 m	Distance from C.G to the rear suspension	1.166 m
Distance from C.G to left suspension	0.335 m	Distance from C.G to right suspension	0.335 m

The vertical acceleration, suspension deflection, and tire deflection responses of a normal vehicle are shown in Figure 13. Two cases were investigated and compared during the simulation. First, the vehicle in CarSim was simulated without an external force, that is, passive suspension. Second, the external force from the HARMRD, controlled by the skyhook controller, was supplied to the vehicle. It is evident that the vehicle controlled by the proposed damper suppressed undesirable vibrations compared with the passive suspension. In addition, the proposed model exhibited a fast settling time after the vehicle hit the bump, which confirms the improvement in ride comfort achieved by the proposed model. To statistically verify the performance, the root mean square (RMS) for both the passive and controlled responses was calculated. The RMS values for passive responses decreased by 66.9%, 69.18%, and 18.12% for vertical acceleration, body deflection, and tire deflection, respectively. Figure 14 shows the time responses for a vehicle powered by IWM. Evidently, the proposed HARMD attenuates the vibrations for acceleration, suspension deflection, and tire deflection. The difference in responses between a vehicle powered by an engine and another powered by IWMs was insignificant. However, the RMS values for the IWM slightly increased compared with the RMS values for a normal vehicle. The passive RMS values for vertical acceleration, suspension deflection, and tire deflection reduced by 69.76%, 73.7%, and 24.03%, respectively.

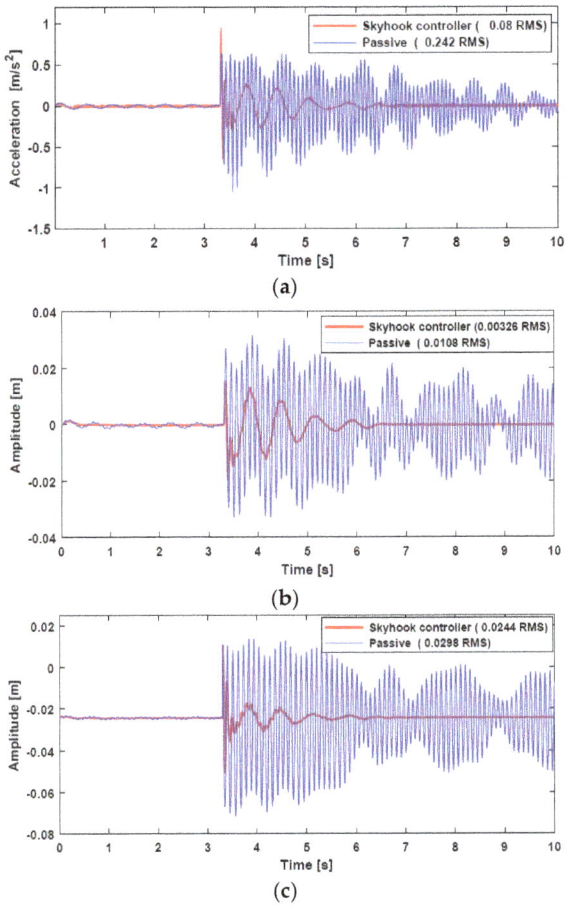

Figure 13. Time responses for a normal vehicle under bump road excitation: (**a**) vertical acceleration from the center of gravity; (**b**) suspension deflection; (**c**) tire deflection.

According to ISO 2631-1, the ride comfort can be mathematically identified and quantified based on the relative discomfort associated with vehicle motion. The frequency weighted RMS Acceleration calculates how intensive mechanical vibrations affect the human body in vehicles. To quantify the relative discomfort, the frequency weighted RMS acceleration for vertical accelerations was calculated as follows.

$$A_W = \left[\frac{1}{T}\int_0^T A(w)^2\right]^{\frac{1}{2}} \tag{19}$$

where $A(w)$ is the frequency weighted acceleration and T is the exposure time. According to ISO 2631-1, when the frequency weighted RMS acceleration is lower than 0.32, the Discomfort level is "Not uncomfortable", and when it is greater than 0.32 and less than 0.5, the Discomfort level is "A little uncomfortable". In the case of a normal vehicle, when the passive damper is used, the frequency weighted RMS acceleration value is 0.42, and when the proposed model is applied, it decreases to a value of 0.18, which satisfies the lowest Discomfort level. Even in the case of the IWM vehicle, when the passive damper is used, the frequency weighted RMS acceleration value is 0.46, and when the proposed model is applied, it can be confirmed that the lowest discomfort level is satisfied by decreasing to a value of 0.16.

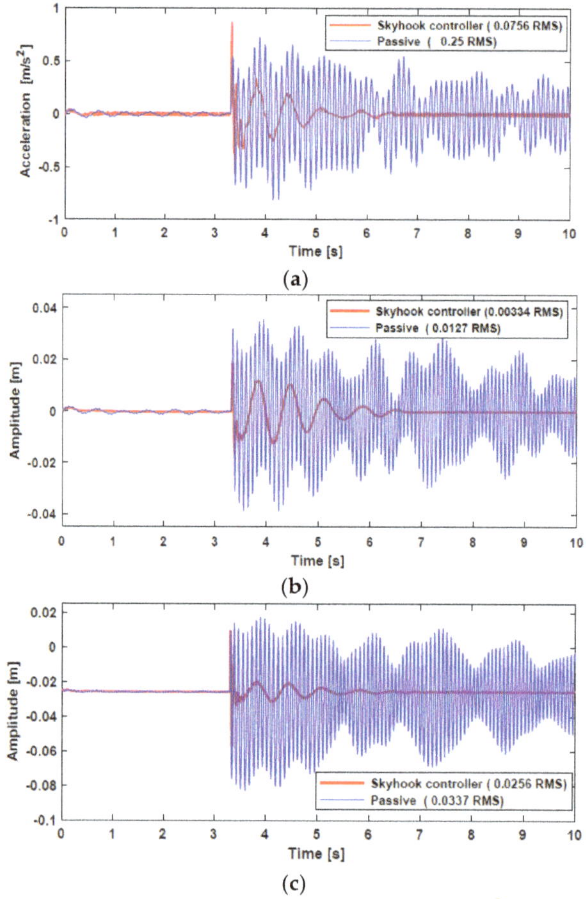

Figure 14. Time responses for IWM under bump road excitation: (**a**) vertical acceleration from the center of gravity; (**b**) suspension deflection; (**c**) tire deflection.

In addition, the proposed model was subjected to a random road to test its capability of attenuating random vibrations. Different grades of a random road are used depending on the purpose and type of vehicles. In this study, a grade B random road was selected as the testing road in CarSim, which can be described as follow [31]

$$\dot{x}(t) = -2\pi f_n v x(t) + 2\pi f_0 \sqrt{s_h(f_0) v} w(t), \tag{20}$$

where $\dot{x}(t)$ represents the road profile, $f_n = 0.01$ is the cut-off frequency, $f_0 = 0.1$ is the spatial frequency, v is the assigned vehicle velocity selected as 60, $s_h(f_0)$ is the road roughness coefficient, and $w(t)$ is the random white noise. During the simulation, a skyhook controller was used to control the input damping force of each tire. The time responses from the random road are not clear to interpret because of the noise; therefore, the frequency responses obtained from the time domain were used to compare controlled and passive responses. Figure 15 shows the frequency responses of a vehicle powered by the IWM. The simulation results indicated that the controlled responses outperformed the passive responses in terms of dissipating energy generated from road unevenness. In addition, all the first peak resonances were in the vicinity of frequencies (1–2 Hz). According to ISO 2631 [32], maximum sensitive vibration to the human body ranges from (4–8 Hz). The RMS values of the body vertical acceleration, suspension deflection, and tire deflection confirm that the proposed damper model has a significant impact on lowering the RMS values compared with the passive values. The power supplied to the semi-active HARMR damper for both bumpy and random road input is illustrated in Figure 16a,b, respectively. Evidently, the proposed model is efficiently able to suppress the vibration for both bumpy and random roads with the generated current of about 1.7 A and 1.2 A, respectively.

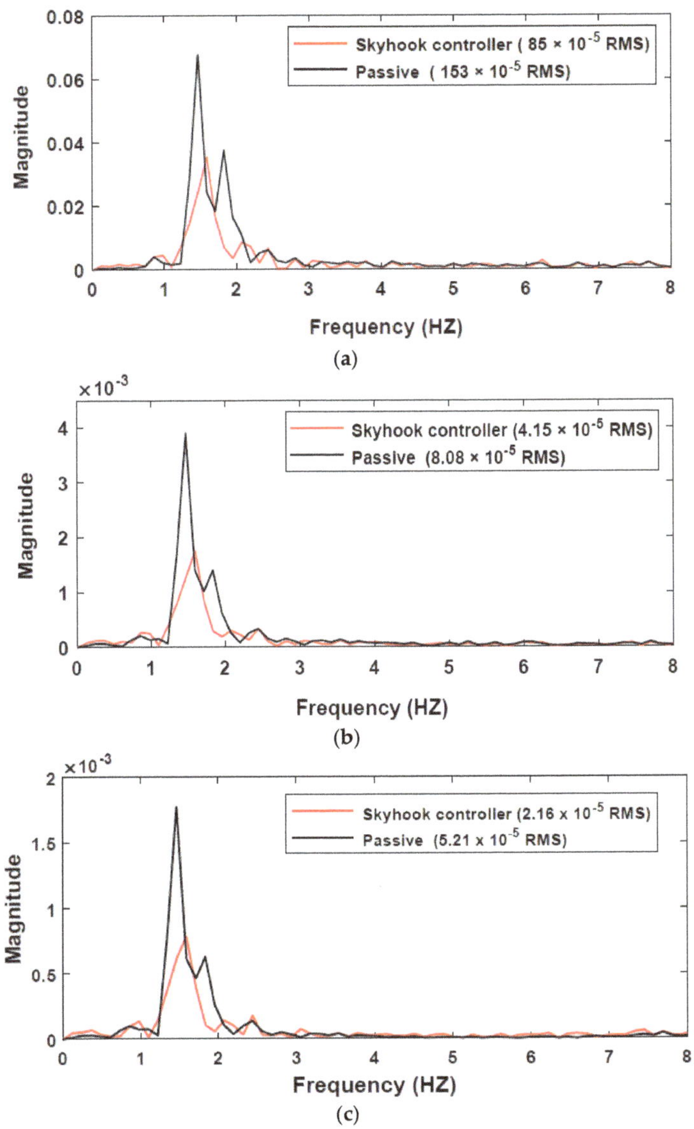

Figure 15. Frequency responses for IWM under random road excitation: (**a**) vertical acceleration from the center of gravity; (**b**) suspension deflection; (**c**) tire deflection.

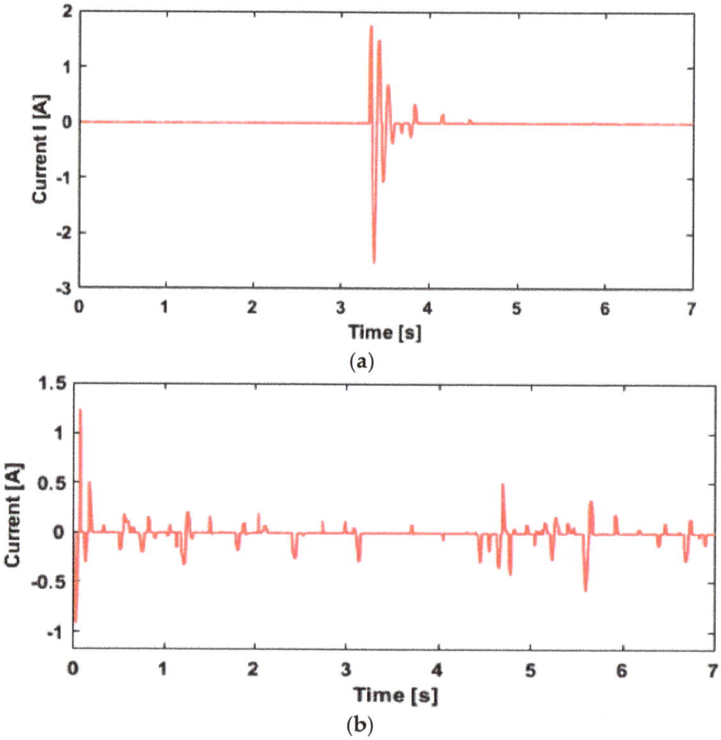

Figure 16. Current input: (**a**) bump road input; (**b**) random road input.

7. Conclusions

In this study, HARMRD was designed and modeled to improve the ride comfort of EVs. The simulation results indicated that HARMRD generates a higher damping force than the conventional MR damper and provides a wider range of vibration controllability. To demonstrate these benefits in terms of the vehicle suspension system, the overall performance of the HARMRD on ride comfort was evaluated via co-simulation using virtual full-vehicle dynamics in CarSim and MATLAB/Simulink. In the simulation, a simple but highly effective skyhook controller was used to control the current supplied to the HARMRD. The simulation results for random road inputs confirmed that the controlled proposed model improved both ride comfort and road holding by significantly suppressing unwanted vibrations compared with passive suspension. It is finally remarked that an experimental validation of the proposed HARMRD needs to be conducted in the near future through a hardware-in-the-loop simulation and real road test.

Author Contributions: Conceptualization, S.-B.C. and J.W.S.; formal analysis, O.M.; investigation, O.M. and P.T.; validation, J.-S.O.; writing—original draft preparation, O.M. and P.T.; writing—review and editing, J.-S.O., S.-B.C. and J.W.S.; supervision, S.-B.C. and J.W.S. All authors have read and agreed to the published version of the manuscript.

Funding: This research was supported by Basic Science Research Program through the National Research Foundation of Korea (NRF) funded by the Ministry of Education (2020R1I1A3074547). This research was supported by the MSIT (Ministry of Science and ICT), Korea, under the Grand Information Technology Research Center support program (IITP-2022-2020-0-01612) supervised by the IITP (Institute for Information and communications Technology Planning and Evaluation).

Data Availability Statement: Not applicable.

Acknowledgments: The computation is implemented under NVIDIA DGX STATION (Future Automotive Intelligent Electronics Core Technology Center, Korea).

Conflicts of Interest: The authors declare no conflict of interest.

eferences

Olabi, A.G.; Grunwald, A. Design and Application of Magneto-Rheological Fluid 'MRF'. *Mater. Des.* **2007**, *28*, 2658–2664. [CrossRef]

Jitesh, S. Magneto Rheological Fluid (MRF). *Int. J. Res. Mech. Eng.* **2014**, *2*, 27–32.

Gołdasz, J. Theoretical study of a twin-tube magnetorheological damper concept. *J. Theor. Appl. Mech.* **2015**, *53*, 885–894. [CrossRef]

Desai, R.M.; Jamadar, M.E.H.; Kumar, H.; Joladarashi, S.; Sekaran, S.C.R. Design and experimental characterization of a twin-tube MR damper for a passenger van. *J. Braz. Soc. Mech. Sci. Eng.* **2019**, *41*, 332. [CrossRef]

Xu, Y.; Liao, G.; Liu, T. Magneto-Sensitive Smart Materials and Magnetorheological Mechanism. In *Nanofluid Flow in Porous Media*; Sheikholeslami, M., Kandelousi, S., Ameen, M., Shaheer, A., Shin, H.-S., Eds.; IntechOpen: London, UK, 2020. [CrossRef]

Seid, S.; Chandramohan, S.; Sujatha, S. Design Evaluation of a Mono-tube Magnetorheological (MR) Damper Valve. In *Innovative Design, Analysis and Development Practices in Aerospace and Automotive Engineering (I-DAD 2018)*; Chandrasekhar, U., Yang, L.-J., Gowthaman, S., Eds.; Springer: Singapore, 2019; pp. 145–151. [CrossRef]

Purandare, S.; Zambare, H.; Razban, A. Analysis of magnetic flux in magneto-rheological damper. *J. Phys. Commun.* **2019**, *3*, 075012. [CrossRef]

Grunwald, A.; Olabi, A.G. Design of magneto-rheological (MR) valve. *Sens. Actuators Phys.* **2008**, *148*, 211–223. [CrossRef]

Gurubasavaraju, T.M.; Kumar, H.; Mahalingam, A. An approach for characterizing twin-tube shear-mode magnetorheological damper through coupled FE and CFD analysis. *J. Braz. Soc. Mech. Sci. Eng.* **2018**, *40*, 139. [CrossRef]

. Jiang, R.; Rui, X.; Zhu, W.; Yang, F.; Zhang, Y.; Gu, J. Design of multi-channel bypass magnetorheological damper with three working modes. *Int. J. Mech. Mater. Des.* **2022**, *18*, 155–167. [CrossRef]

. Olivier, M.; Sohn, J.W. Design and geometric parameter optimization of hybrid magnetorheological fluid damper. *J. Mech. Sci. Technol.* **2020**, *34*, 2953–2960. [CrossRef]

. Jackel, M.; Kloepfer, J.; Matthias, M. *Hybrid Magnetorheological Damper*; LAP Lambert Academic Publishing: Saarbrücken, Germany, 2016.

. Boese, H.; Ehrlich, J. Performance of Magnetorheological Fluids in a Novel Damper with Excellent Fail-safe Behavior. *J. Intell. Mater. Syst. Struct.* **2010**, *21*, 1537–1542. [CrossRef]

. Bai, X.-X.; Wang, D.-H.; Fu, H. Principle, modeling, and testing of an annular-radial-duct magnetorheological damper. *Sens. Actuators Phys.* **2013**, *201*, 302–309. [CrossRef]

. Zhu, S.; Tang, L.; Liu, J.; Tang, X.; Liu, X. A Novel Design of Magnetorheological Damper with Annular Radial Channel. *Shock Vib.* **2016**, *2016*, 8086504. [CrossRef]

. Ai, H.X.; Wang, D.H.; Liao, W.H. Design and Modeling of a Magnetorheological Valve with Both Annular and Radial Flow Paths. *J. Intell. Mater. Syst. Struct.* **2006**, *17*, 327–334. [CrossRef]

. Nguyen, Q.H.; Choi, S.B.; Lee, Y.S.; Han, M.S. Optimal design of high damping force engine mount featuring MR valve structure with both annular and radial flow paths. *Smart Mater. Struct.* **2013**, *22*, 115024. [CrossRef]

. Available online: https://www.danatm4.com/wp-content/uploads/2013/12/Wheel_motor_feature_Electric__Hybrid_July_2013.pdf (accessed on 8 May 2022).

. Tong, W.; Hou, Z. Vertical vibration analysis on electric vehicle with suspended in-wheel motor drives. In Proceedings of the 2013 World Electric Vehicle Symposium and Exhibition (EVS27), Barcelona, Spain, 17–20 November 2013; pp. 1–9. [CrossRef]

. Chen, X.; Yin, J.; Wang, W.; Wu, L.; Tang, F. Approaches to diminish large unsprung mass negative effects of wheel side drive electric vehicles. *J. Adv. Mech. Des. Syst. Manuf.* **2016**, *10*, JAMDSM0064. [CrossRef]

. Jin, X.; Wang, J. Modeling and μ Synthesis Control for Active Suspension System of Electric Vehicle. In Proceedings of the 2021 5th CAA International Conference on Vehicular Control and Intelligence (CVCI), Tianjin, China, 29–31 October 2021; pp. 1–6. [CrossRef]

. Yang, G.; Spencer, B.F.; Carlson, J.D.; Sain, M.K. Large-scale MR fluid dampers: Modeling and dynamic performance considerations. *Eng. Struct.* **2002**, *24*, 309–323. [CrossRef]

. Hu, G.; Zhou, F.; Yu, L. Optimal Design and Performance Analysis of Radial MR Valve with Single Excitation Coil. *Actuators* **2021**, *10*, 34. [CrossRef]

. Hu, G.; Liu, F.; Xie, Z.; Xu, M. Design, Analysis, and Experimental Evaluation of a Double Coil Magnetorheological Fluid Damper. *Shock Vib.* **2016**, *2016*, 1–12. [CrossRef]

. Available online: https://lordfulfillment.com/pdf/44/DS7015_MRF-132DGMRFluid.pdf (accessed on 10 May 2022).

. Nguyen, Q.-H.; Choi, S.-B. Optimal design of a vehicle magnetorheological damper considering the damping force and dynamic range. *Smart Mater. Struct.* **2009**, *18*, 015013. [CrossRef]

. Hong, K.-S.; Sohn, H.-C.; Hedrick, J.J. Modified skyhook control of semi-Active Suspensions: A new model, gain scheduling, and hard ware-in-the-loop tuning. *J. Dyn. Sys. Meas. Control* **2002**, *124*, 158–167. [CrossRef]

28. Shao, X. Active Suspension Control of Electric Vehicle with In-Wheel Motors. Ph.D. Thesis, University of Wollongong, Wollongon Australia, 2018.
29. Wu, H.; Zheng, L.; Li, Y. Coupling effects in hub motor and optimization for active suspension system to improve the vehicle an the motor performance. *J. Sound Vib.* **2020**, *482*, 115426. [CrossRef]
30. Oh, J.-S.; Kim, K.-S.; Lee, Y.-S.; Choi, S.-B. Dynamic simulation of a full vehicle system featuring magnetorheological dampe with bypass holes. *J. Intell. Mater. Syst. Struct.* **2020**, *31*, 253–262. [CrossRef]
31. Bai, R.; Wang, H.-B. Robust Optimal Control for the Vehicle Suspension System With Uncertainties. *IEEE Trans. Cybern.* **202** 1–11. [CrossRef] [PubMed]
32. Oborne, D.J. Whole-Body Vibration and International Standard ISO 2631: A Critique. *Hum. Factors J. Hum. Factors Ergon. So* **1983**, *25*, 55–69. [CrossRef]

sensors

MDPI

·ticle

)esign, Fabrication and Analysis of Magnetorheological ›oft Gripper

kub Bernat [1,*], Piotr Gajewski [2], Rafał Kapela [1], Agnieszka Marcinkowska [2] and Paulina Superczyńska [1]

[1] Institute of Automatic Control and Robotics, Poznan University of Technology, 60-965 Poznan, Poland; rafal.kapela@put.poznan.pl (R.K.); paulina.superczynska@put.poznan.pl (P.S.)
[2] Institute of Chemical Technology and Engineering, Poznan University of Technology, 60-965 Poznan, Poland; piotr.gajewski@put.poznan.pl (P.G.); agnieszka.marcinkowska@put.poznan.pl (A.M.)
* Correspondence: jakub.bernat@put.poznan.pl

Abstract: The magnetorheological elastomer is promising material for applications in soft robotics. Its properties like reactive to external magnetic field and softness allow to construct an attractive devices. This work presents a construction of soft gripper assembled with magnetorheological elastomers. The work describes the detailed molding process of magnetorheological elastomers. Further, the electromechanical properties of magnetorheological elastomers are shown using a simple beam. Finally, the soft gripper is constructed and analyzed with the series of experiments.

Keywords: magnetorheological elastomers; MRE; soft gripper

1. Introduction

The development of an intelligent materials gained the focus of researchers and practitioners from various disciplines [1]. The interesting features like softness which is not available in traditional materials like steel causes the possibilities to construct devices easier to operate with delicate objects or to cooperate with humans [2]. This is especially visible in the development of soft pneumatic devices based on the silicone and pneumatic actuation [3]. Alternative to pneumatic actuation can be an electrical or electromagnet stimuli. The first one is for instance exploited in the electroactive polymers to create deformation [1,4]. The second one is popular for instance in the ferroelectric polymers [1], magnetorheological fluids [5,6] and magnetorheological elastomers [7].

In this work, we are focused on the the magnetorheological elastomers (MRE) which are composite materials reinforced by ferromagnetic particles with ferromagnetic and visco-elastic properties [8–10]. They are solid analog of magnetorheological fluids with a rubber material instead of a liquid. Different medium does not allow particles to settle with time, increases the flexibility (because the container is not needed) and the permeability (needed to generate larger displacement force). Doping with those particles causes sensitivity to the magnetic fields. Due to the fact that MRE depends on magnetic field it is possible to affect on material by changing magnetic field properties such as amplitude or direction of propagation. Several computational and experimental studies have been carried out in the past of materials with MRE core. These studies were mainly concerned with vibration, damping and responses, and visco-elastic behavior [8–10].

In the recent time, the intelligent materials were applied to build various devices for soft robotics [2,11,12]. One of the most challenging tasks is to build soft gripper allows the handling sensitive objects trying to imitate the solutions from nature [2,13]. The soft grippers can be exploited in many area of industrial and domestic applications for instance in medical applications [14] or plant harvesting [15]. The soft grippers can be dived into groups based on the method of actuation such as pneumatic, hydraulic, magnetic field or electric field [2,13]. One of the most interesting method of building soft grippers is application of soft fluid actuators [16] which can be driven by pneumatic or hydraulic excitation. In recent times, these soft grippers has been extensively developed and had a commercial application like PneuNets [17]. Pneumatic soft gripper are build with material possessing the possibility of high stretching and are driven by varying pressure. They enable us to successfully grasp a variety of objects but require compressed air preparation unit to work [2,17]. Beside the success of the soft fluid actuators, the different types of actuation are also used. For instance, in work [18], the Ionic Polymer Metal Composite (IPMC) was applied to build a EMG controlled micro gripper. Another possibility is the Dielectric Electroactive Polymers which was applied to build a gripped driven by high electric field [19]. In the work [20,21] the magnetic soft actuators are summarized to build the milli-, micro- and nanoscale soft devices. The soft magnetic grippers are developed with application of soft permanent magnets and magnetorheologicial fluids [22–24].

In this work, the magnetorheological elastomers are applied to build a soft gripper activated by external electromagnet field. Firstly, to show the properties of magnetorheological elastomers, the sample of beam with various content of iron particles are prepared. Then, the series of experiments are conducted which expose the features of MRE soft beam and its reaction to external stimuli. Secondly, the soft gripper performed with MRE material is designed. Its unique construction with electromagnets allow to grasp varying objects. Further, its features using image processing techniques are analyzed.

2. Magnetoreohogical Materials

2.1. Preparation of Magnetorheological Elastomers

The magnetorheological elastomers were prepared in the following way: an appropriate amount of RTV-2 silicone (OTT-S825 from OTTSilicone) was weighed into a glass vessel, and then an appropriate amount of an iron powder with an average particle size of 63 µm was added. The ingredients were thoroughly mixed to obtain a homogeneous mixture. Then the OTT-S825 catalyst was introduced in the amount of 2% by weight in relation to the amount of silicone used. After thorough mixing, the prepared system was degassed by vacuum treatment to remove any air bubbles introduced into the mixture during stirring the ingredients. This allowed avoiding foaming of the finished composite MRE sample. Then, the mixture was poured into the prepared molds of beam and gripper finger (Figure 1) and allowed to harden by crosslinking reaction for another 12 h. The MRE composites with three different concentrations: 30%, 50% and 70% by weight of the filler were prepared. As a supplementary material, a document is included containing all dimensions of the gripper finger as well as information about the shape.

2.2. Permeability Estimation

The crucial property of the magnetorheological elastomers from electromagnetic point of view is the permeability. It describes the reaction of magnetization on the magnetic field. The higher value of permeability provides that the material stores more magnetic energy or it is more reactive with external magnetic sources.

In this section, we show a simple and fast method of permeability analysis which is suitable for magnetorheological elastomers. It is based on the inductance measurement and it allows to measure the permeability of different MRE beams quickly.

Firstly, we will show the idea of measurement in Ansys software (Ansys® Academic Research Electronics 2021 R2) where the model of coil was analyzed for MRE beam with varying permeability as a core. The model used in the simulation is visible in Figure 2a and the cross section is shown in Figure 2b. The dimension are defined in Table 1. In the simulation, we have assumed that MRE permeability is linear and there is no hysteresis effect. The sensitivity of inductance was checked for different dimensions of coil. We have checked the influence of longer, shorter, thicker and thinner coil's winding. The list coil's dimensions is given in Table 2. The Ansys software was applied to find the self inductance for all cases. In Figure 2c it is visible that the self inductance is very sensitive to the change of windings dimensions. Therefore, the imperfections for instance in producing the coil, can have strong influence on relative permeability estimation. However, the solution of this problem is to calculate the relative inductance

$$k_L(\mu_r) = \frac{L(\mu_r)}{L(\mu_0)} \tag{1}$$

where L is the inductance which depends on the permeability of core material and μ_0 is the relative permeability of coil without core (the relative permeability of vacuum $\mu_0 = 1$). It is visible in Figure 2d that the varying geometry influences the relative inductance much less than absolute inductance presented in Figure 2c.

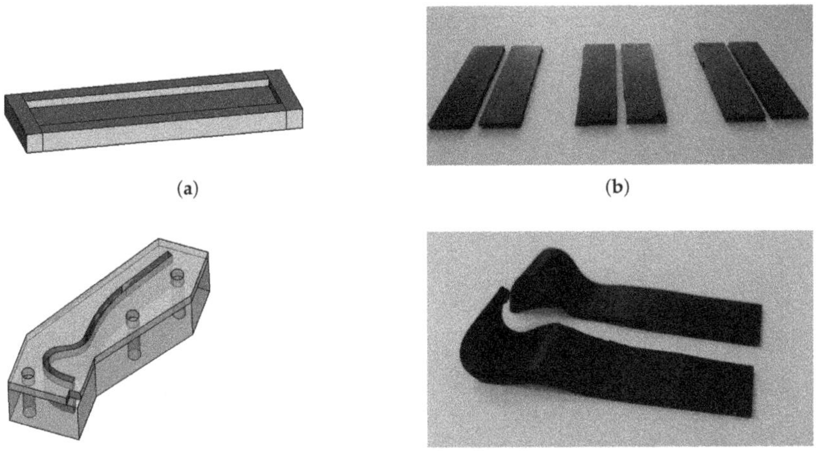

(c) (d)

Figure 1. The mold for preparing sample beams (**a**). Example of performed beams (**b**). The gripper finger mold (**c**). The example of gripper finger (**d**).

Table 1. The dimensions of beam and coil.

Name	Value
beam length	60 mm
beam width	15 mm
beam thickness	2 mm
winding length L	40 mm
winding radius	2 mm
winding thickness z	0.6 mm
winding width	36 mm

Table 2. The various coil models in the analysis.

Name	Length (L)	Thickness (z)
nominal	40 mm	0.6 mm
longer	42 mm	0.6 mm
shorter	38 mm	0.6 mm
thinner	40 mm	0.5 mm
thicker	40 mm	0.7 mm

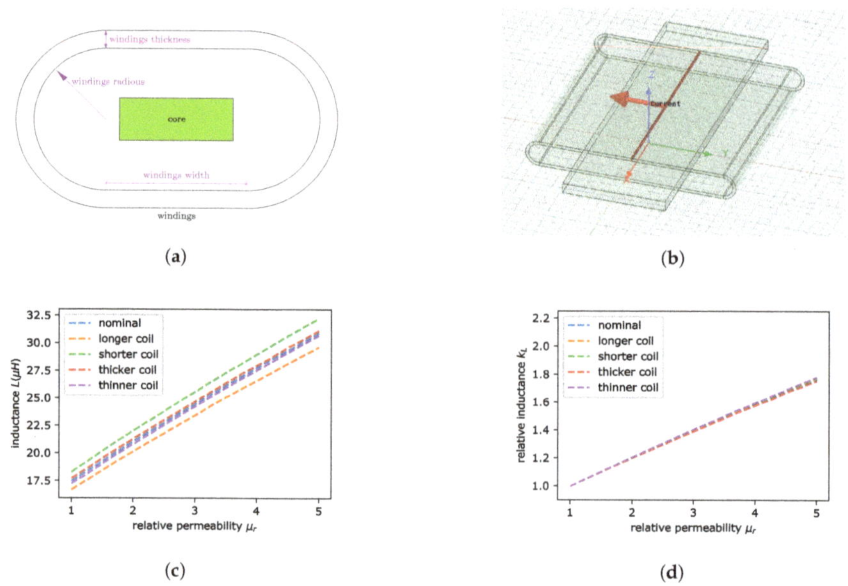

Figure 2. The cross section of coil windings and beam (core) (**a**). The coil model. (**b**) The absolute value of inductance for coil with varying permeability of beam and various geometry (**c**). The relative value of inductance for analogous example (**d**).

To estimate the permeability of magnetorheological elastomers, the two coils were build (one of them is presented in Figure 3). Next, the inductance of MRE beams with various iron powder content was measured by accurate LCR meter (Sanwa LCR700). Based on the measured inductance and characteristics presented in Figure 2d, the permeability of magnetorheological elastomer is estimated by calculating:

$$\hat{\mu}_r = k_L^{-1}\left(\frac{L_s}{L_0}\right) \tag{2}$$

where $\hat{\mu}_r$ is the estimated magnetic permeability, k_L^{-1} is the inverse of relationship between relative inductance and permeability, L_s is the inductance of coil with MRE beam and L_0 is the inductance of coil without core. The results are presented in Table 3. The measurement of relative inductance is much less disturbed by the coil type than the absolute inductance. Further, the permeability of magnetorheological elastomer is low (for instance in compare to electromagnetic steel). It is worth to point that similar range of permeability for MRE is also found in the work [7].

Figure 3. The example of coil built to perform estimate of relative permeability.

Table 3. The results of the inductance measurements for varying iron powder content with the calculated relative inductance and the estimated magnetic permeability.

Coil	Fe Content (%)	L (μH)	$k_L \times 100$	$\hat{\mu}_r$
A	-	22.78	1.0	1.0
B	-	25.10	1.0	1.0
A	30	23.52	3.2	1.12
B	30	26.10	4.0	1.16
A	50	25.05	10.0	1.47
B	50	27.58	9.9	1.46
A	70	28.73	26.1	2.31
B	70	31.50	25.5	2.28

2.3. Damping Properties

In the case of magnetorheological elastomers, the mechanical properties are also very important. As an example of the mechanical behaviour of MRE, the free damped oscillations of MRE beams were analyzed by performing series of experiments. The laboratory kit was assembled as shown in Figure 4 and its main parts are as follows: MRE beam, electromagnet and laser as distance sensor. In the first step, the electromagnet was activated with linearly increasing current and hence it was pulling the beam. This step is visible in Figure 5 between 0 and around 3 s. The second step is active, if the beam was moved by 1 mm, the electromagnet current was set to 0. Because the pulling force stopped working on the beam, it returns to initial position by a free oscillatory movement. The beam come back to initial position by long damped oscillations. Taking into account the described procedures, all samples with various content of iron powder was analyzed (two samples of each content of iron powder, i.e., 30%, 50% and 70% by weight giving a total number of 6 samples). The example result obtained in the experiments are shown in Figure 5. The input signal expressed as PWM duty cycle is presented in Figure 5a,b and the movement of beam measured by laser is visible in Figure 5c,d.

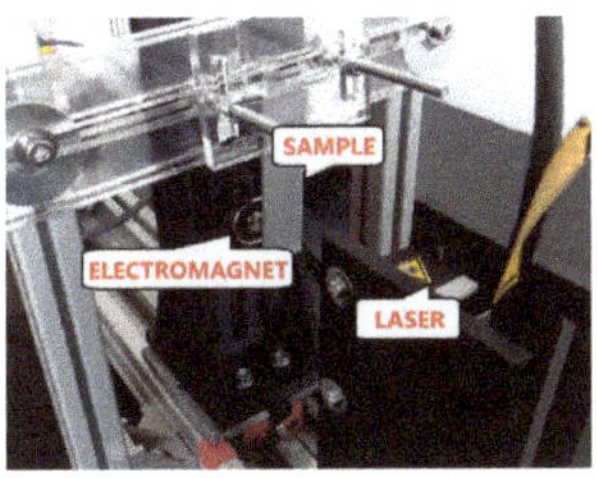

Figure 4. The laboratory set up for analysing the free oscillations of MRE beam.

To analyze the performed experiments, the following model of beam response is considered:

$$y(t) = Ae^{-\alpha t} sin(\omega t) \tag{3}$$

where A is the amplitude of oscillations, α is the decay rate and ω is the angular frequency. The above equation represents a response of second order system. To find the values of A, α and ω, the optimization process is carried out. The goal function is defined as follows:

$$J(A, \alpha, \omega) = \frac{1}{N} \sum_{k=0}^{N} [y_m(t_k) - y(t_k)]^2 \tag{4}$$

where $y_m(t_k)$ is the measured value at time t_k, $y(t_k)$ is the estimated value found from Equation (3) and N is the number of measured points. The optimization was performed by Nelder-Mead algorithm implemented in scipy [25]. The example of estimated output obtained by identification is presented in Figure 6. It is visible that good agreement between the model and measurement is obtained.

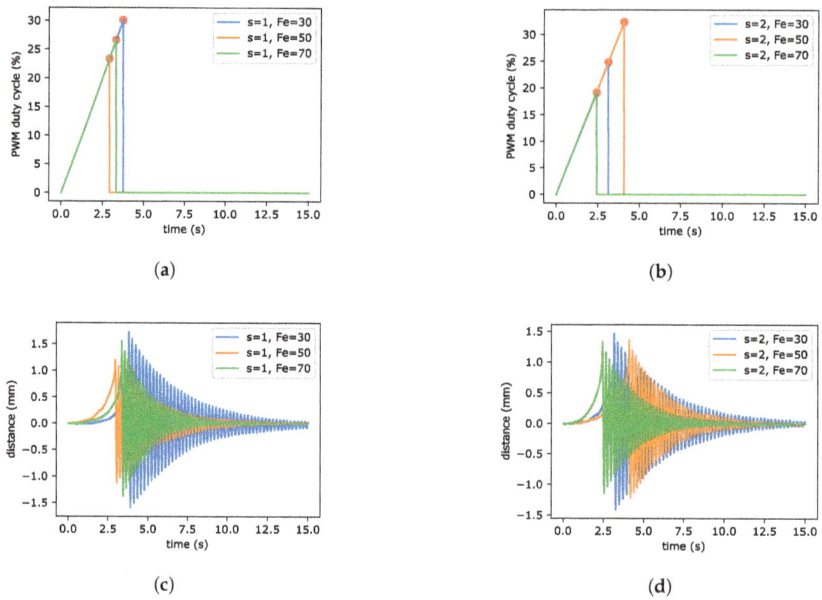

Figure 5. The free vibrations of MRE beam. The duty cycle of PWM and distance response for sample 1 (**a**,**c**). The duty cycle of PWM and distance response for sample 2 (**b**,**d**).

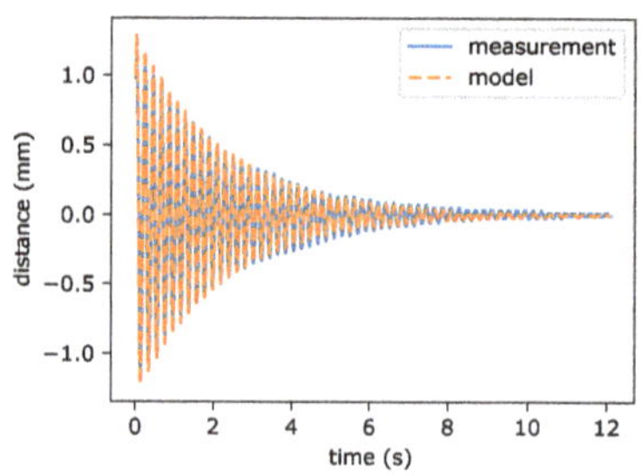

Figure 6. The example of model and measurement transients.

Relying on the above procedure, all responses of MRE beam were analyzed. The obtained results are presented in Table 4. It can be seen that the content of iron in silicone matrix has strong influence on the decay rate and settling time (these parameters are correlated). It can be easily stated that increasing content of iron powder in the silicone matrix causes the decreasing value of the settling time and increasing value of the decay rate. It means that the higher content of iron powder in MRE implies more damped responses. Further, the angular frequency is much less sensitive to the varying Fe content.

Table 4. The settling time of free oscillations for samples of beams with different iron content in MRE.

Sample	Fe Content (%)	Settling Time (s)	α (1/s)	ω (rad/s)
1	30	11.2	0.29	28.5
2	30	11.2	0.29	27.6
1	50	8.9	0.45	30.4
2	50	8.0	0.48	31.6
1	70	6.5	0.63	33.9
2	70	6.6	0.56	30.2

2.4. *Magnetic Attraction Force Experiment*

Experiment presented in this section was to check the difference in the reaction of created beams with different iron content to the magnet approaching them. The laboratory kit was assembled as presented in Figure 7. The main parts are: MRE beam and the permanent magnet placed on the distance sensor. The beam had fastened both ends to avoid any vibrations resulting from free hanging. The permanent magnet with the force sensor approached and moved away from the test object on a distance between 60 and 15 mm as it is illustrated in Figure 8. Experimental results for the magnetic attraction force of the MRE beam are visible in Figure 9. As the magnet moved closer the beam was leaning towards the magnet, what is possible to observe in Figure 9a,b between 750 and 1250 sample. At the point of maximum deflection, and also the strongest force of attraction, the tested element did not touch the magnet, it was tilted to its maximum. As the magnet was receded the MRE returns to its initial position. As in the previous section, all beams samples with various content of Fe powder were analyzed (doubled samples of 30%, 50% and 70% iron powder giving a total number of 6 samples). The example of the input signal expressed as magnet position is presented in Figure 8. Figure 9c,d presents the force as a function of the position for sample 1 and 2, respectively. The maximum force is connected with the maximum content of Fe in the sample. It is around 0.1 N for both tested beams at a distance of around 15 mm.

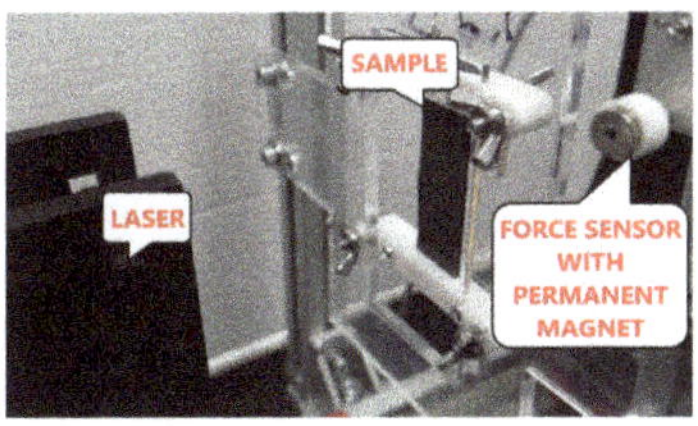

Figure 7. The laboratory set up for analysing the force of MRE beam.

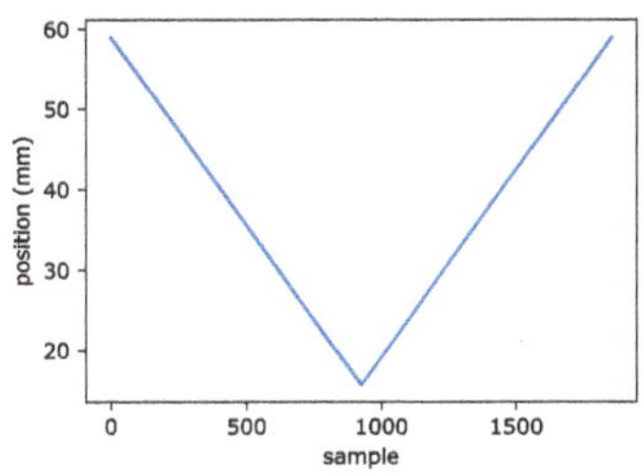

Figure 8. The permanent magnet position in the measurement of force.

2.5. *Tensile Strength Studies*

Mechanical properties of magnetorheological elastomers are very important characteristics of material for industrial applications. Thus, to characterize the mechanical properties of the investigated materials the tensile strength was studied. The tests of prepared beams samples were made most of all to estimate the influence of the filler on silicone matrix stiffness. The measurements were performed at universal testing machine Zwick/Roell model Z020 (Zwick GmbH & Co. KG, Ulm, Germany) at 25 °C according to PN-EN ISO 527-1:1998 (crosshead speed of 5 mm min^{-1}). Results are presented in Table 5.

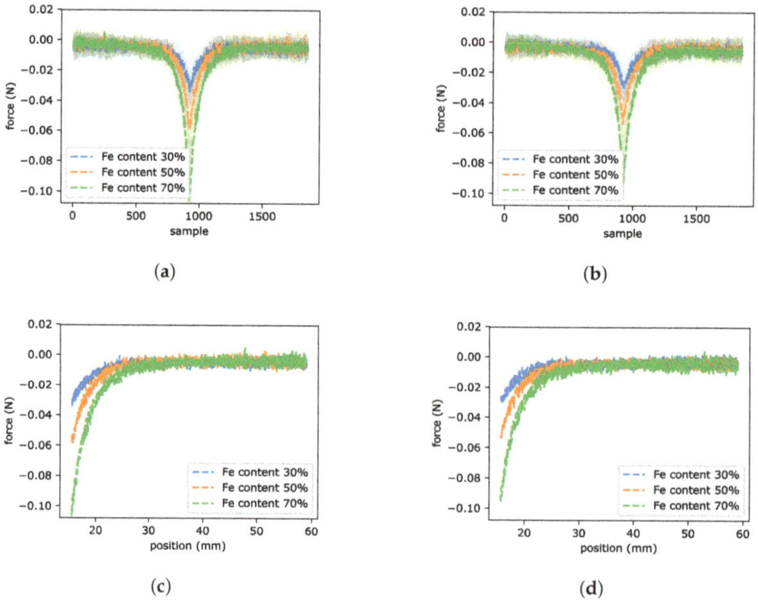

Figure 9. The force measurement between soft beam and permanent magnet. The force sensor for sample 1 (**a**) and sample 2 (**b**). The representation of force as a function of the position for sample 1 (**c**) and sample 2 (**d**).

Table 5. The results of the mechanical strength measurements (Young modulus E_{mod}, tensile strength σ_{max} and elongation at break ε_{max}) for magnetorheological elastomers with various iron powder content.

Fe Content (%)	E_{mod} (MPa)	σ_{max} (MPa)	ε_{max} (%)
0	0.31 ± 0.029	1.55 ± 0.13	255 ± 19
30	0.42 ± 0.045	1.46 ± 0.15	227 ± 17
50	0.62 ± 0.052	1.43 ± 0.17	179 ± 9.0
70	1.17 ± 0.18	1.71 ± 0.10	103 ± 2.5

The stiffness (Young modulus, E_{mod}) of the tested materials increases nonlinearly with the increase of iron content in the composites. This is consistent with the literature [26,27], because inorganic fillers introduced to polymer matrix cause stiffening of the polymer matrix. For materials containing up to 50% by weight, this increase is slight, while 70% by weight of the filler has a greater impact on this parameter—it rises almost four times. Contrary elongation at break decreases with increasing content of Fe in the polymer matrix. Tensile strength is not so sensitive on filler content, and slightly decrease for samples containing 30% and 50% by weight of Fe and slightly increase for sample with 70% by weight of Fe is observed.

3. Soft Magnetorheological Gripper

In this section, the idea of the gripper from prepared MRE materials is presented. The main features of that kind of device are softness and reaction on the magnetic field. Relying on these attributes the gripper construction was proposed.

3.1. Construction

The construction of gripper is based on two MRE fingers. The shape of single finger is presented in Figure 1c and it was performed based on the form presented in Figure 1d. The curved shape of fingers causes an initial force to be presented between them in the closed position. This force is caused by the elasticity of MRE material. To open the gripper, the electromagnets are installed and activated. The electromagnet creates magnetic field which produce a force acting on MRE finger. The force pulls the fingers to electromagnet and gripper is open. It is worth noting that an important aspect of the work of the MRE fingers is the location of the electromagnets. As shown in Figure 9c,d, the force between the source of the magnetic field and the MRE increases rapidly over short distances. Therefore, in the presented gripper, the position of the electromagnet is crucial. This can be seen in Figure 10 where two configurations of the electromagnet are considered. If the upper electromagnet (which is very close to the finger) is turned on, then MRE is pulled to the electromagnet as shown in Figure 10a. Then, the distance between the lower electromagnets is closer and hence turning on the lower electromagnets causes that fingers are pulled by all electromagnets as is visible in Figure 10b. If only the lower electromagnets are turned on, the force acting on fingers is too low for pulling them, so the fingers stay in the close position.

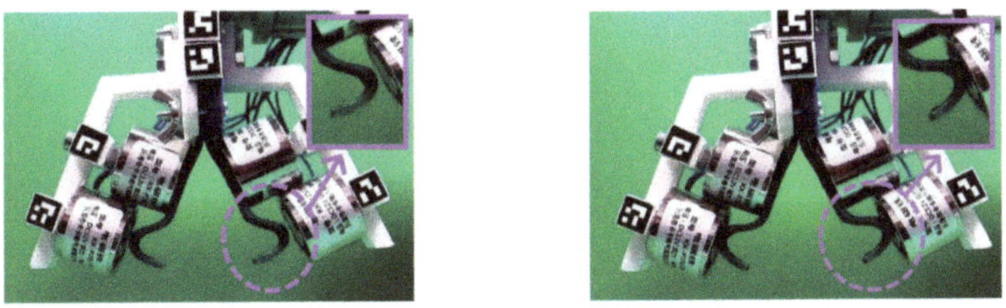

(**a**) (**b**)

Figure 10. The soft gripper reaction on electromagnet. The upper pair of electromagnets turned on (**a**). The both pairs of electromagnets turned on (**b**).

3.2. Experiments

The series of experiments with gripper were performed to show its ability to handling a mass. The set of loads with cylinder shape with different mass were prepared. The steps of experiments were as follows:

- Gripper raised with closed fingers;
- Gripper fingers opened by turning on electromagnet;
- Gripper lowered to load level;
- Gripper fingers closed by turning off electromagnet;
- Gripper raised with mass;
- Gripper fingers opened to release mass.

The procedure was tested for loads with mass: 2.62 g, 2.95 g, 3.28 g, 3.45 g, 3.54 g, 3.62 g, 3.75 g and 3.89 g. In the case of masses 4 g and more, the gripper cannot hold the mass.

3.2.1. Camera System

Figure 11 shows the camera system implemented for our mass holding experiments. It consists of the pan–tilt–zoom camera system [28], custom camera mount, Raspberry Pi 3 embedded computer and a PC computer. Raspberry computer acts as a central server in this architecture that is responsible for control of the PTZ camera system movement and image capturing. It is also capable of computing of some preprocessing steps like finding AruCo markers [29] or basic light/contrast image enhancements. For image capturing and sharing processes a file server is used. Entire communication takes place by http GET/PUT requests that embed JSON format messages sent after the message header shown below:

```
{
  "doAF" : 0,
  "y"  : 0,
  "x"  : 0,
  "focus"  : 0,
  "zoom"   : 0,
  "imgNo" : 5,
  "imgName" : "test_"
}
```

where particular fields denote the following functionality:

- doAF—perform autofocus algorithm;
- x,y—set pan/tilt position;
- focus—set focus position;
- zoom—set zoom position;
- imgNo—set experiment number (included in image names in file server).

(a) (b)

Figure 11. Camera system. Pan–tilt–zoom camera system mount (**a**). The view of the gripper framework (**b**).

3.2.2. Gripper Visual Analysis Algorithm

The series of photos during experiments include the most significant situations during the gripper performance analysis. These include (Figure 12):

- Closed gripper in the base position (Figure 12a);
- Opened gripper in the base position (Figure 12b);
- Opened gripper in the catching position (Figure 12c);
- Closed gripper in the catching position (Figure 12d);
- Closed gripper in the initial lifting position (Figure 12e);
- Closed gripper in the goal lifting position (Figure 12f).

From the visual analysis point of view, there were two major experiments conducted to calculate gripper characteristics: the maximum gap between the gripper fingers and the flexibility analysis of the displacements under variable load. To achieve that, a custom computer vision algorithm has been built. In order to perform the analysis in a robust way it consists of the preprocessing step where the important features are extracted and the actual analysis of the displacements between two analyzed gripper positions.

By controlling the background color and the lighting conditions (Figure 11) it is easier to perform the preprocessing analysis. In addition, the task is to analyze the displacement of the gripper fingers so assuming the camera position does not change during the experiment all the important features can be extracted from the difference of the two images. Under these assumption it is possible to extract gripper fingers envelope shape for each gripper position knowing the work area and the background color. This is done by detecting the edges in the given image (Figure 13a,b).

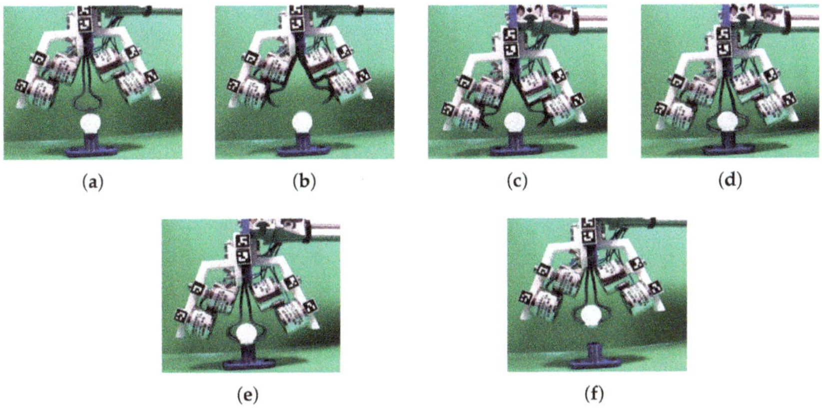

Figure 12. Gripper positions during weight lift experiment. Base position, closed gripper (a). Base position, opened gripper (b). Catching position, opened gripper (c). Catching position, closed gripper (d). Initial lifting position, closed gripper (e). Goal lifting position, closed gripper (f).

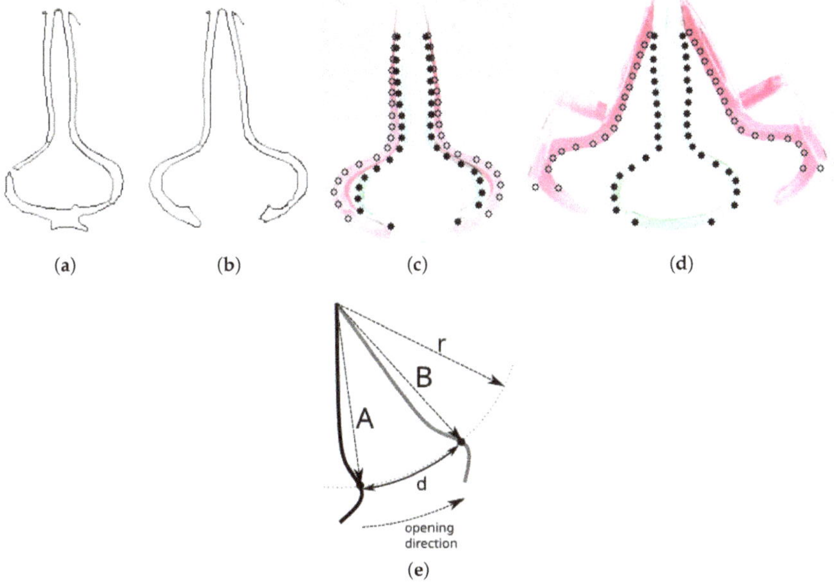

Figure 13. Preprocessing stages of the gripper displacement analysis. Edge image of closed gripper in the base closed position (a). Edge image of the closed gripper holding an object (b). Difference image between the closed empty gripper and closed gripper holding an object (analysis points marked) (c). Difference image between the closed empty gripper and opened gripper (analysis points marked) (d). The idea of identification of displacement analysis points (e).

We have wanted to perform analysis of the fingers deformations with respect to the distance to the center of the gripper that is located in its mount point. For this reason a number of consecutive points is proposed along the major axis of the gripper (Figure 13c) and (Figure 13d). The way the points locations are extracted is that the algorithm finds a cross-section of the gripper outer envelope and a circle with given radius (r) and the center located at the mount point of the gripper. Then the deformation at the given point is measured as a difference between the arc lengths in the two fingers positions accordingly to the Formula (5).

$$d = \|\vec{r}\| \arccos\left(\vec{A} \cdot \vec{B}\right) \tag{5}$$

3.2.3. Gripper Visual Analysis Algorithm Result Analysis

The major experiment conducted with the gripper was the analysis of weight lifting capabilities. It was performed by the series of single weight lifting experiments with a weight as a parameter. The visual algorithm described in the previous paragraph allowed to measure the deformations along the major axis of the gripper.

In order to measure the maximum opening gap of the gripper the last pair of the envelope points is calculated at the outer edge of the opened gripper (Figure 13d). The arc distance measured by the application was 44 mm and the chord distance was 29 mm calculated using the Formula (6).

$$d = 2\|\vec{r}\| \sin\left(0.5\vec{A} \cdot \vec{B}\right) \tag{6}$$

In the next experiment the flexibility of the gripper was tested. In order to conduct an analysis, the gripper was loaded with variable mass and visual analysis was performed to detect the displacements between the base position and the final position in which gripper holds the weight. Figure 13c,d show the idea behind the analysis. Filled dots are the ones that denote gripper shape envelope in the base position—the empty dots denote gripper envelope in mass holding position. For each r there are two pairs of points (left/right) for displacement calculations. Note, that since the fingers of the gripper are rotary the displacement is calculated along the arc of the given r (5).

Figure 14 shows the results of the experiment. The following curves denote the displacements d for the same weight along the radius r. As it can be seen, there is a tendency that bigger masses introduce bigger displacements since all the curves tend to increase values for heavier masses. In addition the displacement increases for increasing r which is obvious for rotary flexible gripper since the flexibility increases over the length of the material. It is also worth to mention that the last displacement for all the weights is always significantly bigger since this is the place where most of the mass is being hold by the gripper.

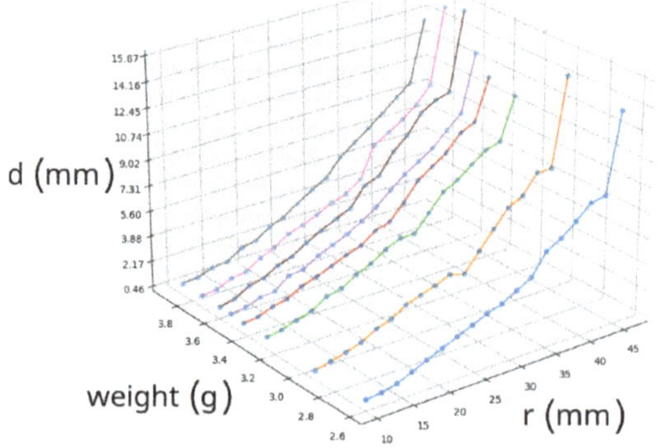

Figure 14. Results from visual analysis. Each line represents the displacement d caused by the particular weight along with the radius r millimeters from the gripper center.

Figure 15 presents standard deviation of the displacements for each measurement point with respect to the radius. Furthermore, positive and negative error bands are included. As can be observed the measurement point with the longest distance from the gripper center is featured with the highest standard deviation. This is most likely due to the accumulation of the flexibility along the gripper fingers. The incline trend in the rest of the observations also confirms this effect. Another observation made is that for increasing weight the positive error of standard deviation also increases. This might be related to two factors: weight disturbances (uneven weight base position or mass placement insight weight) or bending effect of the gripper fingers which occurs frequently at some specific stress for the fingers. It is also worth to mention that positive error is usually bigger than the negative one since the gripper is stretching by the weight.

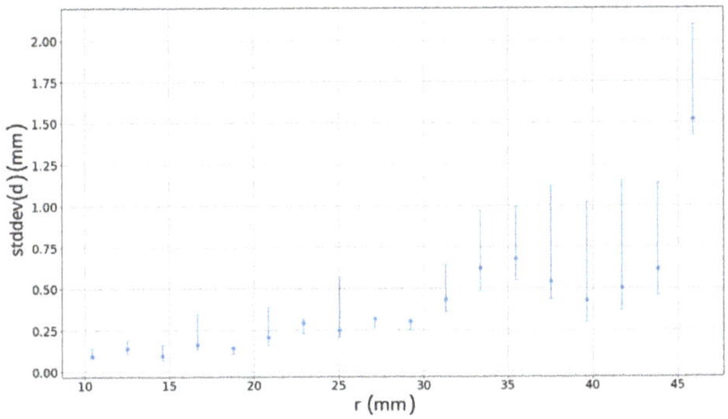

Figure 15. Results from visual analysis. *X* axis represents radius *r*. *Y* axis represents average standard deviation for all the weights with upper and lower error bands.

3.3. Gripping Objects with Varying Shape

To observe possible applications for invented gripper, we have performed gripping test for objects with different shapes, weights, sizes and textures. Table below (Table 6) presents characteristic of chosen objects.

Table 6. Characteristic of gripped objects.

Dimensions	Wood Ball	Game Pawn	Popmpon	Pencil Sharpener
Weight (g)	2.79	0.8	0.18	2.06
Diameter (mm)	20	25.5	16	26
Height (mm)	-	6.5	-	12

The results of the gripping test are shown in Figure 16. All varying shape objects, were lifted up and did not slip out what proves that invented gripper may be used in different applications and easily grasp desired target. It is also possible to observe that the gripper's flexibility allows to deflect and fits to gripped object in order to lift it up. The stiffness of the used materials and forces resulting from a specific shape, affect on the object to hold it the during lifting process.

(a) (b)

Figure 16. *Cont.*

(c)

(d)

Figure 16. The soft gripper holding various objects: wood ball (**a**), game pawn (**b**), pompon (**c**), pencil sharpener (**d**).

4. Conclusions

This article describes a proposal of a soft actuator, presenting its manufacturing process, analysis of permeability and damping properties as well as an experiment of the magnetic attraction force. In the first part of the work, the influence of iron powder content on the properties of MRE was investigated. For this purpose, beams containing 30%, 50% and 70% by weight of iron powder were prepared. On the basis of the performed studies, it was determined that the magnetic properties of investigated samples depend on the iron powder content in MREs. The material with the highest magnetic permeability contains the highest amount of filler (70 wt%) and this MRE was used as a soft actuator—the gripper. In the last section, the operation of the gripper was verified by checking the maximum lifting capacity and deflection of the material during the performed task. In summary, the proposed soft gripper can grasp varying shape objects. The gripper fingers have a moderate possibility to fit external shape. One of the advantages of gripper is the absence of required external stimuli in the closed position. The proposed shape of finger can also be modified to catch different group of objects. In the future, a numerical approach will be used for further evaluation and optimization of the final shape of the MRE actuator.

Supplementary Materials: The following supporting information can be downloaded at: https://www.mdpi.com/article/10.3390/s22072757/s1.

Author Contributions: J.B., design the concept of experiments, FEM modelling, project administration; P.G., analysis and preparation of magnetorheological elastomers, writing manuscript; R.K., software implementation of image processing algorithm, camera setup, writing manuscript; A.M., analysis and preparation of magnetorheological elastomers, writing manuscript; P.S., prepare of experiments, data acquisition, writing manuscript. All authors have read and agreed to the published version of the manuscript.

Funding: This research was funded by Ministry of Education and Science, grant number 0211/SIGR/6434.

Institutional Review Board Statement: Not applicable.

Informed Consent Statement: Not applicable.

Conflicts of Interest: The authors declare no conflict of interest.

References

1. Kim, K.J.; Tadokoro, S. Electroactive Polymers for Robotic Applications. In *Artificial Muscles and Sensors*; Springer: London, UK 2007. [CrossRef]
2. Rus, D.; Tolley, M.T. Design, fabrication and control of soft robots. *Nature* **2015**, *521*, 467–475. [CrossRef]
3. Chen, F.; Wang, M.Y. Design Optimization of Soft Robots: A Review of the State of the Art. *IEEE Robot. Autom. Mag* **2020**, *27*, 27–43. [CrossRef]
4. Rosset, S.; Araromi, O.A.; Schlatter, S.; Shea, H.R. Fabrication Process of Silicone-based Dielectric Elastomer Actuators. *JoVE* **2016**, *108*, e53423. [CrossRef] [PubMed]
5. De Vicente, J.; Klingenberg, D.; Hidalgo-Alvarez, R. Magnetorheological fluids: A review. *Soft Matter* **2011**, *7*, 3701–3710 [CrossRef]
6. Carlson, J.D.; Jolly, M.R. MR fluid, foam and elastomer devices. *Mechatronics* **2000**, *10*, 555–569. [CrossRef]

Böse, H.; Gerlach, T.; Ehrlich, J. Magnetorheological elastomers—An underestimated class of soft actuator materials. *J. Intell. Mater. Syst. Struct.* **2021**, *32*, 1550–1564. [CrossRef]

Yarali, E.; Farajzadeh, M.A.; Noroozi, R.; Dabbagh, A.; Khoshgoftar, M.J.; Mirzaali, M.J. Magnetorheological elastomer composites: Modeling and dynamic finite element analysis. *Compos. Struct.* **2020**, *254*, 112881. [CrossRef]

Kashima, S.; Miyasaka, F.; Hirata, K. Novel soft actuator using magnetorheological elastomer. *IEEE Trans. Magn.* **2012**, *48*, 1649–1652. [CrossRef]

Guan, X.; Dong, X.; Ou, J. Magnetostrictive effect of magnetorheological elastomer. *J. Magn. Magn. Mater.* **2008**, *320*, 158–163. [CrossRef]

Bar-Cohen, Y. *Electroactive Polymer (EAP) Actuators as Artificial Muscles: Reality, Potential, and Challenges*; SPIE Press: Bellingham, WA, USA, 2001.

Carpi, F.; Anderson, I.; Bauer, S.; Frediani, G.; Gallone, G.; Gei, M.; Graaf, C.; Jean-Mistral, C.; Kaal, W.; Kofod, G.; et al. Standards for dielectric elastomer transducers. *Smart Mater. Struct.* **2015**, *24*, 105025. [CrossRef]

Shintake, J.; Cacucciolo, V.; Floreano, D.; Shea, H. Soft Robotic Grippers. *Adv. Mater.* **2018**, *30*, 1707035. [CrossRef] [PubMed]

Rateni, G.; Cianchetti, M.; Ciuti, G.; Menciassi, A.; Laschi, C. Design and development of a soft robotic gripper for manipulation in minimally invasive surgery: a proof of concept. *Meccanica* **2015**, *50*, 2855–2863. [CrossRef]

Navas, E.; Fernández, R.; Sepúlveda, D.; Armada, M.; Gonzalez-de Santos, P. Soft Grippers for Automatic Crop Harvesting: A Review. *Sensors* **2021**, *21*, 2689. [CrossRef] [PubMed]

Pagoli, A.; Chapelle, F.; Corrales-Ramon, J.A.; Mezouar, Y.; Lapusta, Y. Review of soft fluidic actuators: classification and materials modeling analysis. *Smart Mater. Struct.* **2021**, *31*, 013001. [CrossRef]

Mosadegh, B.; Polygerinos, P.; Keplinger, C.; Wennstedt, S.; Shepherd, R.F.; Gupta, U.; Shim, J.; Bertoldi, K.; Walsh, C.J.; Whitesides, G.M. Pneumatic Networks for Soft Robotics that Actuate Rapidly. *Adv. Funct. Mater.* **2014**, *24*, 2163–2170. [CrossRef]

Jain, R.; Datta, S.; Majumder, S. Design and control of an IPMC artificial muscle finger for micro gripper using EMG signal. *Mechatronics* **2013**, *23*, 381–394. [CrossRef]

Acome, E.; Mitchell, S.K.; Morrissey, T.G.; Emmett, M.B.; Benjamin, C.; King, M.; Radakovitz, M.; Keplinger, C. Hydraulically amplified self-healing electrostatic actuators with muscle-like performance. *Science* **2018**, *359*, 61–65. [CrossRef]

Eshaghi, M.; Ghasemi, M.; Khorshidi, K. Design, manufacturing and applications of small-scale magnetic soft robots. *Extrem. Mech. Lett.* **2021**, *44*, 101268. [CrossRef]

Erb, R.M.; Martin, J.J.; Soheilian, R.; Pan, C.; Barber, J.R. Actuating Soft Matter with Magnetic Torque. *Adv. Funct. Mater.* **2016**, *26*, 3859–3880. [CrossRef]

Do, T.N.; Phan, H.; Nguyen, T.Q.; Visell, Y. Miniature Soft Electromagnetic Actuators for Robotic Applications. *Adv. Funct. Mater.* **2018**, *28*, 1800244. [CrossRef]

Ullrich, F.; Dheman, K.S.; Schuerle, S.; Nelson, B.J. Magnetically actuated and guided milli-gripper for medical applications. In Proceedings of the 2015 IEEE International Conference on Robotics and Automation (ICRA), Seattle, WA, USA, 26–30 May 2015; pp. 1751–1756. [CrossRef]

Pettersson, A.; Davis, S.; Gray, J.; Dodd, T.; Ohlsson, T. Design of a magnetorheological robot gripper for handling of delicate food products with varying shapes. *J. Food Eng.* **2010**, *98*, 332–338. [CrossRef]

Virtanen, P.; Gommers, R.; Oliphant, T.E.; Haberland, M.; Reddy, T.; Cournapeau, D.; Burovski, E.; Peterson, P.; Weckesser, W.; Bright, J.; et al. SciPy 1.0: Fundamental Algorithms for Scientific Computing in Python. *Nat. Methods* **2020**, *17*, 261–272. [CrossRef] [PubMed]

Kang, S.S.; Choi, K.; Nam, J.D.; Choi, H.J. Magnetorheological Elastomers: Fabrication, Characteristics, and Applications. *Materials* **2020**, *13*, 4597. [CrossRef] [PubMed]

Zhou, Y.; Li, L.; Li, W.; Wen, S.; Jiang, L.; Jerrams, S.; Ma, J.; Chen, S. The fabrication and properties of magnetorheological elastomers employing bio-inspired dopamine modified carbonyl iron particles. *Smart Mater. Struct.* **2020**, *29*, 055005. [CrossRef]

Arducam PTZ Camera. Available online: https://www.arducam.com/product/arducam-8mp-pan-tilt-zoom-ptz-camera-with-base-for-raspberry-pi-4-3b-3-2/ (accessed on 3 February 2022).

Romero-Ramirez, F.; Muñoz-Salinas, R.; Medina-Carnicer, R. Speeded up Detection of Squared Fiducial Markers. *Image Vis. Comput.* **2018**, *76*, 38–47. [CrossRef]

sensors

MDPI

Communication

Dilute Polymerization of Aniline on PDMS Substrate via Surface Modification Using (3-Aminopropyl)Triethoxysilane for Stretchable Strain Sensor

Chang-Joo Yim, Ji-Yeon Choy, Hae-Kyung Youi, Jung-Hoon Hwang, Eun-Bee Jo, Jun-Ho Lee and Hyun-Seok Kim *

Division of Electronics and Electrical Engineering, Dongguk University-Seoul, Seoul 04620, Korea; akreal@dongguk.edu (C.-J.Y.); jiyeon9711@dongguk.edu (J.-Y.C.); yeahwooong@dgu.ac.kr (H.-K.Y.); dk2323@dgu.ac.kr (J.-H.H.); 2021120265@dgu.ac.kr (E.-B.J.); steve1211@dgu.ac.kr (J.-H.L.)
* Correspondence: hyunseokk@dongguk.edu; Tel.: +82-2-2260-3996

Abstract: Stretchable strain sensors are capable of acquiring data when in contact with human skin or equipment and are widely used in wearable applications. Most strain sensors have tensile properties of less than 20% and have limitations regarding body motion linkage, complex sensor structure, and motion nonreliability. To address these problems, we developed a high tension and high sensitivity sensor with a gauge factor over 40 and tensile stress about 50%. Polydimethylsiloxane (PDMS) was selected as the flexible substrate to ensure tensile strength, and polyaniline (PANI) was used to measure the resistance changes in the sensor. In particular, problems regarding poor uniformity of PANI on PDMS were resolved by surface treatment of the PDMS, wherein PANI polymerization was performed sequentially after forming a self-assembled monolayer (SAM) on the PDMS substrate. O_2 plasma and (3-aminopropyl)triethoxysilane were used to form the SAM. It is expected that this sensor can obtain stable characteristics even under high tensile stress through the evenly formed PANI films on the surface-treated PDMS substrate and may be used in various flexible sensor applications.

Keywords: strain sensor; conducting polymer; self-assembled monolayer; dilute polymerization; gauge factor

1. Introduction

Conductive polymers have the electrical and optical properties of metals as well as mechanical properties of original polymers; hence, extensive research is currently underway on their potential applications [1–5]. Polyaniline (PANI) has good electrical properties and high mechanical strength, thereby rendering it suitable for application to flexible-substrate-based wearable sensors [6–8]. Existing flexible substrate-based wearable sensors have been studied for various applications, such as electronic skin, flexible displays, health monitors, and energy harvesters; however, they are limited by several restrictions regarding use in practical applications owing to their lack of tensile properties [9–11]. Therefore, using PANI to develop stretchable sensors with excellent tensile properties and conductivity, such as strain-sensor-based electronic skin, is expected to have highly valuable applications [12,13].

PANI films can be fabricated easily on substrates by various chemical, electrochemical, template, and interfacial synthesis methods [14–17]. The chemical polymerization method, which uses redox reactions, is widely utilized owing to simplicity of synthesis [18,19]. However, variables such as the type of oxidant, synthesis temperature and time, and surface treatment must be considered carefully to control PANI's shape, length, electrical characteristics, and yield [20,21]. In particular, the surface functional groups on a substrate are crucial for achieving high adhesive force, uniform coverage, and good electrical properties of PANI during the chemical polymerization process [22–24]. The present study is aimed at manufacturing a crack-based strain sensor by synthesizing a PANI film on a polydimethylsiloxane (PDMS) substrate via surface treatment through self-assembled monolayer (SAM) formation [22,24,25].

The gauge factor (GF) is an essential indicator for evaluating the performance of a strain sensor [26]. GF is defined as

$$GF = \frac{R/R_0}{L/L_0} \quad (1)$$

GF was expressed as the change in relative resistance with relative length (Equation (1)). R_0 and L_0 denote the resistance and the length in the initial state. R and L represent the resistance and length in the strain state. Generally, for metal thin films, the GF is as low as 2–5. Low GF indicates low sensor sensitivity, which means that a material with high GF is required for precise sensing.

Previously, several researchers have attempted PANI polymerization on various substrates with insufficient tensile properties, such as Si, PEN and PET [27,28]. In addition, a few studies have attempted PANI polymerization on tensile substrates such as PDMS [8,29]. It is not difficult to synthesize PANI films on PDMS substrates, but uniform and stable characteristics in manufacturing sensors are the greater concerns. Therefore, a simple process for synthesizing PANI films on self-assembled monolayer (SAM)-treated PDMS substrate is suggested in this study to ensure high GF and tensile properties. The suggested synthesis process is suitable for manufacturing inexpensive, flexible, sensitive, and reversible sensors.

2. Materials and Methods

2.1. Materials

PDMS used as the flexible substrate was made using a Si elastomer base and a curing agent (sylgard 184) purchased from Dow Corning Corporation (Midland, MI, USA). Aniline (≥99.5%) for PANI synthesis was purchased from Sigma-Aldrich (St. Louis, MO, USA), perchloric acid (70%) was purchased from Sigma-Aldrich (St. Louis, MO, USA), and ammonium persulfate (≥98%) was purchased from Sigma-Aldrich (St. Louis, MO, USA). Toluene (≥99.9%) for surface treatment and (3-aminopropyl)triethoxysilane (APTES, 99%) were both purchased from Sigma-Aldrich (St. Louis, MO, USA). Acetone and isopropanol (IPA) used for cleaning were purchased from J.T. Baker (Radnor, PA, USA).

2.2. Methods

The PDMS substrate synthesis process is as follows. PDMS is combined in a base to curing agent ratio of 10:1 in a 4-inch Petri dish and mixed thoroughly for 10 min; then, a treatment procedure involving vacuum for 5 min, release for 1 min, and waiting for 5 min was repeated three times in a vacuum desiccator to remove air bubbles. Thereafter, curing was performed at 100 °C for 48 min in an oven to complete the fabrication of bulk PDMS, followed by cooling and removal from the Petri dish [18]. Finally, the PDMS was cut using a mold made according to the ASTM D412 model [30–32].

The PDMS substrate was then subject to the following surface treatment process. First, the surface was cleaned with acetone, IPA, and deionized (DI) water for 1 min each and dried with nitrogen. Then, surface modifications are performed using O_2 plasma under 100 W, 20 sccm, and 100 mTorr for 2 min. Thereafter, the PDMS surface treatment is completed by immersing the substrate in 10 mM APTES in toluene for 30 min. Finally, the substrate is naturally dried for 3 h.

PANI films are deposited on the PDMS surface by the dilute polymerization process. First, the concentration at which PANI can be most evenly formed on the surface was selected. The typical ratio of aniline to APS is 1.5 [18]. However, we tried to find the most suitable synthesis concentration of aniline in the range of 6–14 mM while fixing the APS concentration as shown in Figure 1. The density with 14 mM of aniline is sufficiently ensured and better as compared to other concentrations. Therefore, we prepare 14 mM of aniline solution and 9.33 mM APS solution in 1 M perchloric solution, followed by precooling for 30 min. The surface-treated PDMS substrate is attached to the beaker, and the prepared aniline solution is first added, followed by slow addition of the prepared APS solution. The beaker is then maintained at 0 °C for 24 h and stirred at 200 rpm. After completion of synthesis, the final product is cleaned in DI water and dried with a nitrogen gun. The various prepared samples are listed in Table 1. The sample 1 is a PANI on PDMS without any treatment. The sample 2 is a PANI on PDMS after O_2 plasma treatment. The sample 3 is a PANI on PDMS after O_2 plasma and APTES treatments. Circle and cross indicate whether or not they have gone through any treatment processes.

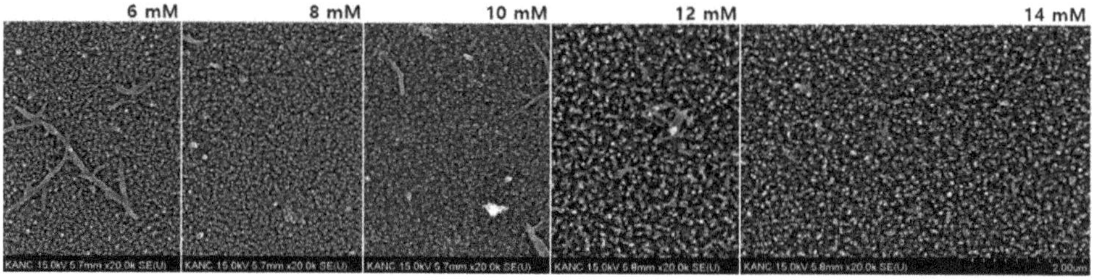

Figure 1. Scanning electron microscopy images of PANI with different mole concentrations of aniline.

Table 1. Three different samples with PANI on PDMS.

	PANI	O_2 Plasma	APTES
Sample 1	O	X	X
Sample 2	O	O	X
Sample 3	O	O	O

3. Results

3.1. SAM on PDMS

A SAM refers to an organic monolayer that is formed spontaneously on the surface of the substrate. The organic molecules of the SAM consist of three parts: the first part is the head group, which enables chemical adsorption of the molecules on the surface; the second part is the body group that enables ordered molecular film formation; the third part is the tail group that determines the chemical functional groups of the formed molecular film. Here, the amine group (NH_2) is used as the tail group to ensure high deposition rate with PANI during the dilute polymerization process. Synthesis is performed using the same characteristics of the amine group (NH_2) of PANI and the tail group of SAM. Figure 2 shows the process of synthesis of PANI after surface treatment. When PANI is synthesized on a substrate, aniline must be deposited in seed form for PANI growth. If this seed is nonuniformly deposited on the substrate, the final deposited product will also be nonuniform. Therefore, for PANI to grow uniformly on the substrate, it is noted that the surface of the substrate must also be chemically uniform. PDMS has a $-CH_3$ functional group (Figure 2a) that is converted to an –OH group through O_2 plasma treatment (Figure 2b). Thereafter, this functional group is replaced with the NH_2 group to facilitate PANI synthesis through APTES treatment (Figure 2c). Finally, PANI polymerization is performed on the modified PDMS surface (Figure 2d). To grow a uniform PANI layer on the PDMS surface and increase the overall coverage, it is essential to form a SAM. Because the surface is not chemically uniform in the pristine PDMS, it is subjected to O_2 plasma and then silane (here, APTES) treatments to form a chemically uniform surface capable of achieving good adhesion with PANI.

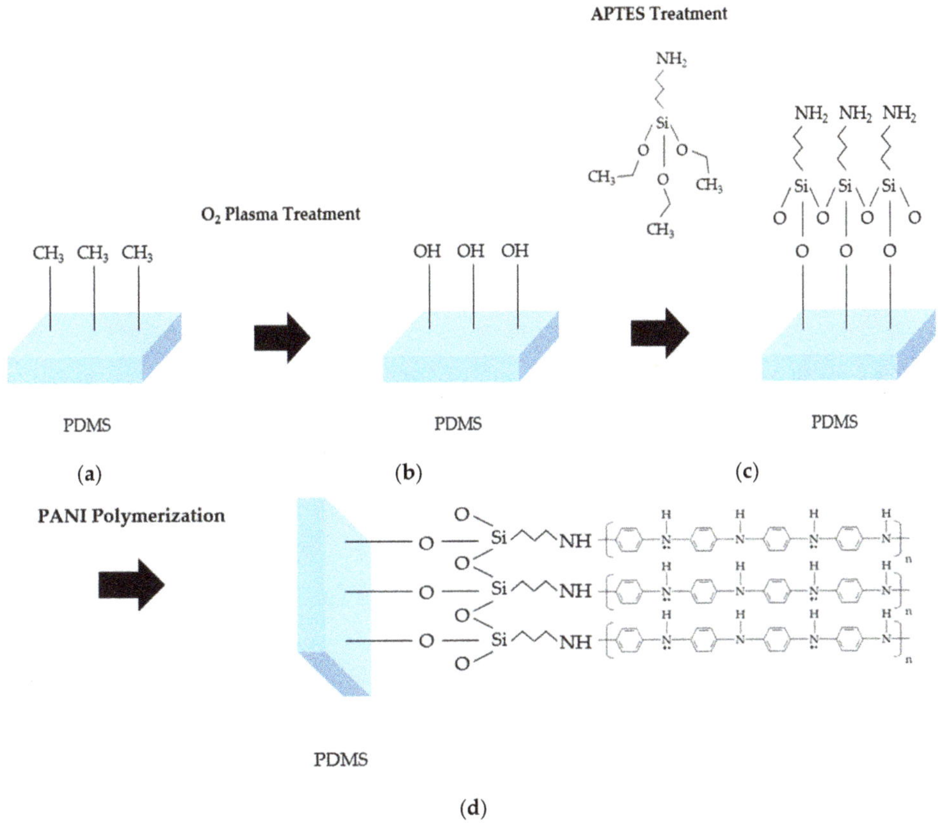

Figure 2. PDMS surface modification: (**a**) pristine PDMS; (**b**) O_2 plasma treatment; (**c**) APTES treatment; (**d**) PANI polymerization after surface modification.

3.2. Scanning Electron Microscopy (SEM)

We selected PDMS as the substrate and confirmed PANI film growth based on surface treatment in the dilute polymerization process of PANI. The surface morphology and elemental composition were studied through SEM-coupled energy-dispersive X-ray spectroscopy (EDS) analysis. Figure 3a,b shows the growth of PANI on PDMS without any surface treatment (sample 1); it is seen that the PANI film was deposited on the PDMS surface such that the density was sufficiently secured. However, Figure 3c,d shows PANI growth after O_2 plasma treatment of the PDMS (sample 2), and it is seen that the density is significantly less than that of PANI grown on pristine PDMS. This shows that synthesis is poor when PANI is grown immediately after O_2 plasma treatment. The formation of –OH group on PDMS substrate alone proves that PANI is difficult to synthesize normally. Figure 3e,f shows that APTES treatment was performed immediately after O_2 plasma treatment of the PDMS, followed by PANI growth (sample 3); here, the shape itself is not much different from the sample with PANI growth on pristine PDMS, but the density is sufficiently ensured, thickness is greater, and linkage is better. The EDS pattern of the sample 3 (Figure 3g) show several peaks for the PANI film, with carbon (C) and nitrogen (N) signals representing PANI, silicon (Si), and oxygen (O) signals derived from APTES surface treatment, and chloride (Cl) signals from perchloric acid.

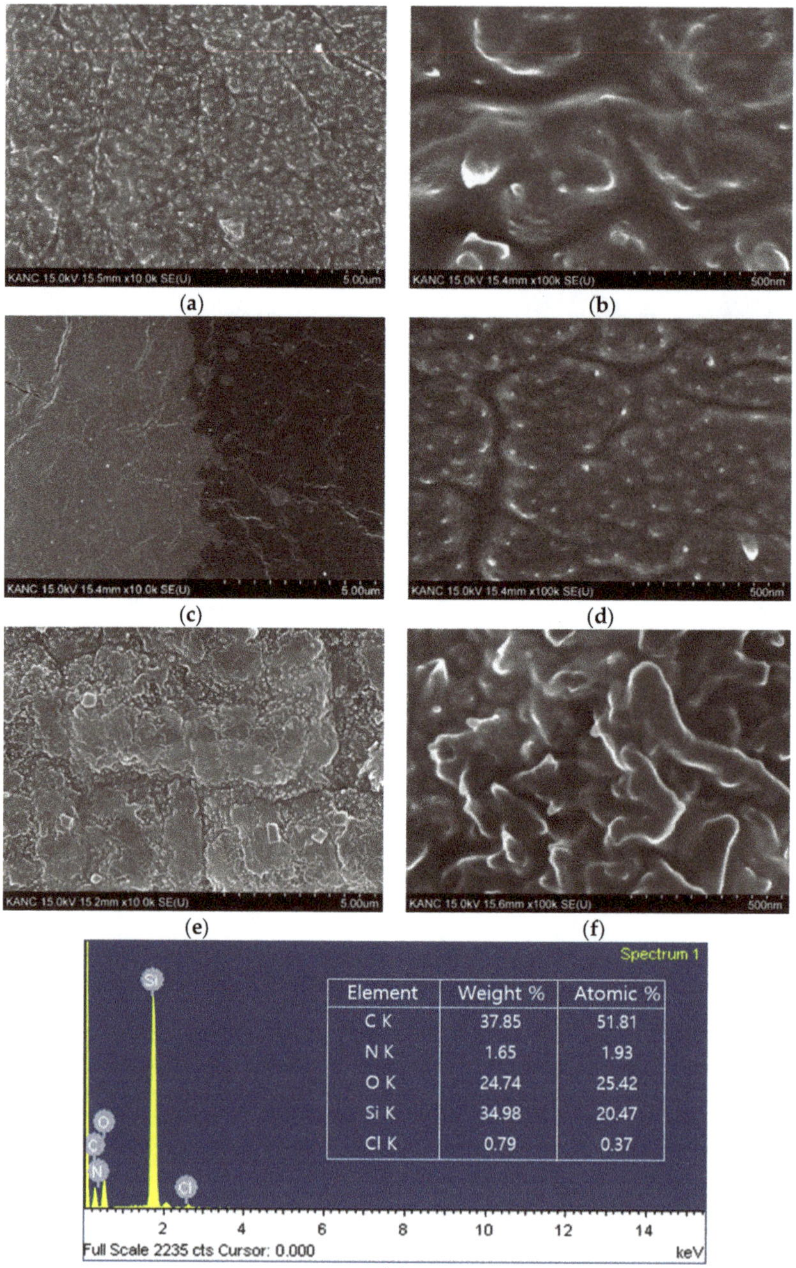

Figure 3. Scanning electron microscopy images of sample 1, 2 and 3 at 10 k and 100 k magnifications: (**a**,**b**) PANI on pristine PDMS; (**c**,**d**) PANI on PDMS after O_2 plasma treatment; (**e**,**f**) PANI on PDMS after O_2 plasma and APTES treatments; (**g**) energy-dispersive X-ray spectroscopy pattern and element composition of the PANI film (sample 3).

3.3. Contact Angle

The contact angle (CA) is a good indicator of the hydrophilicity or hydrophobicity of the surface. Figure 4 shows the contact angles of DI water between the PDMS and PANI surfaces. Since the pristine PDMS surface has $-CH_3$ functional groups, it is predicted that the surface is hydrophobic; thus, it is confirmed from the CA of over 110° that the pristine PDMS is hydrophobic (Figure 4a). From the O_2 plasma treatment of the PDMS surface, it is seen that the PDMS surface becomes hydrophilic with a CA of less than 20° (Figure 4b). However, after O_2 plasma treatment and PANI polymerization, the surface returns to the hydrophobic state with a CA over 110°, similar to that of pristine PDMS (Figure 4c). Thus, it is seen that the surface of PANI is also hydrophobic like that of PDMS. The O_2 plasma treatment process is hence required for good APTES deposition on the PDMS surface; when the CA is measured after completing APTES treatment and PANI synthesis, it is confirmed that the PANI surface becomes hydrophobic with a CA over 90° (Figure 4d).

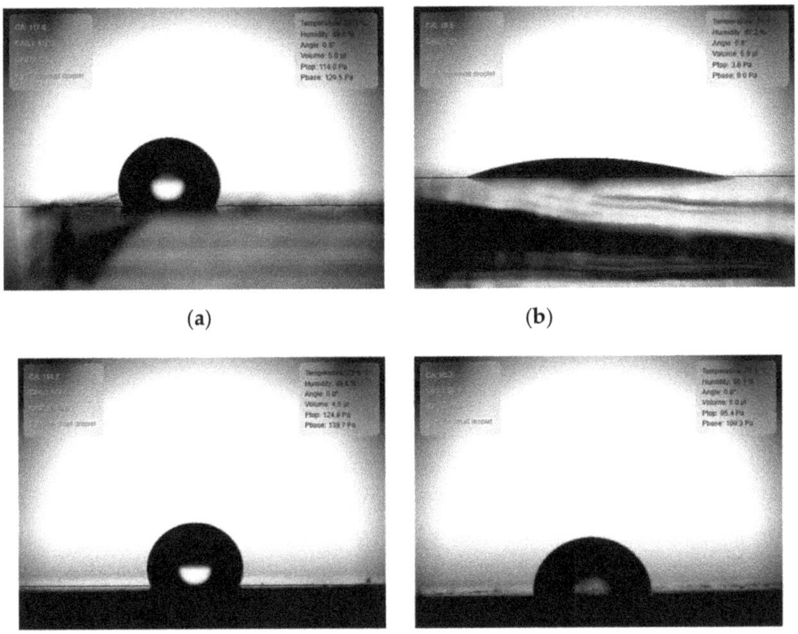

(a) (b) (c) (d)

Figure 4. Comparison of contact angles: (**a**) pristine PDMS; (**b**) PDMS after O_2 plasma treatment; (**c**) PANI growth on PDMS after O_2 plasma treatment; and (**d**) PANI growth on PDMS after O_2 plasma and APTES treatments.

3.4. Strain and I–V Characteristics

PANI can be formed in three types as leucoemeraldine, emeraldine, and pernigraniline through the synthesis processes and represents the base and salt structures via redox reactions in each form. One of the characteristics of PANI is that it shows different colors depending on the synthesized form. Among these, the emeraldine salt structure having conductivity exhibits a green color. Figure 5 illustrates the synthesis of PANI on PDMS substrate, and all three illustrations above show a green color and conductivity; thus, it is expected that PANI would have an emeraldine salt structure. Figure 5a shows PANI synthesized on PDMS without any treatment, and Figure 5b shows PANI synthesis after O_2 plasma treatment. Compared with synthesis on pristine PDMS, when a plasma treatment is performed, the coverage is significantly lowered by the PDMS surface property, which become hydrophilic owing to the –OH groups. However, as in Figure 5c, when PANI synthesis is performed after O_2 plasma and APTES treatment with $-NH_2$ functional groups, the coverage is confirmed to be better than that after plasma treatment of the pristine surface. The $-NH_2$ functional group is hydrophilic similar to the –OH group, but both PANI and APTES have the same $-NH_2$ functional groups, which offer advantages in the synthesis process.

(**a**) (**b**) (**c**)

Figure 5. PANI growth on PDMS: (**a**) PANI on pristine PDMS; (**b**) PANI on PDMS after O_2 plasma treatment; (**c**) PANI on PDMS after O_2 plasma and APTES treatments.

We conducted I–V measurements according to the strain for PANI on pristine PDMS and PANI on PDMS after O_2 plasma and APTES treatments. PANI on PDMS after O_2 plasma treatment is judged to have not been polymerized on the substrate because of significantly lower coverage than those of the other samples, thereby showing the characteristics of poor conductor; hence, I–V measurements were not shown here. Figure 6a shows the current measurements performed by attaching nickel tape and varying the voltage from −2 V to 2 V in steps of 0.1 V with dual sweeping. The strain was assessed for 0%, 10%, 20%, 30%, 40% and 50%, and Figure 6b–d shows the I–V values according to each strain. Figure 6b shows I–V trends according to strain of the PANI polymerization sample for pristine PDMS. When strain was 0%, it showed about 5×10^{-9} A at 2 V, and as seen from the graph, it was not possible to ensure stable current values based on the voltage and strain values. Figure 6c shows the I–V measurements for PANI on PDMS after O_2 plasma and APTES treatments, which has a value of 6×10^{-7} A at 2 V with 0% strain, and it was possible to obtain stable measurements depending on the voltage. In addition, when the strain was gradually increased from 0% to 50%, the current decreased, and when the maximum tensile value was 50%, a current of about 3×10^{-8} A was obtained. It should be noted here that a stable current value was seen in accordance with the voltage value for voltages of 0 V to 2 V, and the current value tended to decrease in accordance with 0% to 50% strain.

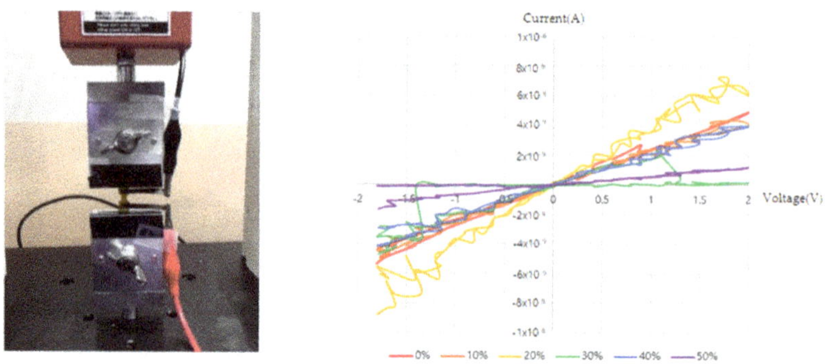

(**a**) (**b**)

Figure 6. *Cont.*

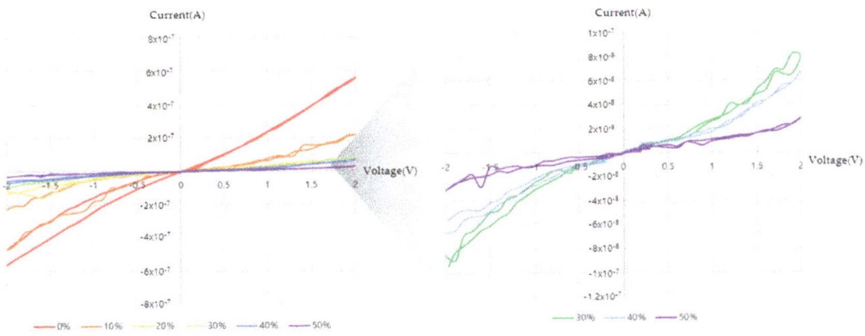

(**c**) (**d**)

Figure 6. I–V characteristics: (**a**) connection to the tensile machine; (**b**) PANI growth on pristine PDMS; (**c**) PANI growth on PDMS after O_2 plasma and APTES treatments; (**d**) magnified view of PANI growth characteristics on PDMS after O_2 plasma and APTES treatments.

Figure 7 shows the GF values. In Figure 7a, the GF is in the range of 6–14 when the strain is 50%; this is not only a low value for a strain sensor but also shows poor stability of the sample measuring at different strains. However, from Figure 7b, a GF of 12–30 is confirmed when the strain is 10%, and a high GF of 38–49 is available when the strain is 50%. Relatively high and stable GF values can be secured. Additionally, we prepared three more samples after O_2 plasma and APTES treatments. Although the GF value at 50% strain is not able to measure, average GF values with the strain from 10 to 40% are more stable according to the strain as shown in Figure 7c. Thus, the PANI synthesis on PDMS after APTES treatment of the strain sensor produces higher sensitivity and reliability than pristine PDMS.

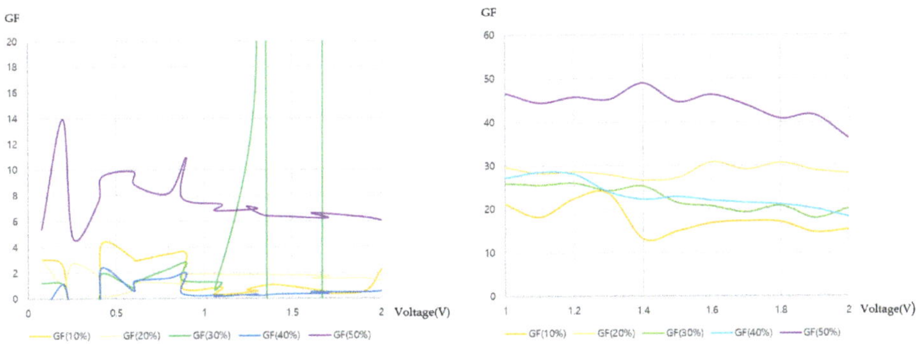

(**a**) (**b**)

Figure 7. *Cont.*

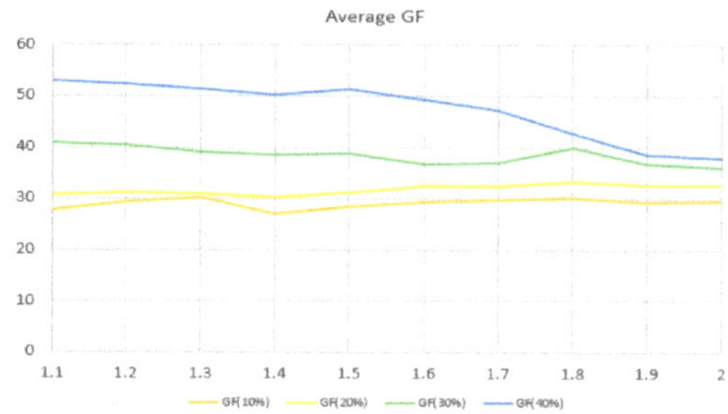

(c)

Figure 7. Gauge factor: (**a**) PANI growth on pristine PDMS; (**b**) PANI growth on PDMS after O_2 plasma and APTES treatments; (**c**) average of three PANI samples on PDMS after O_2 plasma and APTES treatments.

4. Discussion

Various applications using flexible substrates are being commercialized and various materials are being developed. Among them, there are different process methods to utilize the good characteristics of PANI. However, the synthesis of PANI on PDMS is a well-known process, but it is not easy to obtain stable properties by directly synthesizing PANI on PDMS substrates with high uniformity. The process of bonding two different substances without any intermediate treatment is not easy when considering intermolecular interconnections. In order to obtain stable properties by combining heterogeneous substances, a surface treatment process must be accompanied. In this study, we propose a process to obtain higher uniformity and reliable operational characteristics than previously known PANI synthesis on PDMS through chemical surface treatment. In particular, the surface treatment method using a SAM is a very important part of synthesizing a conductive polymer on a flexible substrate. The result of PANI on PDMS synthesis without any surface treatment and the result of PANI on PDMS synthesis through surface treatment were compared. Finally, the PANI synthesis on the PDMS strain sensor after O_2 plasma and APTES treatments produced a maximum GF of 49 at 50% strain, which shows that this configuration obtained higher coverage and better electrical characteristics than PANI synthesis on a pristine PDMS strain sensor. High GF at a high strain rate was secured, which is considered to be a good indicator in using PANI and PDMS in the future. We believe that strain sensors manufactured using surface treatments to synthesize conductive polymers can be used in various applications, such as wearables, electronic skin, and robotics.

Author Contributions: Methodology, conceptualization, formal analysis, and writing—original draft, C.-J.Y.; conceptualization and formal analysis, J.-Y.C.; data curation and validation, H.-K.Y. and J.-H.H.; software and visualization, E.-B.J. and J.-H.L.; resources, funding acquisition, supervision, and writing—review and editing, H.-S.K. All authors have read and agreed to the published version of the manuscript.

Funding: This work was supported by the Mid-career Researcher Program through the National Research Foundation of Korea (NRF) funded by the Ministry of Science and ICT (No. 2019R1A2C2086747).

Conflicts of Interest: The authors declare no conflict of interest.

References

1. Akheel, A.S.; Maravattickal, K. Review: Polyaniline a novel polymeric material. *Talanta* **1991**, *38*, 815–837.
2. Huang, Z.; Wang, P.-C.; MacDiarmid, A.G. Selective deposition of conducting polymers on hydroxyl-terminated surfaces with printed monolayers of alkylsiloxanes as templates. *Langmuir* **1997**, *13*, 6480–6484. [CrossRef]
3. Gangopadhyay, R.; Dea, A. Conducting polymer nanocomposites: A brief overview. *Chem. Mater.* **2000**, *12*, 608–622. [CrossRef]

Unsworth, J.; Lunn, B.A.; Innis, P.C.; Jin, Z.; Kaynak, A.; Booth, N.G. Conducting polymer electronics. *J. Intell. Mater. Syst. Struct.* **1992**, *3*, 380–395. [CrossRef]

Choy, J.-Y.; Jo, E.-B.; Yim, C.-J.; Youi, H.-K.; Hwang, J.-H.; Lee, J.-H.; Kim, H.-S. Improvement in Strain Sensor Stability by Adapting the Metal Contact Layer. *Sensors* **2022**, *22*, 630. [CrossRef]

Jia, X.-S.; Tang, C.-C.; Yan, X.; Yu, G.-F.; Li, J.-T.; Zhang, H.-D.; Li, J.-J.; Gu, C.-Z.; Long, Y.-Z. Flexible polyaniline/poly(methyl methacrylate) composite fibers via electrospinning and in situ polymerization for ammonia gas sensing and strain sensing. *J. Nanomater.* **2016**, *2016*, 9102828. [CrossRef]

Gong, S.; Lai, D.T.; Wang, Y.; Yap, L.W.; Si, K.J.; Shi, Q.; Jason, N.N.; Sridhar, T.; Uddin, H.; Cheng, W. Tattoolike polyaniline microparticle-doped gold nanowire patches as highly durable wearable sensors. *ACS Appl. Mater. Interfaces* **2015**, *7*, 19700–19708. [CrossRef]

Hong, S.Y.; Lee, Y.H.; Park, H.; Jin, S.W.; Jeong, Y.R.; Yun, J.; You, I.; Zi, G.; Ha, J.S. Stretchable active matrix temperature sensor array of polyaniline nanofibers for electronic skin. *Adv. Mater.* **2016**, *28*, 930–935. [CrossRef]

Wagner, S.; Lacour, S.P.; Jones, J.; Pai-hui, I.H.; Sturm, J.C.; Li, T.; Suo, Z. Electronic skin: Architecture and components. *Phys. E* **2004**, *25*, 326–334. [CrossRef]

Yeo, J.C.; Yap, H.K.; Xi, W.; Wang, Z.; Yeow, C.-H.; Lim, C.T. Flexible and stretchable strain sensing actuator for wearable soft robotic applications. *Adv. Mater. Technol.* **2016**, *1*, 1600018. [CrossRef]

Pantelopoulos, A.; Bourbakisa, N.G. A survey on wearable sensor-based systems for health monitoring and prognosis. *IEEE Trans. Syst. Man Cybern. Part C* **2009**, *40*, 1–12. [CrossRef]

Huang, Y.; Gao, L.; Zhao, Y.; Guo, X.; Liu, C.; Liu, P. Highly flexible fabric strain sensor based on graphene nanoplatelet–polyaniline nanocomposites for human gesture recognition. *J. Appl. Polym. Sci.* **2017**, *134*, 45340. [CrossRef]

Jiang, J.-K.; Liang, Q.-H.; Men, R.-S.; Yang, Q.; Sun, X.; Yang, D.-G.; Zhang, G.-Q.; Chen, X.-P. The influence of tensile stress on polyaniline as strain sensor. *IEEE Electron. Device Lett.* **2016**, *37*, 1636–1638. [CrossRef]

Cao, Y.; Andreatta, A.; Heeger, A.J.; Smith, P. Influence of chemical polymerization conditions on the properties of polyaniline. *Polymer* **1989**, *30*, 2305–2311. [CrossRef]

Hussain, A.M.; Kumar, A. Electrochemical synthesis and characterization of chloride doped polyaniline. *Bull. Mater. Sci.* **2003**, *26*, 329–334. [CrossRef]

Chen, W.; Rakhi, R.B.; Alshareef, H.N. Facile synthesis of polyaniline nanotubes using reactive oxide templates for high energy density pseudocapacitors. *J. Mater. Chem. A* **2013**, *1*, 3315–3324. [CrossRef]

Zhang, X.; Chan-Yu-King, R.; Jose, A.; Manohara, S.K. Nanofibers of polyaniline synthesized by interfacial polymerization. *Synth. Met.* **2004**, *145*, 23–29. [CrossRef]

Chiou, N.R.; Lu, C.; Guan, J.; Lee, L.J.; Epstein, A.J. Growth and alignment of polyaniline nanofibers with superhydrophobic, superhydrophilic and other properties. *Nat. Nanotechnol.* **2007**, *2*, 354–357. [CrossRef] [PubMed]

Chiou, N.-R.; Epstein, A.J. Polyaniline nanofibers prepared by dilute polymerization. *Adv. Mater.* **2005**, *17*, 1679–1683. [CrossRef]

Zhang, D.; Wang, Y. Synthesis and applications of one-dimensional nano-structured polyaniline: An overview. *Mater. Sci. Eng. B* **2006**, *134*, 9–19. [CrossRef]

Stejskal, J.; Spirkova, M.; Riede, A.; Helmstedt, M.; Mokreva, P.; Prokes, J. Polyaniline dispersions 8. The control of particle morphology. *Polymer* **1999**, *40*, 2487–2492. [CrossRef]

Niu, F.; Wang, Y.; Zhang, Y.; Xi, S.; Ma, L.; Wang, C.; Mao, Y. A hierarchical architecture of PANI/APTES/SiC nanocomposites with tunable dielectric for lightweight and strong microwave absorption. *J. Mater. Sci.* **2019**, *54*, 2181–2192. [CrossRef]

Lee, Y.B.; Park, C.H.; Jang, S.K.; Kim, F.S. Mechanical Hardness and Adhesion Property of Surface-Polymerized Nanocomposites based on Polyaniline and Sol-Gel Silica. *J. Korean Phys. Soc.* **2019**, *75*, 248–253. [CrossRef]

Zhang, P.; Cao, L.; Wang, X.; Cui, J.; Lin, Z.; Ngai, S.; Vogel, F.; Wang, H.; Li, W.; Li, S.; et al. Improvement of electrochemical performance of titania nanowires for supercapacitor electrodes by in-situ growth of polyaniline nanoparticles. *Ceram. Int.* **2022**, *48*, 1731–1739. [CrossRef]

Wang, H.; Ma, L.; Gan, M.; Zhou, T.; Sun, X.; Dai, W.; Wang, H.; Wang, S. Fabrication of polyaniline/urchin-like mesoporous TiO_2 spheres nanocomposite and its application in supercapacitors. *Electrochim. Acta* **2015**, *163*, 232–237. [CrossRef]

Gong, X.X.; Fei, G.T.; Fu, W.B.; Fang, M.; Gao, X.D.; Zhong, B.N.; Zhang, L.D. Flexible strain sensor with high performance based on PANI/PDMS films. *Org. Electron.* **2017**, *47*, 51–56. [CrossRef]

Flavel, B.S.; Yu, J.; Shapter, J.G.; Quinton, J.S. Patterned polyaniline & carbon nanotube–polyaniline composites on silicon. *Soft Matter* **2009**, *5*, 164–172.

Zou, W.; Quan, B.; Wang, K.; Xia, L.; Yao, J.; Wei, Z. Patterned growth of polyaniline nanowire arrays on a flexible substrate for high-performance gas sensing. *Small* **2011**, *7*, 3287–3291. [CrossRef] [PubMed]

Cai, G.; Wang, J.; Lin, M.-F.; Chen, J.; Cui, M.; Qian, K.; Li, S.; Cui, P.; Lee, P.S. A semitransparent snake-like tactile and olfactory bionic sensor with reversibly stretchable properties. *NPG Asia Mater.* **2017**, *9*, e437. [CrossRef]

Johnstan, I.D.; McCluskey, D.K.; Tan, C.K.L.; Tracey, M.C. Mechanical characterization of bulk Sylgard 184 for microfluidics and microengineering. *J. Micromech. Microeng.* **2014**, *24*, 035017. [CrossRef]

Mazurek, P.; Hvilsted, S.; Skov, A.L. Green silicone elastomer obtained from a counterintuitively stable mixture of glycerol and PDMS. *Polymer* **2016**, *87*, 1–7. [CrossRef]

Jeronimo, K.; Koutsos, V.; Cheung, R.; Mastropaolo, E. PDMS-ZnO Piezoelectric Nanocomposites for pressure sensors. *Sensors* **2021**, *21*, 5873. [CrossRef]

sensors

MDPI

Article

Design and Control of Multi-Plate MR Clutch Featuring Friction and Magnetic Field Control Modes

Jin-Young Park [1], Jong-Seok Oh [2,*] and Young-Choon Kim [2,*]

[1] Department of Mechanical Engineering, Kongju National University, Cheonan 31080, Korea; pjy0186@kongju.ac.kr

[2] Department of Future Automotive Engineering, Kongju National University, Cheonan 31080, Korea

* Correspondence: jongseok@kongju.ac.kr (J.-S.O.); yckim59@kongju.ac.kr (Y.-C.K.)

Abstract: A magnetorheological (MR) multi-plate clutch was proposed with both mechanical friction mode and magnetic field control modes. The magnetic field control mode was based on an MR fluid coupler that changed its viscous properties according to the density of an applied magnetic field. This mode was used in the early stage of clutch operation to reduce the impact of friction between the disc and plate, and eliminate to the extent possible the difference in their relative speeds when contacting each other in later stages. Once the rotational speed difference between the disc and plate was reduced, the clutch was operated in mechanical friction mode by compressing the friction surfaces together. A torque modeling equation was then derived for each mode based on the Bingham model of the MR fluid, and the transmission torque of the proposed multi-plate clutch was derived using these equations as well as magnetic field analysis results obtained using ANSYS Maxwell. A multi-plate MR clutch was then fabricated, and its torque transmission characteristics were evaluated in the magnetic field control and mechanical friction modes. The results confirmed that the model-based torque calculations were consistent with the observed transmission torque. Finally, control algorithms for mechanical friction only and mixed mechanical friction/magnetic field control torque tracking of the proposed MR multi-plate clutch were designed, and their performances were evaluated when applying unit step command, half-sine-wave command, and rotational speed changes. The results indicated that the torque tracking control was performed smoothly, demonstrating the advantages of the proposed clutch.

Keywords: magnetorheological (MR) fluid; multi-plate MR clutch; torque transmission; magnetic field control mode

1. Introduction

Interest in high-performance power transmission components, such as clutches, has recently been increasing owing to the demands for high-quality power transmission systems. Most power transmission components use multiplate friction clutches to provide high power and smooth transmission characteristics; however, noise and vibration are generated by the friction between the clutch disc and plate, potentially decreasing rider comfort. Research and development of new friction clutch systems is therefore actively underway to address this shortcoming, and has to date primarily focused on improving the control methods and reducing friction noise and vibration using algorithms. For example, Vasca et al. proposed a torque transmission model for a dry clutch that applied the effect of clutch spring characteristics [1]; Jibin et al. analyzed the effects of contact angle and drag torque according to the friction plate pattern and wet clutch spacing [2]; Agostino et al. applied various slip control methods to realize the soft combination of dry clutch and power [3]; and Desmidt et al. analyzed a power transmission system to propose a weight optimization method for a multiplate clutch in a two-speed rotorcraft transmission [4]. As demonstrated by these examples, many studies have attempted to realize smooth clutch fastening, but it remains impossible to ideally reduce the shock and vibration that occur during the clutch fastening process. New control methods and materials to address this issue are therefore the subject of intense ongoing research.

One such material, magnetorheological (MR) fluid increases its resistance to flow under a magnetic field load. As the yield stress of the MR fluid changes according to the applied magnetic field, efforts have been undertaken to apply it in a wide variety of systems, including dampers, valves, mounts, clutches, and brakes. Two MR fluid modes have been widely used to date according to the fluid stimulation mechanism and relative motion. The first is the flow mode, in which the fluid moves through fixed stimuli (such as in a damper, valve, or mount); in this mode, changes in the pressure and flow rate of the fluid in a pipe occur simultaneously. The second is the shear mode or rotational shear mode, in which the fluid passes between two parallel plates, one of which is fixed and the other of which rotates or moves (such as in clutch or brake systems). Notably, Mukhopadhyay et al. investigated the effect of plate orientation on MR fluid behavior in shear mode under a magnetic field [5].

Ongoing research on MR fluids has improved their performance to the extent that they have been successfully applied in various clutches. Neelakantan and Washington proposed a method applying the MR sponge concept to solve problems related to the centrifuging of the MR fluid owing to clutch rotation [6]. Benetti and Dragoni performed design and electromagnetic field analyses to develop a 100 Nm-class multiplate MR brake [7]. Gratzer et al. suggested an efficient MR clutch that combined the clutch design for an all-wheel drive vehicle with fluid development and magnetic circuit optimization [8]. Kikuchi et al. suggested a design method for an MR multiplate clutch with a 50 μm MR fluid layer between the input and output plates, and presented the results of an experimental evaluation [9]. Zhang et al. proposed a design method using nondimensional parameters for a large-capacity MR single-plate clutch [10]. Güth et al. suggested a new design concept combining a variable-speed electric actuator and MR clutch [11]. Törőcsik studied the transmission torque and corresponding power consumption according to the number and thickness of plates used in an MR multiplate clutch [12]. Wang et al. proposed a mechanical method for increasing the number of plates to generate a high transmission torque with an MR multiplate clutch [13]. Kavlicoglu et al. analyzed the output torque of a single-plate clutch employing an MR grease with six different iron particle sizes [14]; they also developed a double-plate prototype of an MR clutch after conducting a performance design using a variable torque range according to changes in input current [15]. Thakur and Sarkar proposed a simulation-based method to estimate the maximum transmission torque and response time of an MR clutch using the shear mode of the MR fluid [16]. Yadmellat et al. presented simulation and experimental results describing the performance of an MR clutch using the Preisach model, constituting a hysteresis model approach [17]. Bucchi et al. suggested a method for improving torque characteristics by applying a neural network to an MR clutch controller [18]. Pilch proposed an MR fluid for an optimal clutch design based on simulation results to realize variable controllability according to temperature [19]. Leong et al. proposed a method for the design of a magnetic coupler to amplify axial speed using magnetic attraction [20]. Fernando suggested the use of magnetic field modulation and an EM clutch slip controller through d–q current control [21]. Kavlicoglu et al. proposed a design and slip controller for a multiplate limited-slip differential clutch to achieve high torque transmission using an MR fluid [22]. Rizzo et al. proposed a method to improve the transmission torque of an MR fluid clutch using the eddy current induced in the conduction region by relative rotation [23]. Latha et al. proposed MR fluid coupler type multiplate clutch to overcome mechanical friction [24]. Olszak et al. designed a hydrodynamic clutch that incorporated an MR fluid and presented the results of simulations and experiments considering friction loss [25]. Zhang et al. designed the magnetic field structure of a new dual clutch and suggested a relationship between torque and current using a finite element analysis [26]. Finally, Kluszczynski and Pilch compared the clutch torque per total volume from the perspectives of torque, diameter, external length, and ratio according to the number of plates in an MR clutch [27].

Recently, Park et al. developed an MR fluid-based multiplate clutch that provided the advantages of both magnetic field control mode and mechanical friction mode, a notable advance beyond the conventional MR multiplate clutch, which only supports magnetic field control mode [28]. An optimal design of this clutch was previously achieved using torque modeling of mechanical friction mode and an electromagnetic field analysis of magnetic field control mode, and its performance was evaluated in each mode. However, torque tracking control employing both the magnetic field control and mechanical friction modes has yet to be evaluated for the proposed clutch.

Indeed, the following outstanding research issues related to MR clutches were identified by the authors:

1. There is no existing MR clutch capable of transmitting power by compression force;
2. There is a lack of research on hybrid MR clutches that employ two operating modes;
3. The MR fluid clutch can transmit high power and provide smooth engagement but requires a large size or high current; however, studies on the minimum size or current have been lacking;
4. Existing studies have thus far failed to verify the torque transmission and control performance of the MR fluid clutch in both magnetic field control and mechanical friction modes.

Therefore, in this study, the rotational shear mode—a characteristic of the MR fluid—was used to realize the magnetic field control of a multi-plate MR clutch, while the operation mode of a conventional friction clutch was used to realize the mechanical friction mode. To do so, torque modeling and electromagnetic field analysis of the MR single-plate clutch were performed, and the transmission torque was identified by fabricating a prototype multi-plate clutch. The applied current and compression forces were then controlled to investigate its torque tracking performance using an algorithm designed to realize smooth multi-plate MR clutch operation.

2. Torque Modeling of the Multi-Plate MR Clutch

The MR multi-plate clutch proposed in this study was able to operate in one of two modes—mechanical friction or magnetic field control—by moving the output shaft in the axial direction, as shown in Figure 1. In mechanical friction mode, shown in Figure 1a, power is transmitted from the input shaft to the output shaft through the friction between the input and output plates, similar to the operation of a conventional general clutch. In magnetic field control mode, shown in Figure 1b, power is transmitted between the input and output plates through the shear stress induced in the MR fluid by sliding in its shear mode according to its properties as dictated by the applied magnetic field. In this section, the properties of the MR fluid are first briefly discussed, then the torque modeling in each mode is described based on these properties.

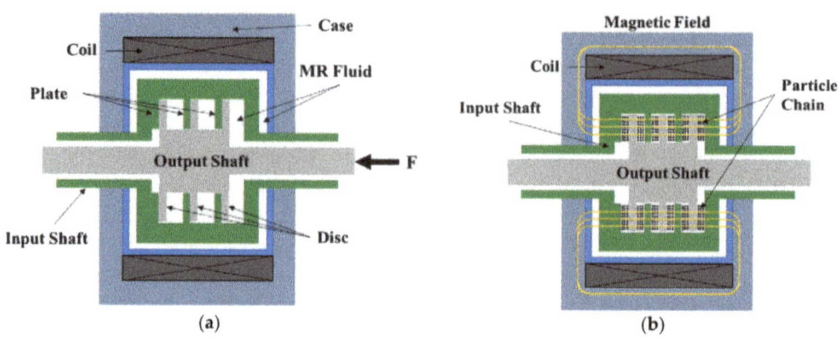

Figure 1. Operational concept of the multi-plate MR clutch: (**a**) mechanical friction mode; (**b**) magnetic field control mode.

2.1. Modeling of MR Fluid

Generally, MR fluids consist of magnetic (typically iron) particles suspended in a low-permeability solvent carrier fluid; the particles used in this study were micron-sized iron particles. The resistance of the MR fluid to flow increases under an applied magnetic field load. When there is no magnetic field load, an MR fluid exhibits the same behavior as a Newtonian fluid, in which the suspended particles move freely. However, in the presence of a magnetic field load, an MR fluid exhibits Bingham fluid behavior, presenting a yield strength as the particles are charged to form a chain structure within the fluid. Thus, the rheological properties of an MR fluid change with the magnetic field strength. The shear stress in the MR fluid τ can therefore be predicted using the Bingham model as follows [29,30]:

$$\tau = \tau_y + \eta\dot{\gamma} \quad (1)$$

where τ_y and η denote the yield stress and viscosity coefficient after MR fluid yielding, respectively, and $\dot{\gamma}$ is the shear rate of the MR fluid at its interface with the discs and plates. Because the gap between each torque input plate and torque output disc is very small, a linear distribution of shear can be assumed. Thus, the shear rate of the MR fluid at its interfaces can be determined as follows:

$$\dot{\gamma} = \frac{r\omega}{d} \quad (2)$$

where r and d are the radius of the smallest area on the disc and the gap between the disc and plate, respectively, and ω is the difference in angular velocity between the disc and plate. The yield shear stress in the MR fluid represents the shear stress at the moment the fluid begins to flow when under an applied magnetic field. Because the fluid flow begins in a very short time and exhibits complex behaviors, it is difficult to accurately measure the corresponding yield shear stress. Thus, a linear equation for shear rate and shear stress was derived, and the shear stress when the shear rate was zero was defined as the dynamic yield stress. In general, the dynamic yield stress can be treated as the yield shear stress of the MR fluid and is expressed as a function of the magnetic field intensity.

Previous studies have ignored the claim that the viscosity of the MR fluid has a significant effect on the applied magnetic field. To consider this issue in the present study, the approximate constant was used as a unique characteristic value determined by the type of solvent and particles in the MR fluid, their composition ratio, and the composition environment [31,32] as follows:

$$\eta = \eta_\infty + (\eta_0 - \eta_\infty)(2e^{-B\sigma_{S\eta}} - e^{-2B\sigma_{S\eta}}) \quad (3)$$

$$\tau_y = \tau_{y\infty} + (\tau_{y0} - \tau_{y\infty})(2e^{-B\sigma_{S\tau_y}} - e^{-2B\sigma_{S\tau_y}}) \quad (4)$$

where η_∞ and η_0 respectively denote the saturated and natural states of the viscosity coefficient, $\tau_{y\infty}$ and τ_{y0} respectively denote the saturated and natural states of the yield stress, $\sigma_{S\eta}$ is the saturated moment for the viscosity coefficient, $\sigma_{S\tau_y}$ is the saturated moment for the yield stress, and B is the magnetic field density applied to the MR fluid.

Compared to Ferrofluids [33], MR fluids consist of magnetic (typically iron) particles in a carrier fluid. In the presence of a magnetic field, the micron-sized particles link and change the fluid to a semi-solid in milliseconds. When the magnetic field is removed, the fluid just as quickly reverts back to its natural free-flowing state.

The unique characteristics of the MRF-132DG MR fluid (Lord Cop.) used in this study are listed in Table 1. For more detailed information, please refer to technical data from Lord Corp. [34].

Table 1. Rheological properties of the MRF-132DG MR fluid.

Parameter	Value
Natural state of viscosity coefficient, η_0	0.1 [pa·s]
Saturated state of viscosity coefficient, η_∞	3.8 [pa·s]
Saturated moment for viscosity coefficient, $\sigma_{S\eta}$	4.5 [T^{-1}]
Natural state of yield stress, τ_{y0}	12 [pa]
Saturated state of yield stress, $\tau_{y\infty}$	40,000 [pa]
Saturated moment for yield stress, $\sigma_{S\tau_y}$	2.9 [T^{-1}]
Density	2.95–3.15 [g/cm^3]
Solids content by weight	80.98%
Operating temperature	−40 to 130 [°C]

2.2. Torque Modeling of Mechanical Friction Mode

A multiplate clutch has N friction surfaces, and its size and weight are different from those of a single-plate clutch. However, torque modeling of a single-plate clutch was performed in this study to design the proposed multi-plate MR clutch, assuming that the size and weight were the same to simplify the mathematics. It was assumed that the single-plate clutch is composed of input and output shafts, with a friction pad in between.

The transmission torque of the clutch can be obtained through torque modeling, and the axial compression force F applied to the cross section of the clutch plate can be expressed as follows:

$$F = \int_{r_i}^{r_o} 2\pi rpdr = \pi p(r_o^2 - r_i^2) \tag{5}$$

where p is the pressure acting on the friction pad cross section when the clutch is connected, assumed to act uniformly per unit area, and r_i and r_o are the inner and outer diameters of the clutch friction pad, respectively.

The transmission torque T_M generated by the compression between clutch disc and plate in a single-plate clutch is described as follows:

$$T_M = \int\int_{r_i}^{r_o} r\mu \mathrm{sgn}(\omega) dF dr = \frac{2\pi\mu p}{3}(r_o^3 - r_i^3)\mathrm{sgn}(\omega) = \frac{3F\mu(r_o^3 - r_i^3)}{3(r_o^2 - r_i^2)}\mathrm{sgn}(\omega) \tag{6}$$

where μ is the friction coefficient of the clutch friction pad, defined as a function of the plate's angular speed ω and pressure p. For a more detailed explanation of torque modeling of mechanical friction mode, please refer to the authors' previous study [28].

2.3. Torque Modeling of Magnetic Field Control Mode

In magnetic field control mode, the torque generated by the magnetic field induced viscosity of the MR fluid is transmitted from the friction surface of the input shaft to the friction surface of the output shaft. The transmitted torque is generated by the friction of the MRF acting on its interfaces with the disc and plate as well as the outer annular enclosure. The magnetic field control mode torque T_{MR} of an MR single-plate clutch can be determined by the yield shear stress τ_y and friction area of the MR fluid and can therefore be derived from the properties of the MR fluid as follows:

$$T_{MR} = \int_A \tau r dA = 2\pi \int_{r_i}^{r_o} \tau r^2 dr \tag{7}$$

Substituting Equations (1) and (2) into Equation (7) and rearranging, the following equation can be obtained:

$$T_{MR} = 2\pi \int_{r_i}^{r_o} \tau_y r^2 dr + 2\pi \int_{r_i}^{r_o} r^2 \eta\left(\frac{r\omega}{d}\right) dr \tag{8}$$

Shear stress is also generated in the MR fluid in the annular enclosure between the ends of the disc and plate of the multi-plate MR clutch. Hence, similar to the MRF fluid in the disc–plate gap, using the average magnetic density obtained from numerical integration and assuming a linear distribution of the shear rate in the gap, the following expression can be obtained [32]:

$$T_{MR} = 2\pi r^2 l\left(\tau_y + \eta\frac{r\omega}{d}\right) \tag{9}$$

where r and l denote the radius and length of the annular enclosure, respectively.

When no magnetic field is applied to the MR fluid, the torque transmission equation of the MR single-plate clutch is given as follows [35–37]:

$$T_0 = \frac{\pi\eta_0}{h}(r_o^4 - r_i^4)\omega + \frac{4\pi\tau_{y0}}{3}(r_o^3 - r_i^3) + 2\pi r_o^2 t_d(\tau_{y0} + \eta_0\frac{r_o\omega}{h}) \tag{10}$$

The torque Equations (6)–(10) for a single-plate clutch were used to determine the torque model of the multiplate clutch designed in this study. Because the multi-plate MR clutch has N friction surfaces, the mechanical friction mode transmission torque T_{M_m} of the multiplate clutch can be expressed as:

$$\begin{aligned} T_{M\_m} &= T_M + T_0 = \tfrac{3F\mu(r_o^3 - r_i^3)}{3(r_o^2 - r_i^2)}\mathrm{sgn}(\omega)N \\ &+ \sum_{I=0}^{n} \tfrac{\pi\eta_{0I}}{h}(r_o^4 - r_i^4)\omega + \tfrac{4\pi\tau_{y0I}}{3}(r_o^3 - r_i^3) + 2\pi r_o^2 t_d\left(\tau_{y0I} + \eta_{0I}\tfrac{r_o\omega}{h}\right) \end{aligned} \tag{11}$$

where h is the gap between the disc and plate, t_d is the thickness of the disc and plate, τ_{y0I} and η_{0I} are the ith values of τ_{y0} and η_0, respectively, and n is the total number of gaps between discs and plates.

However, because torque transmission between the disc and plate is performed by the MR fluid according to the applied magnetic field, the torque must be calculated at every location in the fluid. Thus, the magnetic field control mode transmission torque T_{MR_m} is expressed as the sum of all torques acting in the gap between the disc and plate and can be expressed as:

$$\begin{aligned} T_{MR\_m} &= \sum_{I=0}^{n}\{T_{MR} + T_{th} + T_0\} \\ &= \sum_{I=0}^{n}\left\{ \begin{array}{l} \tfrac{2\pi\tau_{yI}}{3}(r_o^3 - r_i^3) + \tfrac{\pi\eta_I\left(\frac{\omega}{d}\right)}{2}(r_o^4 - r_i^4) \\ +2\pi r_o^2 t_d\left(\tau_{yI} + \mu_I\tfrac{r_o\omega}{h}\right) \\ +\tfrac{\pi\eta_{0I}}{h}(r_o^4 - r_i^4)\omega + \tfrac{4\pi\tau_{y0I}}{3}(r_o^3 - r_i^3) + 2\pi r_o^2 t_d\left(\tau_{y0I} + \eta_{0I}\tfrac{r_o\omega}{h}\right) \end{array} \right\} \end{aligned} \tag{12}$$

where η_I is the Ith value of η.

The value of t_d in this study was 23 mm, and h ranged from 0 to 2 mm. Note that the transmission torque of the MR friction clutch was obtained by ignoring the torque generated by the friction with the seal and in the bearing. In addition, the electromagnetic time constant for arranging the ferromagnetic particles along the magnetic field lines was 0.4–0.8 ms [36,37]. As such a small delay can be neglected, the response delay to reach the steady-state damping force is typically modeled as a first-order dynamic system [38].

3. Design and Magnetic Analysis of Multi-Plate MR Clutch

The size of the proposed multi-plate MR clutch was designed based on a small-capacity multiplate clutch. To predict the torque transmission of the designed multi-plate MR clutch, the torque generated by the mechanical friction in the clutch was calculated using the torque modeling Equation (8). The axial holding force F was evaluated at 0–350 N and the rotation speed was evaluated at 0–200 rpm, as shown in Figure 2. The objective dry torque was at 0–25 Nm, which is sufficient for power transmission. To meet the objective dry torque requirements, two design variables were established: the motor revolution speed was set to 200 rpm and the compression force was set to 320 N. These values resulted in a calculated dry torque of 26.5 Nm. The MR multi-plate clutch was modeled and fabricated as shown in Figure 3. During the design process, the geometric tolerance was set to 0.1 mm.

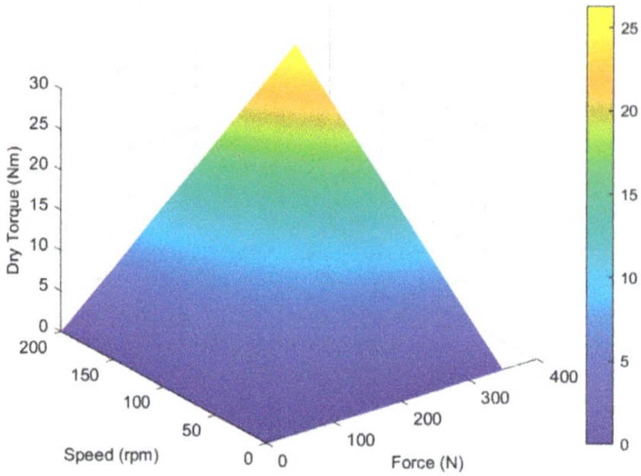

Figure 2. Torque transmission calculation results according to rotational speed and axial compression force.

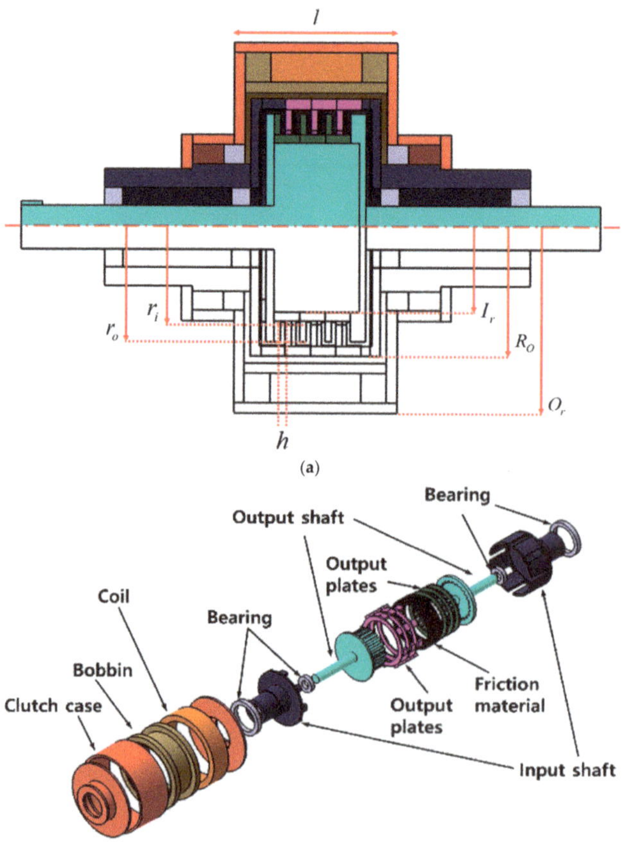

Figure 3. Multi-plate MR clutch (**a**) design parameters and (**b**) model.

The parts of the multi-plate MR clutch were fabricated with a machining tolerance of 0.05 mm, which represents the finest accuracy achievable with common machining tools. During the assembly of the multi-plate MR clutch, a uniform distribution of gaps between the plates was considered for simplicity. The final configuration and dimensions of the designed multi-plate MR clutch are summarized in Figure 3 and Table 2.

Table 2. Geometry of the proposed hybrid multi-plate MR clutch.

Parameter	Description	Value [mm]
O_r	Radius of output shaft	61
I_r	Radius of input shaft	44
r_i	Inner radius of clutch plate and disc	45
r_o	Outer radius of clutch plate and disc	55
h	Disc-to-plate gap	2
R_o	Outer radius of multi-plate MR clutch	88
ℓ	Length of multi-plate MR clutch except shaft	77

Before mathematically evaluating the multi-plate MR clutch torque transmission in magnetic field control mode, an electromagnetic analysis was performed using ANSYS Maxwell to investigate the formation of magnetic fields according to the gap between the disc and plate and the applied current strength. For this analysis, the inner space of the clutch was assumed to be filled by the MR fluid without voids. In addition, the MR fluid mesh was set denser than the clutch mesh to clearly identify the result of the magnetic field acting between the disc and plate, where torque is transmitted. The conditions for this electromagnetic analysis are listed in Table 3.

Table 3. Multi-plate MR clutch electromagnetic field analysis conditions.

Parameter	Value
Number of turns	1000
Coil resistance	22 [Ω]
Current	0–2 [A]
Mesh	200,000

The finite element model of the multi-plate MR clutch was constructed using the eight-vertice 2D-axisymmetric couple element (PLANE 13) included in the commercial ANSYS software to reflect the magnetic circuits. A total of elements 204,132 were employed in the completed model.

In addition, the temperature increase in the MR fluid arises partially from the fluid friction and partially from the Ohmic losses in the magnetic circuit. The characteristics of the damping force owing to viscous damping reduction at high temperatures were revealed by the ANSYS model. Such changes lead to a model mismatch and degradation of the control effect [39,40]. However, the proposed hybrid multi-plate MR clutch was provided with a torque tracking control mode that is capable of overcoming these changes. Accordingly, degradation owing to heating was neglected in this study.

To determine the strength of the magnetic field acting on the MR fluid between the clutch plates and discs using the electromagnetic field analysis, the average value in the MR fluid was derived by establishing four measurement locations at the gaps between the disc and plate along perpendicular lines passing through the center of the clutch shafts, as shown in Figure 4. The gap was varied between 0.02 and 2.0 mm in 0.2-mm increments. The results of the analysis are summarized in Figure 5 and Table 4. The magnetic field density reached a maximum of 0.249 T when the gap between the disc and plate was 0.4 mm, and a maximum of 0.297 T when the gap was 2 mm. However, note that when the gap between the disc and plate was 0.4 mm on one side of the disc, a 3.6 mm gap was generated on the opposite side of the disc. Thus, the magnetic field density decreases as the largest distance to the plate on either side of the disc decreases, with an optimum when the gap is equal on both sides of the disc.

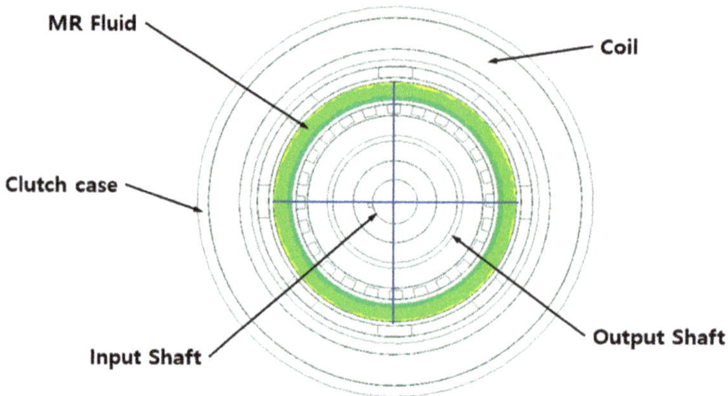

Figure 4. Measurement positions of the average magnetic field density.

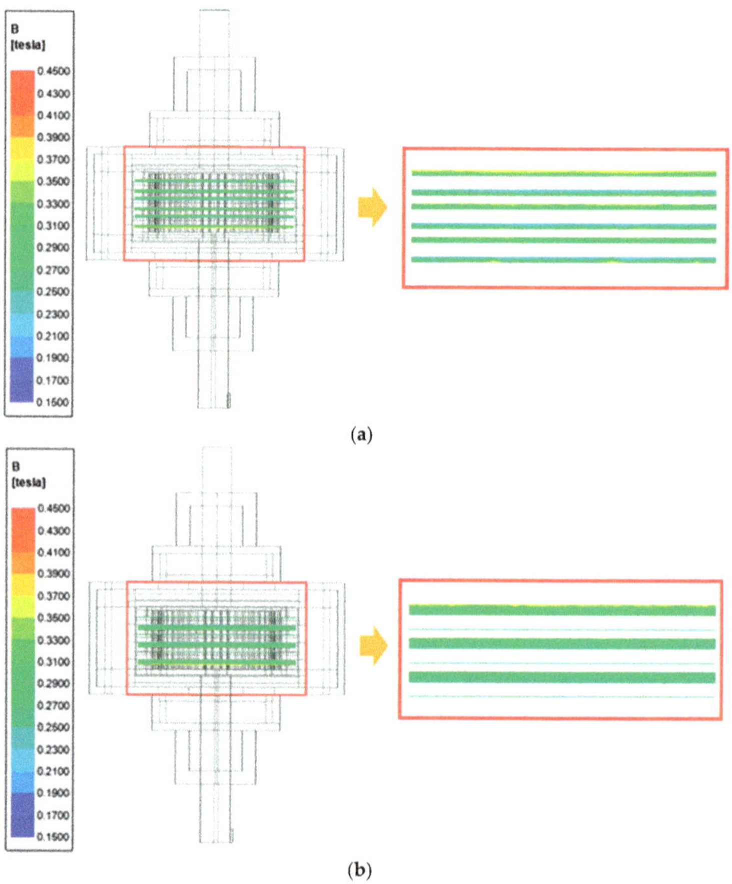

Figure 5. Electromagnetic field analysis results for a gap between the disc and plate of: (**a**) 2.0 and (**b**) 0.4 mm.

Table 4. Average magnetic field density according to the gap between disc and plate.

Gap [mm]	Applied Current [A] 0.2	0.4	0.6	0.8	1.0	1.2	1.4	1.6	1.8	2.0
	Magnetic field density [T]									
2.0	0.035	0.067	0.099	0.129	0.161	0.189	0.216	0.244	0.271	0.297
0.4	0.028	0.056	0.082	0.106	0.133	0.159	0.182	0.204	0.227	0.249

Based on the magnetic field analysis results according to the gap between disc and plate, the *B–h* curve, and the shear stress graph of 132 DG MR fluid, when the inner diameters of disc and plate were changed, the magnetic strength governing the transmission torque in magnetic field control mode increased with increasing current. Hence, when the gap was 2 mm, the transmission torque increased from 0.56 to 17.24 Nm as the current increased from 0.2 to 2 A, and when the gap decreased to 0.4 mm, the transmission torque at 2 A decreased from 17.24 to 15.9 Nm. Furthermore, when the gap was 0.62 mm, a small torque of 15.74 Nm was obtained, and when the gap decreased, the transmission torque tended to increase again. This is illustrated in Figure 6.

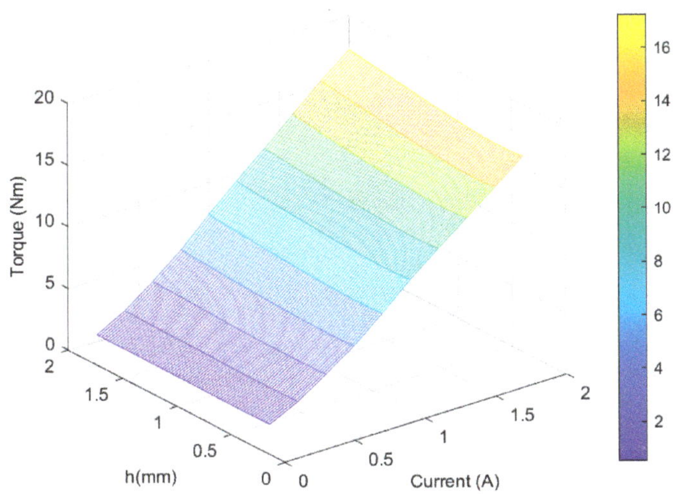

Figure 6. Torque transmission results according to the applied current and gap *h* between disc and plate.

4. Experimental Results and Discussion

4.1. Results

The multi-plate MR clutch proposed in Section 3 was fabricated and used to conduct torque transmission experiments. As shown in Figure 7, the experimental setup for measuring the transmitted torque comprised a motor to rotate the input shaft and a torque sensor to measure the transmission torque of the output shaft. To manage the control signals and sensor measurements, peripherals were designed to transmit and collect digital and analog signals, and fabricated on a single control board. The Cortex M7 was used as the micro control unit (MCU) for torque control, and the IAR Embedded Workbench was used to run the compiled control program. A motor controller was used for the drive motor control, an electronic brake was used to generate a physical load on the output shaft, and an amplifier was used to amplify the torque sensor signals. In addition, a power supply with 350 V, a 10-A output was used to form a magnetic field in the MR fluid using a coil. To collect the experimental data, the desired torque and applied current were measured using LeCroy's four-channel oscilloscope WavePro 604 HD (Future Automotive Intelligent Electronics Core Technology Center, Republic of Korea).

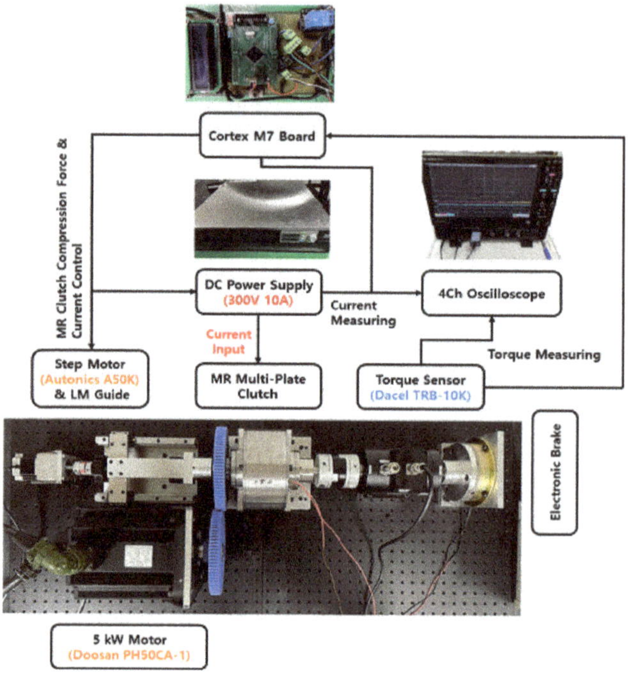

Figure 7. Schematic diagram of the experimental setup.

Clutch operation could be performed normally at rotational speed as high as 1800 rpm, the maximum speed of the motor; however, in this study, experiments were conducted at a lower rotational speed of 200 rpm for safety. In the clutch experiment, the torque transmission characteristics were examined as the compression force, current, and gap between disc and plate were changed to perform basic evaluations of the magnetic field control and mechanical friction modes of the fabricated multi-plate MR clutch. In magnetic field control mode, the torque transmission was evaluated according to the current and gap between disc and plate. The magnetic field was again applied to the MR fluid by adjusting the current flowing in the coil in the range of 0–2 in 0.2-A increments; the gap was also changed in the range of 0.2–2.0 mm in 0.2-mm increments during the experiment. Finally, the ability of the clutch to transmit a torque of up to 30 Nm through the compression force in mechanical friction mode was investigated.

The transmission torque was evaluated by applying a compression force between the disc and plate while applying a current to the gap using the coil during clutch operation at a constant rpm. The experimental protocols used are listed in Table 5.

Figure 8 shows the experimental results for proposed multi-plate MR clutch in mechanical friction mode. Measurements of the torque transmitted through the compression force were performed at a motor rotation of 200 rpm until the compression force was removed. Because the compression force could not be measured in mechanical friction mode, it was back calculated by measuring the transmission torque after moving the output shaft 0.2 mm in the axial direction using a step motor and a ball screw at the point of attachment. Detailed results are shown in Table 6. In the mechanical friction mode experiment, the transmission torque increased up to 23.7 Nm starting when compression was applied, exhibiting a maximum steady-state arrival time of 680 ms. Furthermore, a torque ripple of up to 1 Nm was generated, and approximately 5 s were required until the power transmission was blocked from the maximum friction. This indicates that the torque transmission response was delayed owing to the limited rotational torque of the step motor.

Table 5. Experimental protocols and simulation parameters.

Title of Experiment	Input Condition	Current Input
Mechanical friction mode	Compression force 15–320 [N]	0 [A] (fixed)
Magnetic field control mode	Gap 0.2–2 [mm]	0–2 [A]
Mechanical friction mode for torque tracking control	Determined by control algorithm	0 [A] (fixed)
Mixed mode for torque tracking control	Determined by control algorithm	

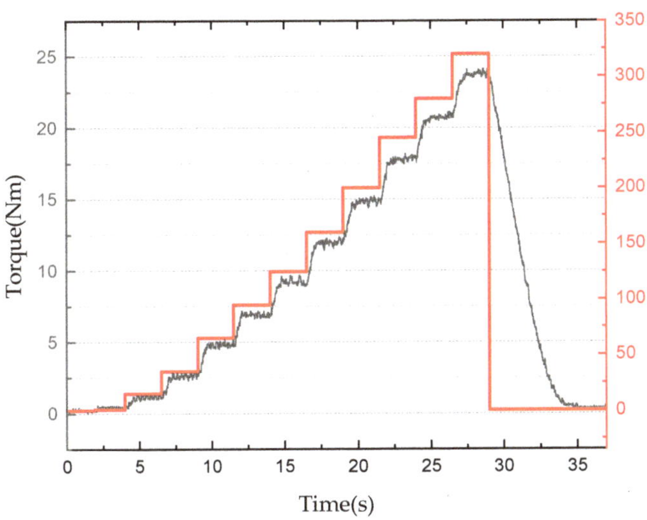

Figure 8. Torque transmission in mechanical friction mode according to applied compression force.

Table 6. Torque transmission experiment results in mechanical friction mode according to applied compression force.

Applied Force [N]	15	35	65	95	125	160	200	245	280	320
Torque [Nm]	1.03	2.43	4.75	6.94	9.2	11.8	14.75	18.01	20.77	23.7
Response time [ms]	639	545	667	657	569	655	680	578	602	576
Torque ripple [Nm]	0.63	0.81	0.7	0.77	0.84	0.92	0.94	0.73	1.03	0.78

Figure 9 shows torque transmission results in the magnetic field control mode experiment according to the applied current using the same parameters applied in the mechanical friction mode. The transmission torque was measured as the current was increased from 0 to 2 A in 0.2-A increments at 1.5-s intervals with a gap of 2 or 0.2 mm. The figure shows that the transmission torque changed according to the applied current and provided gap. Detailed results are provided in Table 7. The experimental results indicate that when the gap was fixed at 2 mm and the current was increased from 0.2 to 2 A, the transmission torque increased to a maximum of 15.43 Nm, and the maximum steady-state arrival time was 220 ms. When the gap was 0.2 mm, the maximum transmission torque was 15.93 Nm and the maximum steady-state arrival time was 200 ms. The transmission torque response in magnetic field control mode was notably faster than that in mechanical friction mode. However, a transmission torque ripple of up to 0.76 Nm occurred owing to the vibration resulting from the friction between the disc and plate at a gap of 0.2 mm.

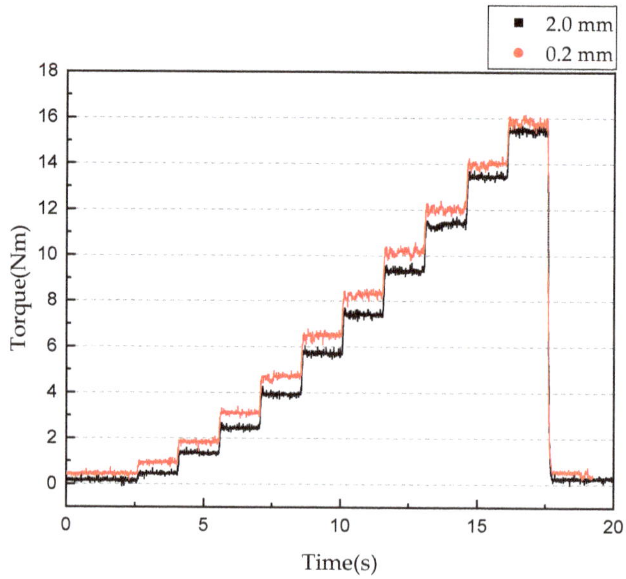

Figure 9. Torque transmission in magnetic field control mode according to applied current.

Table 7. Torque transmission experiment results in magnetic field control mode according to applied current with 2.0- and 0.2-mm gaps.

Gap between Disc and Plate	**Applied Current [A]**									
	0.2	**0.4**	**0.6**	**0.8**	**1.0**	**1.2**	**1.4**	**1.6**	**1.8**	**2.0**
	Transmitted torque [Nm]									
2.0 [mm]	0.41	1.28	2.42	3.95	5.75	7.48	9.44	11.4	13.5	15.4
0.2 [mm]	1.05	1.93	3.17	4.84	6.59	8.25	10.2	12.1	14.1	15.9
	Response time [ms]									
2.0 [mm]	146	137	205	195	162	179	203	220	195	152
0.2 [mm]	148	134	132	134	146	183	186	213	177	200
	Torque ripple [Nm]									
2.0 [mm]	0.43	0.34	0.39	0.51	0.54	0.55	0.55	0.60	0.47	0.51
0.2 [mm]	0.38	0.64	0.50	0.56	0.66	0.76	0.76	0.64	0.66	0.76

Figure 10 shows the transmission torque measured during the magnetic field control mode experiment when the provided gap was decreased from 2.0 to 0.2 mm in 0.2-mm increments at 1.5-s intervals. Similar to Figure 9, Figure 10 shows that the transmission torque changed depending on the current and gap. experimental results describing the change in transmitted torque according to gap and applied current are summarized in Table 8. The transmission torque was 0.7 Nm when the applied current was 2 A and 0.25 Nm when it was 0.2 A. The closer the friction surfaces of the disc and plate, the more the transmission torque tended to increase. At approximately 0.2 mm, immediately before contact, fine friction began to occur and the transmission torque increased further. This friction loss occurred at the oil seal keeping the MR fluid inside the clutch as well as at the thrust bearing used to maintain the gap between the disc and plate and the shaft bearings used to prevent the deflection of the input and output shafts. Thus, it was confirmed that the torque increased as the applied current increased, and increased faster as the gap increased. Furthermore, it was determined that the transmitted torque was affected by friction at low applied currents.

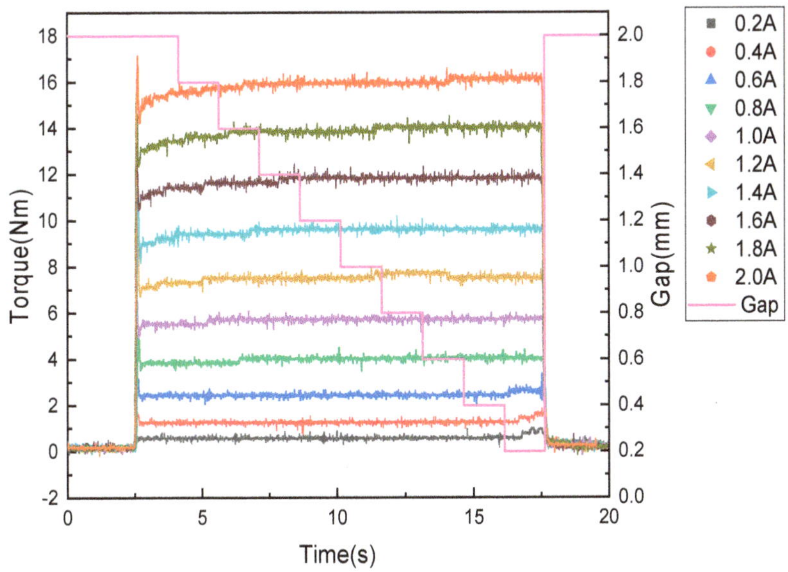

Figure 10. Torque transmission in magnetic field control mode according to changes in gap and current.

Table 8. Torque transmission experiment results in magnetic field control mode according to current at gaps of 2 and 0.2 mm.

Gap between Disc and Plate	Applied Current [A]									
	0.2	**0.4**	**0.6**	**0.8**	**1.0**	**1.2**	**1.4**	**1.6**	**1.8**	**2.0**
	Transmitted torque [Nm]									
2.0 [mm]	0.64	1.3	2.53	3.88	5.58	7.3	9.4	11.4	13.4	15.6
0.2 [mm]	0.89	1.47	2.78	4.2	5.85	7.67	9.8	11.8	14	16.3

The experiments conducted using the proposed multi-plate MR clutch in mechanical friction mode and magnetic field control mode confirmed that in mechanical friction mode, the torque response time was slow and a torque ripple was generated, whereas in magnetic field control mode, the torque response time was approximately three times faster and the torque ripple was small. Therefore, to quickly and smoothly control the torque tracking of the proposed multi-plate MR clutch, a control strategy should be established that prioritizes magnetic field control and assists torque control using mechanical friction when a large torque is necessary.

4.2. Torque Control Strategy

To achieve adequate torque control of the proposed multi-plate MR clutch, torque control performance should be evaluated using the flowcharts for the control algorithms shown in Figures 11 and 12. First, according to the control flowchart when using only the mechanical friction mode (Figure 11), when the program starts, the MCU is initialized, and the parameters are set. Then, a 10-ms interrupt service routine is set with the peripherals. Once all settings are completed, the clutch is in standby mode, and its state is the same as it is under the stop command. When the user gives a start command, tracking control is performed using the desired torque value, and the interrupt service routine is performed every 10 ms. The current torque is then measured, and the error is calculated to control the rotational direction and speed of the step motor.

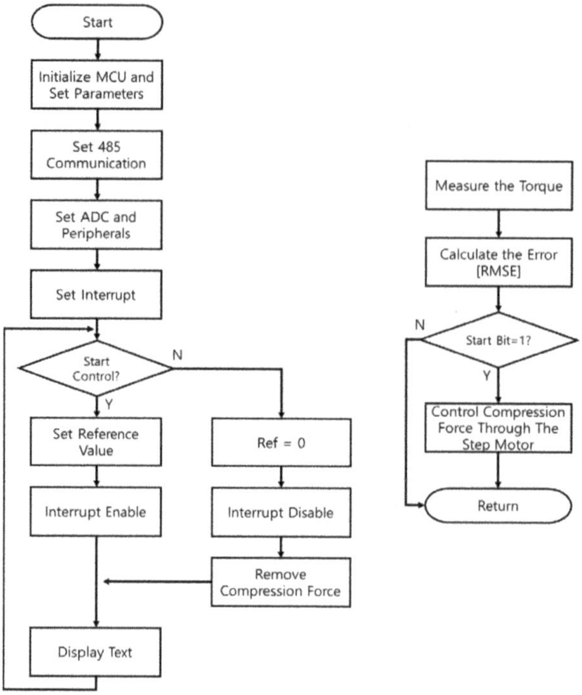

Figure 11. Control algorithm for the multi-plate MR clutch using only mechanical friction mode.

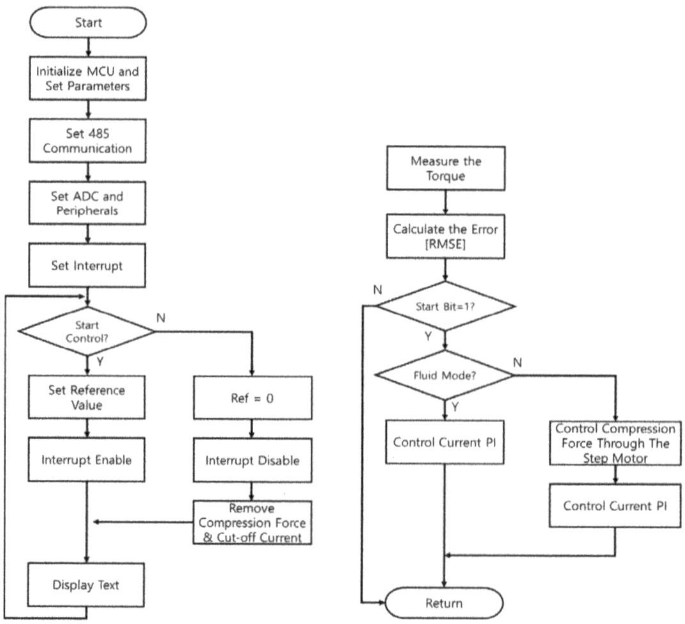

Figure 12. Control algorithm for the multi-plate MR clutch in mixed mechanical friction/magnetic field control mode.

The initialization stage is the same in the control flowchart for the mixed mode (Figure 12). However, during the interrupt service routine, the control mode of the multi-plate MR clutch can be changed according to the desired torque, and the necessary parameters are separately controlled accordingly.

Figures 13 and 14 show the experimentally obtained torque control characteristics when using the control algorithms for mechanical friction mode and mixed mode, respectively. The command was given using the proposed algorithm to increase the torque from 0 to 30 Nm in 5-Nm intervals and then decrease it in the same fashion. Torque tracking using only mechanical friction mode, shown in Figure 13, required a maximum of 1000 ms and minimum of 642 ms after the torque command was issued. In addition, the root mean square error (RMSE) was used to evaluate the control performance as follows:

$$RMSE = \sqrt{\frac{\sum_{i=1}^{n} (\hat{y}_i - y_i)^2}{n}} \quad (13)$$

where $\hat{y}$ is the measured torque, y is the torque command, and n is the number of data points. Thus, an RMSE of approximately 1.192 Nm was obtained when using mechanical friction mode.

Torque tracking in the mixed mode required a maximum of 870 ms and a minimum 310 ms after the torque command was issued. The RMSE for the torque tracking error was approximately 0.649 Nm. According to the control algorithm in Figure 14, the current control was performed first, then from the moment when the 15 Nm torque command was received, the mixed and mechanical friction modes began to run in parallel. The vibration was found to be more severe when using only the mechanical friction mode than when using the mixed mode. However, the actual torque matched well with the desired torque when using either mode. Furthermore, the control performance in the mixed mode was superior in terms of both the tracking control error and steady-state arrival time. Figures 13 and 14 also show enlarged views of the measured change in torque near the 15-Nm value. Even when the current control mode was switched to the mixed mode, the torque tracking error and vibration for the stepped command values remained very small. The tracking response times for each mode are listed in Tables 9 and 10.

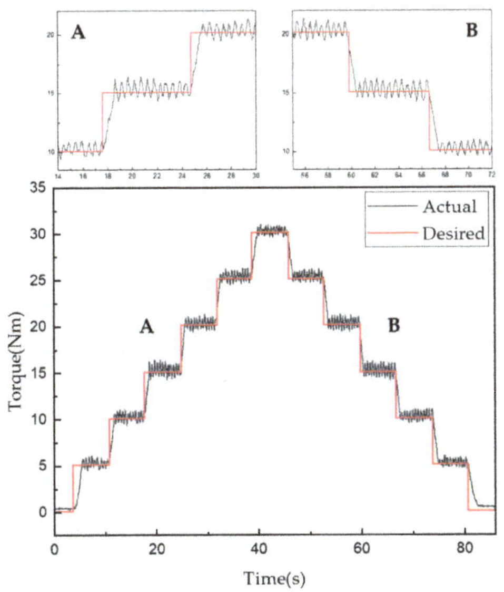

Figure 13. Torque tracking results for step control in mechanical friction mode.

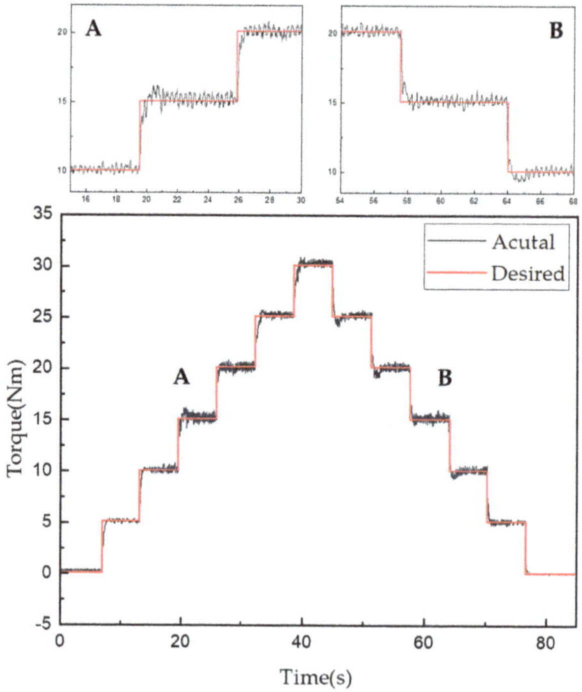

Figure 14. Torque tracking results for step control in mixed mode.

Table 9. Torque tracking response times according to step control in mechanical friction mode.

Desired Torque [Nm]	5	10	15	20	25	30	25	20	15	10	5	0
Response time [ms]	1000	740	680	710	660	874	738	769	722	642	781	854

Table 10. Torque tracking response times according to step control in mixed mode.

Desired Torque [Nm]	5	10	15	20	25	30	25	20	15	10	5	0
Response time [ms]	750	395	538	456	870	780	551	563	601	532	522	310

Figures 15 and 16 depict graphs of the tracking control results for a torque command of 20 Nm when the rotational speed of the input motor was changed from 50 to 100 to 200 to 50 rpm after giving a torque command of 20 Nm. When using mechanical friction mode, a torque ripple was generated by default, and an RMSE for torque tracking of 0.6126 Nm was observed when the rotational speed was changed. Thus, although the torque fluctuated significantly when the rotational speed was changed, torque tracking remained possible. Furthermore, in the mixed mode, an RMSE for torque tracking of approximately 0.437 Nm occurred when the rotational speed was changed. This suggests that, in the mixed mode, the torque can be tracked more smoothly with a smaller error than in the mechanical friction mode, even when the rotational speed is changed.

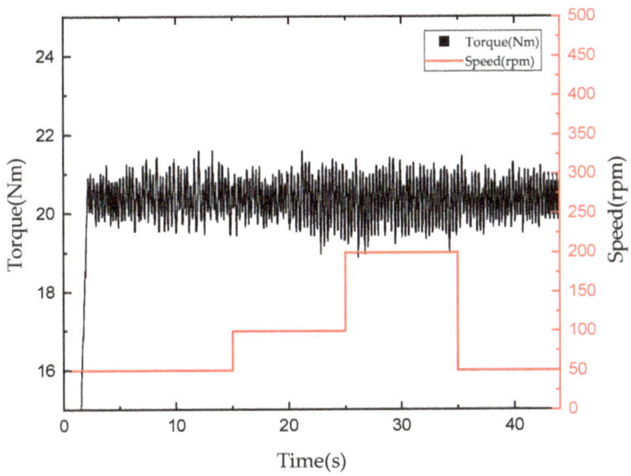

Figure 15. Torque tracking results according to rotational speed in mechanical friction mode.

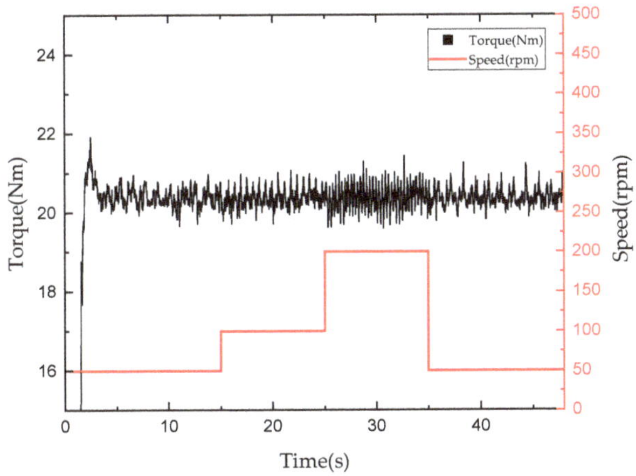

Figure 16. Torque tracking results according to rotational speed in mixed mode.

Figures 17 and 18 show torque tracking graphs when half-sine-wave type torque commands of 0–10, 0–20, and 0–30 Nm were provided using the two control modes. When using only the mechanical friction mode, a torque ripple was generated, and the RMSE for torque tracking was 0.866 Nm; however, the possibility of torque tracking for half-sine-wave type torque commands was verified. In the mixed mode, the RMSE for torque tracking was 0.553 Nm. It was assumed that in this case, some noise also existed in the desired torque, and the error of the proportional–integral (PI) controller was small at the beginning when the torque increased, resulting in a tracking delay and tracking error. However, because the torque tracking error remained less than 1.7% at the maximum torque, the torque tracking control was considered to be smoothly conducted.

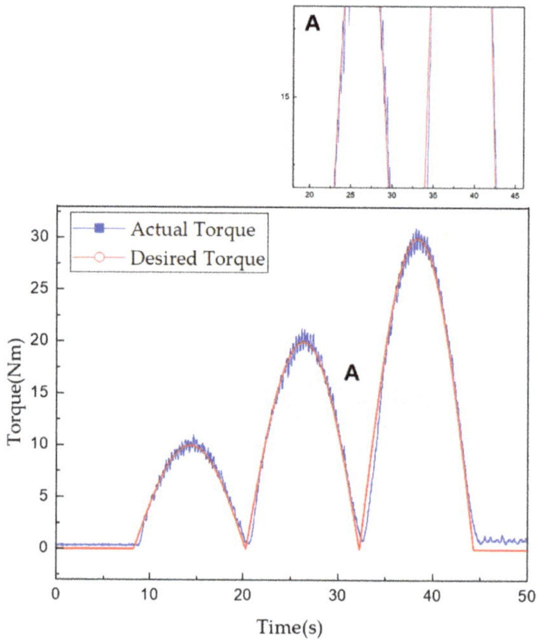

Figure 17. Torque tracking results for half-sine-wave input in mechanical friction mode.

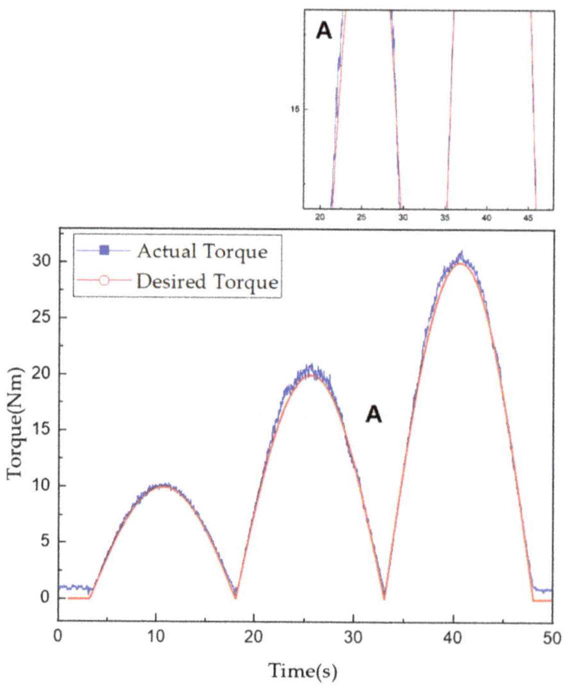

Figure 18. Torque tracking results for half-sine-wave input in mixed mode.

5. Conclusions

This study modeled and evaluated the torque transmission of a multi-plate MR clutch in mechanical friction mode, magnetic field control mode, and a mixed mode. Torque tracking control was then performed using the multi-plate MR clutch based on the derived torque modeling equations and the magnetic field analysis results obtained using ANSYS Maxwell.

The results of the torque performance experiment indicated that the maximum transmitted torque was 24.1 Nm in mechanical friction mode and 15.5 Nm in magnetic field control mode. These values were approximately 2 Nm smaller than their corresponding simulation results of 26 and 17 Nm, respectively. This difference was attributed to the mechanical structure required to create a compression force between the friction surfaces and small errors in the magnetic field analysis. The change in torque according to the applied rotational speed exhibited only a slight change, with a maximum difference of 0.7 Nm observed when compared to the torque model calculation result. The cause of this difference was assumed to be the loss of power owing to friction at the clutch seal and bearings. However, because this loss a small compared to the maximum torque, the torque modeling equation and magnetic field analysis were determined to be reliable.

After establishing a control algorithm and using it to control the torque tracking of the proposed MR multi-plate clutch, the RMSEs in mechanical friction mode for the step command, rotational speed change, and half-sine wave command experiments were 1.192, 0.616, and 0.866 Nm, respectively. Compared with the results in mechanical friction mode, the RMSEs were relatively smaller in the mixed mode. Furthermore, the basic torque ripple was large in mechanical friction mode but smaller in mixed mode. Indeed, in mixed mode, a relatively large RMSE value appeared in the half-sine-wave experiment, but the tracking error was small, except for an initial increase. This indicates that the proposed MR multi-plate clutch can be applied to realize large and accurate power transmission systems.

The proposed multi-plate MR clutch was shown to operate smoothly by capitalizing on the advantages of both dry and wet clutches. In other words, the proposed multi-plate MR clutch reduced vibration by performing smooth shifting operations without torque ripples compared to a purely mechanical friction clutch and exhibited excellent power transmission performance similar to that of a conventional multiplate clutch. These properties are expected to be advantageous in a variety of fields.

In future research, the maximum performance of the clutch could be realized if the drive motor and the step motor controlling the compression force are replaced with equivalents having a higher capacity and rotation speed while producing less noise.

Author Contributions: Conceptualization, J.-S.O.; validation, J.-Y.P.; investigation, J.-Y.P.; writing—original draft preparation, J.-Y.P.; writing—review and editing, J.-S.O.; supervision, Y.-C.K. All authors have read and agreed to the published version of the manuscript.

Funding: This research was supported by the Basic Science Research Program through the National Research Foundation of Korea (NRF) funded by the Ministry of Education (2018R1D1 A3B0705028814) and by a research grant from Kongju National University in 2020.

Institutional Review Board Statement: Not applicable.

Informed Consent Statement: Not applicable.

Conflicts of Interest: The authors declare no conflict of interest.

References

Vasca, F.; Iannelli, L.; Senatore, A.; Scafati, M.T. Modeling torque transmissibility for automotive dry clutch engagement. In Proceedings of the 2008 American Control Conference, Seattle, WA, USA, 11–13 June 2008; pp. 306–311.

Jibin, H.; Zengxiong, P.; Chao, W. Experimental research on drag torque for single-plate wet clutch. *J. Tribol.* **2012**, *134*, 014502. [CrossRef]

D'Agostino, V.; Cappetti, N.; Pisaturo, M.; Senatore, A. Improving the engagement smoothness through multi-variable frictional map in automated dry clutch control. In *ASME International Mechanical Engineering Congress and Exposition*; American Society of Mechanical Engineers: New York, NY, USA, 2012; Volume 45271, pp. 9–19.

4. Desmidt, H.; Smith, E.C.; Bill, R.C.; Rao, S. Multi-Plate Dry Clutch Design and Analysis for Dual Speed Rotorcraft Application In Proceedings of the 56th AIAA/ASCE/AHS/ASC Structures, Structural Dynamics, and Materials Conference, Kissimmee, FL USA, 5–9 January 2015; p. 0953.
5. Arief, I.; Mukhopadhyay, P.K. Magnetorheology in CoNi nanoplatelet-based MRFs: Effect of platelet orientation and oscillator shear. *J. Magn. Magn. Mater.* **2019**, *479*, 326–331. [CrossRef]
6. Neelakantan, V.A.; Washington, G.N. Modeling and reduction of centrifuging in magnetorheological (MR) transmission clutche for automotive applications. *J. Intell. Mater. Syst. Struct.* **2005**, *16*, 703–711. [CrossRef]
7. Benetti, M.; Dragoni, E. Nonlinear magnetic analysis of multi-plate magnetorheological brakes and clutches. *In Proceedings of th COMSOL Users Conference.* 2006. Available online: https://www.researchgate.net/profile/Eugenio-Dragoni/publication/2660950 8_Nonlinear_Magnetic_Analysis_of_Multi-plate_Magnetorheological_Brakes_and_Clutches/links/56e5ab1708aedb4cc8ae7986 Nonlinear-Magnetic-Analysis-of-Multi-plate-Magnetorheological-Brakes-and-Clutches.pdf (accessed on 20 December 2021).
8. Gratzer, F.; Steinwender, H.; Kušej, A. Magnetorheological AWD clutches. *ATZautotechnology* **2008**, *8*, 32–37. [CrossRef]
9. Kikuchi, T.; Ikeda, K.; Otsuki, K.; Kakehashi, T.; Furusho, J. Compact MR fluid clutch device for human-friendly actuator. *J. Phys Conf. Ser.* **2009**, *149*, 012059. [CrossRef]
10. Saifei, Z.; Yong, L. Disc Shaped High-Torque-MRF-Clutch Design. In Proceedings of the International Conference on Compute Application and System Modeling, Taiyuan, China, 22–24 October 2010; Volume 13, pp. 374–377.
11. Güth, D.; Aust, M.; Maas, J. Novel concepts for MRF based clutch systems with integrative functionalities. In Proceedings o the 2010 IEEE/ASME International Conference on Advanced Intelligent Mechatronics, Montreal, QC, Canada, 6–9 July 201(pp. 1191–1196.
12. Törőcsik, D. Some design issues of multi-plate magnetorheological clutches. *Hung. J. Ind. Chem.* **2011**, *39*, 41–44.
13. Wang, D.; Tian, Z.; Meng, Q.; Hou, Y. Development of a novel two-layer multiplate magnetorheological clutch for high-powe applications. *Smart Mater. Struct.* **2013**, *22*, 085018. [CrossRef]
14. Kavlicoglu, B.M.; Gordaninejad, F.; Evrensel, C.A.; Cobanoglu, N.; Liu, Y.; Fuchs, A.; Korol, G. High-torque magnetorheologica fluid clutch. In *Smart Structures and Materials 2002: Damping and Isolation*; International Society for Optics and Photonics Bellingham, WA, USA, 2002; Volume 4697, pp. 393–400.
15. Kavlicoglu, B.M.; Gordaninejad, F.; Wang, X. Study of a magnetorheological grease clutch. *Smart Mater. Struct.* **2013**, *22*, 12503([CrossRef]
16. Thakur, M.K.; Sarkar, C. Development and performance analysis of a Magnetorheological fluid Clutch. In Proceedings of th COMSOL Conference, Boston, MA, USA, 3–5 October 2018.
17. Yadmellat, P.; Kermani, M.R. Adaptive modeling of a magnetorheological clutch. *IEEE/ASME Trans. Mechatron.* **2013**, *1*9 1716–1723. [CrossRef]
18. Bucchi, F.; Forte, P.; Frendo, F. Temperature effect on the torque characteristic of a magnetorheological clutch. *Mech. Adv. Mate Struct.* **2015**, *22*, 150–158. [CrossRef]
19. Pilch, Z. Analysis of Established Thermal Conditions for Magnetorheological Clutch for Different Loading Conditions. In *Analysi and Simulation of Electrical and Computer Systems*; Springer: Cham, Switzerland, 2015; pp. 197–213.
20. Leong, F.H.; Tsai, N.C.; Chiu, H.L. Infinite-stage magnetic clutch for shaft speed amplification. *Proc. Inst. Mech. Eng. Part C J Mech. Eng. Sci.* **2017**, *231*, 193–204. [CrossRef]
21. Fernando, N. A Field Modulated Electromagnetic Clutch with Slip Control. In Proceedings of the 2016 IEEE 2nd Annual Southerr Power Electronics Conference (SPEC), Auckland, New Zealand, 5–8 December 2016; pp. 1–6.
22. Kavlicoglu, B.; Gordaninejad, F.; Evrensel, C.; Fuchs, A.; Korol, G. A semi-active, high-torque, magnetorheological fluid limitec slip differential clutch. *J. Vib. Acoust.* **2006**, *128*, 604–610. [CrossRef]
23. Rizzo, R.; Musolino, A.; Bucchi, F.; Forte, P.; Frendo, F. A multi-gap magnetorheological clutch with permanent magnet. *Smar Mater. Struct.* **2015**, *24*, 075012. [CrossRef]
24. Latha, K.H.; Sri, P.U.; Seetharamaiah, N. Design and manufacturing aspects of magneto-rheological fluid (MRF) clutch. *Mate Today Proc.* **2017**, *4*, 1525–1534. [CrossRef]
25. Olszak, A.; Osowski, K.; Kesy, Z.; Kesy, A. Investigation of hydrodynamic clutch with a magnetorheological fluid. *J. Intell. Mate Syst. Struct.* **2019**, *30*, 155–168. [CrossRef]
26. Zhang, H.; Du, H.; Sun, S.; Li, W.; Wang, Y. Design and Analysis of a Novel Magnetorheological Fluid Dual Clutch for Electri Vehicle Transmission (No. 2019-01-5014). *SAE Tech. Paper* **2019**. [CrossRef]
27. Kluszczyński, K.; Pilch, Z. Basic features of MR clutches-resulting from different number of discs. In Proceedings of the 2019 15th Selected Issues of Electrical Engineering and Electronics (WZEE), Zakopane, Poland, 8–10 December 2019; pp. 1–4.
28. Park, J.Y.; Kim, G.W.; Oh, J.S.; Kim, Y.C. Hybrid multi-plate magnetorheological clutch featuring two operating modes: Fluic coupling and mechanical friction. *J. Intell. Mater. Syst. Struct.* **2021**, *32*, 1537–1549. [CrossRef]
29. Oh, J.S.; Choi, S.H.; Choi, S.B. Design of a 4-DOF MR haptic master for application to robot surgery: Virtual environment work *Smart Mater. Struct.* **2014**, *23*, 095032. [CrossRef]
30. Han, Y.M.; Oh, J.S.; Kim, S.; Choi, S.B. Design of multi-degree motion haptic mechanisms using smart fluid-based devices. *Mech Based Des. Struct. Mach.* **2017**, *45*, 135–144. [CrossRef]
31. Nguyen, Q.H.; Choi, S.B. Optimal design of a novel hybrid MR brake for motorcycles considering axial and radial magnetic flux *Smart Mater. Struct.* **2012**, *21*, 055003. [CrossRef]

2. Quoc, N.V.; Tuan, L.D.; Hiep, L.D.; Quoc, H.N.; Choi, S.B. Material characterization of MR fluid on performance of MRF based brake. *Front. Mater.* **2019**, *6*, 125. [CrossRef]
3. Vékás, L. Ferrofluids and magnetorheological fluids. In *Advances in Science and Technology*; Trans Tech Publications Ltd.: Bäch, Switzerland, 2008; Volume 54, pp. 127–136.
4. Available online: https://www.lord.com/sites/default/files/Documents/TechnicalDataSheet/DS7015_MRF-132DGMRFluid.pdf (accessed on 20 December 2021).
5. Nguyen, P.B.; Choi, S.B. Accurate torque control of a bi-directional magneto-rheological actuator considering hysteresis and friction effects. *Smart Mater. Struct.* **2013**, *22*, 055002. [CrossRef]
6. Yoon, D.S.; Kim, G.W.; Choi, S.B. Response time of magnetorheological dampers to current inputs in a semi-active suspension system: Modeling, control and sensitivity analysis. *Mech. Syst. Signal Process.* **2021**, *146*, 106999. [CrossRef]
7. Yoon, D.S.; Park, Y.J.; Choi, S.B. An eddy current effect on the response time of a magnetorheological damper: Analysis and experimental validation. *Mech. Syst. Signal Process.* **2019**, *127*, 136–158. [CrossRef]
8. Yang, G.; Spencer, B.F., Jr.; Carlson, J.D.; Sain, M.K. Large-scale MR fluid dampers: Modeling and dynamic performance considerations. *Eng. Struct.* **2002**, *24*, 309–323. [CrossRef]
9. Priya, C.B.; Gopalakrishnan, N. Temperature dependent modelling of magnetorheological (MR) dampers using support vector regression. *Smart Mater. Struct.* **2019**, *28*, 025021. [CrossRef]
0. Sahin, I.; Cesmeci, S.; Wereley, N.M. Sensitivity of magnetorheological damper behavior to perturbations in temperature via Bouc-Wen model. In *Electro-Rheological Fluids and Magneto-Rheological Suspensions*; World Scientific Publishing Co Pte Ltd.: Singapore, 2011; pp. 665–671.

sensors

MDPI

Article

Improvement in Strain Sensor Stability by Adapting the Metal Contact Layer

Ji-Yeon Choy †, Eun-Bee Jo †, Chang-Joo Yim, Hae-Kyung Youi, Jung-Hoon Hwang, Jun-Ho Lee and Hyun-Seok Kim *

Division of Electronics and Electrical Engineering, Dongguk University, Seoul 04620, Korea; jiyeon9711@naver.com (J.-Y.C.); eunbee.j123@gmail.com (E.-B.J.); akreal@naver.com (C.-J.Y.); yeahwooong@gmail.com (H.-K.Y.); junghoonh0326@gmail.com (J.-H.H.); junho_1211@naver.com (J.-H.L.)

* Correspondence: hyunseokk@dongguk.edu; Tel.: +82-2-2260-3996

† These authors contributed equally to this work.

Abstract: Research on stretchable strain sensors is actively conducted due to increasing interest in wearable devices. However, typical studies have focused on improving the elasticity of the electrode. Therefore, methods of directly connecting wire or attaching conductive tape to materials to detect deformation have been used to evaluate the performance of strain sensors. Polyaniline (PANI), a p-type semiconductive polymer, has been widely used for stretchable electrodes. However, conventional procedures have limitations in determining an appropriate metal for ohmic contact with PANI. Materials that are generally used for connection with PANI form an undesirable metal-semiconductor junction and have significant contact resistance. Hence, they degrade sensor performance. This study secured ohmic contact by adapting Au thin film as the metal contact layer (the MCL), with lower contact resistance and a larger work function than PANI. Additionally, we presented a buffer layer using hard polydimethylsiloxane (PDMS) and structured it into a dumbbell shape to protect the metal from deformation. As a result, we enhanced steadiness and repeatability up to 50% strain by comparing the gauge factors and the relative resistance changes. Consequently, adapting structural methods (the MCL and the dumbbell shape) to a device can result in strain sensors with promising stability, as well as high stretchability.

Keywords: stretchable sensor; strain sensor; wearable device; polyaniline; PDMS

1. Introduction

Highly flexible and stretchable next-generation wearable devices are anticipated for applications in various fields, such as flexible robotics, displays, smart watches for healthcare, and functional clothing. Accordingly, it is appropriate to conduct in-depth research aimed at developing a high-sensitivity strain sensor that can operate with stability despite deformation, while in contact with human skin, in light of the increasing demand for body-mounted devices.

Studies on strain sensors that can convert mechanical strain, such as body movement, into electrical signals have been actively pursued [1]. Novel electrodes and substitutes for typical metal or inorganic films have been studied to ensure conductivity and stretchability under strain, including metal nanowires [2], conductive polymers [3,4], the carbon nanotube (CNT) [5], graphene [6], and composites of these devices.

In particular, polyaniline (PANI), one of the conductive polymers, has attracted attention as an appropriate substance, due to its environmental stability, cost-effectiveness, and tunable conductivity [7], which provides scalability in a multi-functional strain sensor capable of sensing various stimuli, such as chemical gases, temperatures, and humidity [8]. However, strain sensors made by spin-coating or printing PANI dispersed in a solvent are limited because it is difficult to secure high elasticity due to their rigid chain structure.

Cai et al. recently demonstrated stretchability of up to 150% with high sensitivity, and excellent mechanical adaptability via the island-bridge morphology of PANI film using dilute-chemical polymerization, resulting in the expansion of PANI's versatility as a stretchable electrode [9].

Studies on stretchable electrodes have been actively conducted. However, studies on packaging, integration, and measurement of strain sensors have not been sufficiently conducted [10]. Silver paste for wire connections, conductive tapes such as copper tape, and conductive fabric have been directly attached to stretchable electrodes for estimating their performance [11]. However, such methods have several problems. First, it is difficult to detect minute strain because the deformed region of the stretchable electrode cannot be precisely controlled. Second, contact resistance, based on the unstable attachment and increased instability under external stress, degrades a device's efficiency. Moreover, the effectiveness of materials that have been used for measurement, including copper and silver, are limited.

There is a restriction in establishing a stable metal semiconductor junction with the various stretchable electrodes. In particular, PANI is a p-type semiconductor whose conductivity results from delocalization by the π-conjugation structure, filling the local electron-deficient part of the cation radical. Thus, its work function is dependent on synthetic methods, such as chemical, electrochemical, and interfacial methods, synthetic conditions, such as the conditions related to dopants, acids, and solvents, and the ambient environment.

The typical work function of PANI is in the range of 4.2–4.7 eV. It should have ohmic contact with metal within this range to achieve high sensitivity [12–14]. However, it was not possible to form ohmic contact or to reduce contact resistance by simply employing a metal tape. An additional step, such as using silver paste, has been required [7,9]. In addition, most devices do not come with a separate contact area or appropriate connecting methods [15–17]. Therefore, adapting a metal contact layer to sensors could assure ohmic contact.

Silicone-based elastomers, such as polydimethylsiloxane (PDMS), Ecoflex, Dragon Skin, and hydrogel, are commonly used as stretchable substrates for body-attached devices. In particular, PDMS is frequently used due to its high elasticity, chemical stability, biological compatibility, transparency, and thermal stability [18]. Additionally, adjusting the mixing ratio of its components and curing conditions can easily control physical properties [19,20]. However, most studies on PDMS have focused on surface treatment, molding, or adhesion improvement. Hence, fabrication methods for a single substrate that is composed of more than two types of PDMS with different physical rigidities, using their advantages, have not been suggested.

Therefore, only a substrate with a single PDMS is provided for stretchable sensors or their applications. Hence, when a metal thin film that is susceptible to deformation is applied to sensors configured with PDMS, polymers such as polyimide and parylene have been required as a protective layer [21]. Furthermore, although studies on the fabrication of metal patterning on PDMS have advanced, their low surface energy hinders the application of photoresist, due to non-uniformity and low adhesion, preventing precise pattern formation. Additionally, its coefficient of thermal expansion, swelling, and flexibility acts as a hindrance [22,23]. Consequently, the patterning on PDMS via conventional photolithography is difficult and restricted.

This paper introduces a study that ultimately aims to improve the operational stability of strain sensors by adapting the metal contact layer (the MCL). We fabricated a strain sensor that included PANI as a sensing material to detect deformation, the MCL to retain ohmic contact with PANI, and a buffer layer to prevent breakage of the MCL by external stress. Additionally, we cut the sensor in the shape of a dumbbell to maximize the protective effect of the MCL and the buffer layer. Furthermore, the possibility of patterning PANI, the MCL, and the buffer layer is suggested, to provide potential for a diverse selection of materials on stretchable electrodes or the MCL (Au and Ag), indicating the versatility of fabrication of stretchable strain sensors.

The buffer layer formation was established from the facile controllability of PDMS's physical stiffness. We fabricated a single substrate composed of two different types of PDMS. As a result, it has high elasticity, but partial regions which have low elasticity can be protected from deformation. These partial regions, the buffer layers, can prevent breakage or peel-off of the MCL during the fabrication or application. The buffer layer is PDMS with a Young's modulus greater than 3 MPa (more than 6 MPa in this study). It is called hard PDMS (H-PDMS) because it is larger than the general Young's modulus. The substrate that determines the overall elasticity of the sensor has a Young's modulus less than 1 MPa (less than 700 kPa in this study); it is called soft PDMS (S-PDMS) because its modulus is smaller than the general Young's modulus. Furthermore, we carried out the PDMS patterning process to formulate PDMS with different physical properties within a single substrate. We introduced the transfer with a sacrificial layer and an auxiliary substrate to minimize damage during the process.

PANI, the MCL, and the H-PDMS can be formed on a single plane with a facile procedure. We demonstrated promising stability, reproducibility, and classification of novel structured sensors based on a comparison of the gauge factors (GFs) of three samples, depending on the MCL and the buffer layer and their transient measurement of relative resistance change under repeated tensile stress. Moreover, current-voltage (I−V) graphs of three types of sensors with different positions of the MCL according to PANI lengths were compared to identify the effect of the dumbbell-shape that was designed in accordance with the standards of the American Society for Testing of Materials (ASTM) [24–26].

2. Materials and Methods

2.1. Materials

Aniline (>99.5%, ACS reagent, 62-53-3), sulfuric acid (98%, for analysis EMSURE, 7664-93-9), hydrochloric acid (37%, ACE reagent, 7647-01-0), (3-Aminoprophyl) triethoxysilane (99%, 2554-06-05), and 2,4,6,8-tetramethyl-2,4,6,8-tetravinylcyclotetrasiloxane (2554-06-05) were purchased from Sigma-Aldrich, Seoul, Korea. A graphite rod (GR002H) was purchased from QRINS, Seoul, Korea. ITO-coated glass was purchased from iTASCO, Seoul, Korea. A silicone elastomer base and a curing agent (Sylgard 184 kit) were purchased from Dow Corning, Midland, MI, USA. VDT-731 (67762-94-1), SIT-6831.2 (68478-92-2), and HMS-301 (68037-59-2) were purchased from Gelest, Morrisville, PA, USA. Copper tape was purchased from 3M, Seoul, Korea. All reagents were used as received without further purification.

2.2. Characterization

The thickness of PDMS and PANI was analyzed using a surface profiler (Alpha Step 200, KLA/TENCOR). The alignment of PANI, Au thin film (the MCL), and the H-PDMS were observed using a microscope (MX51, Olympus, Tokyo, Japan). The surface morphology of transferred PANI was analyzed via FE-SEM (S-4800, Hitachi, Tokyo, Japan) at the Korea Advanced Nano Fab Center.

2.3. Fabrication of Strain Sensor

2.3.1. PANI/Au/H-PDMS Strain Sensor

We introduced a strain sensor with the MCL and the buffer layer (PANI/Au/H-PDMS strain sensor). Its fabrication process is shown in Figure 1. First, we prepared the ITO-coated glass. It was cleaned with acetone, IPA, and deionized water prior to processing. Patterning directly on PDMS may cause alignment difficulties. A transfer process using ITO (185 nm, 10 Ω/cm^2) as a sacrificial layer and glass as an auxiliary substrate was used (Figure 1a). ITO was chosen as a sacrificial layer because it can act as a working electrode during subsequent PANI synthesis. Typical photolithography was performed to establish the MCL. Photoresist (AZ5214E, AZ Electronics Materials, Charlotte, NC, USA) was spin-coated at 3000 rpm for 30 s. After the exposure, it was immersed in developer (CD-30, AZ Electronics Materials, Charlotte, NC, USA) for 1 min 10 s. Au, which has a significantly larger work function than PANI, was selected as the MCL to form a stable ohmic contact. Thereafter, 150 nm of Au thin film was deposited via an e-beam evaporator (KVE-T5560, Korea Vacuum Tech, Gimpo, Korea), using Au pellets (purity 99.999%) as a source. The deposition conditions were as follows: 2.8×10^{-5} Torr, 140 mA, and an 0.8 kÅ/s of deposition rate. It was lifted into a rectangular shape (3000 μm × 1010 μm) and served as the MCL (Figure 1b).

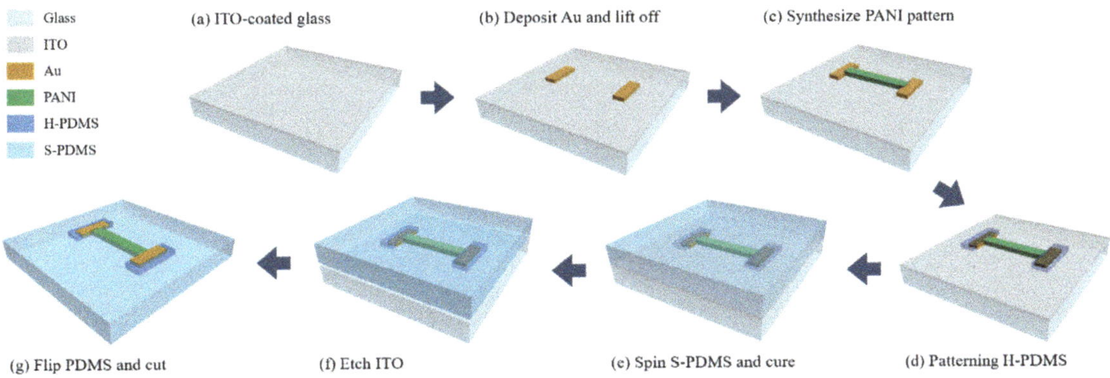

Figure 1. Fabrication of PANI/Au/H-PDMS strain sensor: (**a**) preparing the auxiliary substrate with a sacrificial layer; (**b**) formation of MCL; (**c**) formation of sensing material; (**d**) formulation of the buffer layer; (**e**) formulation of the stretchable substrate; (**f**) separating the stretchable substrate and the auxiliary substrate; (**g**) cutting PDMS into a dumbbell shape.

Next, photolithography was implemented to define PANI capable of detecting mechanical strain. A PANI mask was aligned to overlap with 20 μm of the MCL. A thicker photoresist (KPRO-15, KemLab, Woburn, MA, USA) was used to ensure PANI thickness and spin-coated at 1000 rpm for 45 s. PANI was synthesized electrochemically. Photoresist acted as a deposition mask for patterning (width of 1500 μm and length of 8270 μm). The polymerization conditions were as follows: 0.3 M of aniline and 0.7 M aqueous solution of sulfuric acid as an electrolyte, with the ITO-coated glass as a working electrode and the graphite rod as a counter electrode. The electrodeposition was performed for 1 min 10 s at a fixed voltage of 1.1 V at room temperature [7]. Thereafter, it was baked on the hot plate at 100 °C for 10 min to enhance adhesion to the substrate and dried at room temperature for 24 h to completely remove moisture. It was then soaked in acetone for 10 min to remove the photoresist and the residue was cleaned with IPA and deionized water (Figure 1c).

Subsequently, the H-PDMS patterning was conducted, as described in detail in Section 2.4 (Figure 1d). The S-PDMS determines the stretchability of a strain sensor. It was manufactured by mixing the base and a curing agent (Sylgard 184) with a weight ratio of 15:1. Prior to spin-coating the mixture, the curing agent was spin-coated at 3000 rpm for 30 s to ensure adhesion to the H-PDMS. The mixture was then spin-coated at 500 rpm for 30 s and sufficiently cured at room temperature for 72 h (Figure 1e). Finally, the ITO film was removed by the etchant, which is a diluent of HCl (37%), and deionized water with a weight ratio of 1:3 (Figure 1f). In this step, partial PANI that lost conductivity during the process was oxidized to emeraldine salt form and restored in an acidic etchant having low pH. The PDMS film separated from the glass as the ITO was etched, flipped, and washed with deionized water. After that, it was cut into a dumbbell shape based on ASTM D412 (Figure 1g) [24–26]. Additionally, three samples with different PANI lengths (5020 μm, 7020 μm, and 8270 μm) were fabricated under the same conditions to optimize the strain sensor according to the position of the MCL and the buffer layer within the same dumbbell structure.

2.3.2. PANI and PANI/Au Strain Sensor

We compared the strain sensor consisting of only PANI (PANI strain sensor) and the PANI/Au/H-PDMS strain sensor to demonstrate the effectiveness of adapting MCL. Additionally, a strain sensor without the buffer layer (PANI/Au strain sensor) and PANI/Au/H-PDMS strain sensor were compared to verify the efficiency of the buffer layer under deformation. The conditions of fabricating PANI, the MCL, and the S-PDMS in these samples were the same as those for the PANI/Au/H-PDMS strain sensors. In the case of the PANI strain sensor, the S-PDMS was applied immediately after defining the PANI on the ITO-coated glass. The sacrificial layer was then etched. PDMS film separated from the glass was cleaned with deionized water and cut into a rectangular shape (7 mm × 30 mm). The dumbbell shape was not required because it was composed only of PANI, excluding the MCL and the buffer layer that require structural protection.

The PANI/Au strain sensor was fabricated by coating the S-PDMS on the patterned MCL and PANI. Similarly, the MCL and PANI were aligned with 20 μm of the overlapping area. The surface treatment with (3-Aminoprophyl) triethoxysilane (APTES) was processed to enhance the adhesion between the S-PDMS and the MCL just before the S-PDMS coating. An aqueous solution diluted APTES with a weight ratio of 1:100 was used. It was dried with an N_2 gun without a further washing process after immersion of 20 min [27]. After that, the S-PDMS was coated and cured for a sufficient time. The sacrificial layer was removed using HCl diluent to isolate it from the glass. It was washed by deionized water and cut into the dumbbell shape to protect the MCL structurally.

2.4. *Patterning the H-PDMS*

We applied the H-PDMS with a high Young's modulus as the buffer layer to protect the MCL from deformation. We introduced its patterning process as shown in Figure 2. First, the thick photoresist was applied to the sample to complete the step shown in Figure 1c, to ensure sufficient thickness that is fundamental for a durable buffer layer. We spin-coated the photoresist (KPRO-15, KemLab, Woburn, MA, USA) at 1000 rpm for 45 s and soft-baked it to form the first photoresist layer. After cooling, a sufficient thickness of 35 $\sim$ 40 μm was obtained by repeating the same process under identical conditions. Thereafter, a pattern with a size of 4000 μm × 1500 μm was aligned with PANI overlapped 10 μm and the bulk MCL via conventional photolithography (Figure 2a). To develop the photoresist to a thickness greater than 35 μm, it was immersed in developer (CD-30, AZ Electronics Materials, Charlotte, NC, USA) for 20 min. APTES surface treatment was performed prior to applying the H-PDMS to improve adhesion between the MCL and the H-PDMS. After immersing in APTES aqueous solution diluted at a weight ratio of 1:100 for 20 min, it was taken out and dried with an N_2 gun.

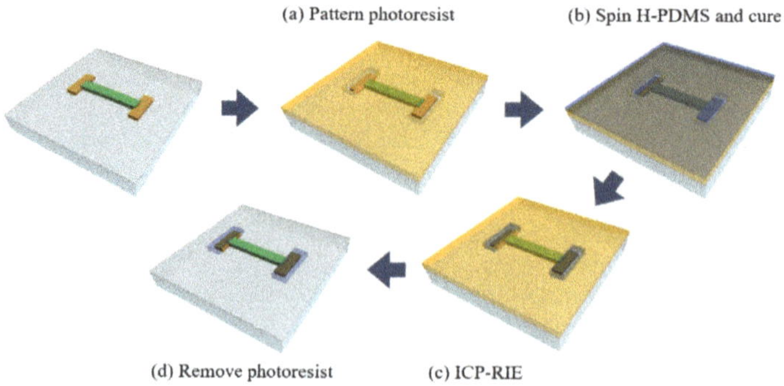

Figure 2. The process of patterning the H-PDMS: (**a**) photoresist as an etching mask; (**b**) formation of H-PDMS; (**c**) etching the top of H-PDMS to expose photoresist surface; (**d**) remained H-PDMS as the buffer layer.

The H-PDMS was established in the following steps. First, 3.7 g of VDT-731 and 50 μL of Pt catalyst (SIP-6831.2) were mixed. Then, 0.1 g of modulator (2,4,6,8-tetramethyl-2,4,6,8-tetravinylcyclotetrasiloxane) and 1 g of HMS-301 were added and mixed. The mixture was degassed whenever each substance was added [28,29]. The mixture was spin-coated at 1000 rpm for 30 s. Self-leveling for 10 min on a flat place at room temperature was performed immediately after the preceding step. It was then cured in an oven at 60 °C for 4 h (Figure 2b). ICP-RIE was performed to expose the top surface of the photoresist by etching the H-PDMS. The selectivity of the H-PDMS photoresist was 1.5:1. The following were the ICP-RIE conditions: 500 W of RF power, 200 W of RF bias power, 10 mTorr of pressure with 90 sccm of SF_6 and 6 sccm of O_2, which lasted 32 min (Figure 2c) [30]. It was then immersed in an acetone solution to remove the photoresist. Notably, the exposed top surface facilitated its removal. The thickness of the remaining H-PDMS was 12 μm. The H-PDMS containing 10 μm of PANI and the bulk MCL and an overlapped area with 10 μm of the MCL and bulk PANI, situated externally to the H-PDMS, guaranteed the interconnection of each part under strain. Additionally, we manufactured Ag/H-PDMS samples and observed the variation of the surface conditions and the resistance based on strain to verify the effectiveness of the H-PDMS. Ag (150 nm) as the MCL and the H-PDMS (25 μm) as the buffer layer were progressed under the same conditions as mentioned above. The S-PDMS was also formed identically using the mixture of a Sylgard 184 kit with a 15:1 weight ratio, and cut into the dumbbell shape.

2.5. Measurement Conditions

We prepared three different types of samples with and without the Au thin film as the MCL and the H-PDMS as the buffer layer: Sample A (PANI strain sensor), Sample B (PANI/Au strain sensor), and Sample C (PANI/Au/H-PDMS strain sensor). Sample A was cut into a rectangular shape because it lacked metallic thin film and was vulnerable to external stress. Samples B and C were cut into the shape of the dumbbell based on the ASTM D412 model to maximize the efficiency of the MCL protection. Each sample had a PANI pattern, 1500 μm of width and 8270 μm of length, which can detect mechanical deformation. We also prepared three different lengths (5020, 7020, and 8270 μm) of PANI in Sample C to explore the effect of the dumbbell structure. The copper tape was prepared in size 0.125 cm × 3 cm and attached directly to PANI or the MCL to measure the electrical properties (Figure 3). The measurement was repeated consecutively for all samples under the same conditions to obtain I−V characteristics. Measurement conditions were as follows: dual sweep from −10 V to 10 V with 0.1 V step voltage in a fixed state with deformation at 0%, 10%, 20%, 30%, 40%, and 50%. Transient characteristics were measured at speeds of 50 mm/min and 6 mm/min. The former had a waiting time of 10 s, and the latter had a waiting time of 0.1 s. Ag/H-PDMS samples were cut into the ASTM D412 model, and their surface condition was observed using a microscope at 0%, 20%, and 60% strains. Resistance was measured under the 0%, 20%, 40%, and 60% strains using the Au probe.

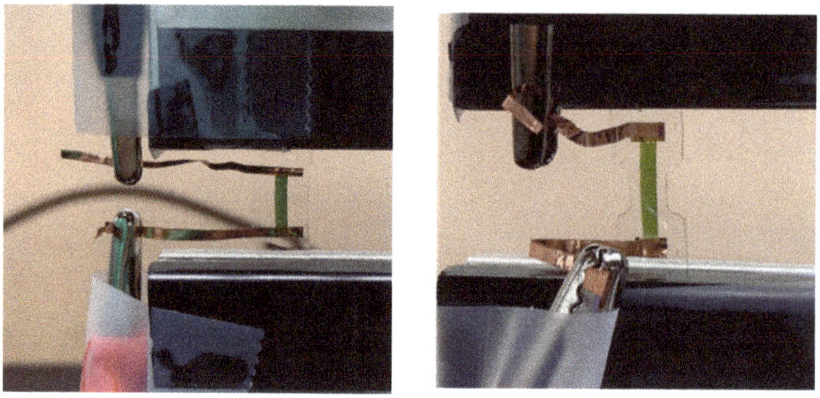

(a) (b)

Figure 3. Measuring electrical properties. (**a**) Sample A (PANI strain sensor), (**b**) Sample B (PANI/Au strain sensor), and Sample C (PANI/Au/H-PDMS strain sensor).

3. Results

3.1. PANI Morphology

It was important to maintain the interconnection of PANI under high tension to have reversible tensile strength. Therefore, the construction of an island-bridge structure was significant because the previous study showed its effectiveness of up to 150% [9]. In the earlier study, we found that electrochemically synthesized PANI formed the initial thin film that served as an island, and the nanowires that grew on it acted as a bridge. Figure 4a shows the scanning electron microscope (SEM) image of PANI taken after transfer with PDMS, depicting island-bridge constitutions. Figure 4b is the enlarged image, revealing the organization with small dots instead of nanowires. These small dots, the reverse side of the PANI film, appeared because the transfer process caused PDMS to embed it. Therefore, the nanowires of the island-bridge buried in this way were able to tolerate up to 50% strain.

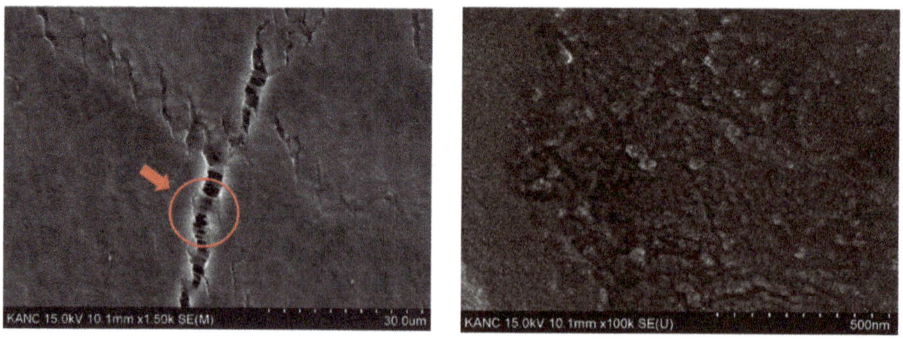

(a) (b)

Figure 4. SEM image of transferred PANI film: (**a**) the island-bridge structure; (**b**) enlarged image of PANI film.

3.2. The MCL Protection Effect of the Dumbbell Structure and the H-PDMS

Figure 5 is an equivalent circuit model of the dumbbell shape of Sample C. Part 1 is the narrowest area, Part 2 is the curved area, and Part 3 is the widest area. The resistance of PANI formed in each area is denoted by R_1, R_2, R_3. The equation for the resistance of PANI (R_{pani}) is expressed in Equation (1), and a Young's modulus of the S-PDMS (E) is presented in Equation (2). ρ is the resistivity of PANI, l is the length of PANI in the corresponding part (l_1 = 2.4 mm, l_2 = 1.6 mm, l_3 = 1.335 mm), and A_{pani} is the cross-sectional area of PANI. Since E is related to the S-PDMS, F is the force applied to the S-PDMS, and A_{pdms} is the cross-sectional area of the S-PDMS (A_{pdms1} = 3t mm, $A_{pdms2} \approx$ 4.6t mm, A_{pdms3} = 6.26t mm, t = thickness of the S-PDMS).

$$R_{pani} = \frac{\rho l}{A_{pani}} \tag{1}$$

$$E = \frac{stress}{strain} = \frac{F/A_{pdms}}{\Delta l/l} \tag{2}$$

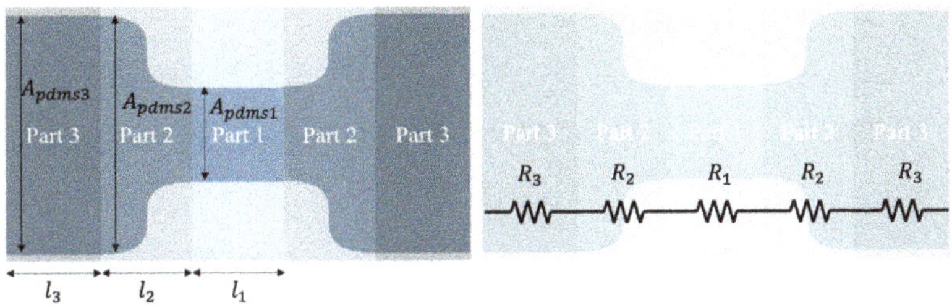

Figure 5. An equivalent circuit model of dumbbell shape.

The strain (S) of Part 1 can be expressed as S_1 (Equation (3)). Then, the changes in S, when A_{pdms} are different, can be expressed in Equations (4) and (5).

$$\Delta l_1/l_1 = S_1 \tag{3}$$

$$S_2 = S_1\frac{A_{pdms1}}{A_{pdms2}} \tag{4}$$

$$S_3 = S_1\frac{A_{pdms1}}{A_{pdms3}} \tag{5}$$

The change in resistance (ΔR_{pani}) can be obtained based on the change in PANI's length (Δl) (Equations (6)–(8)). By referring to Equations (3)–(5), Δl can be expressed by the equations related to S_1 (Equations (9) and (10)). As a result, the dumbbell structure is protective of the MCL because Δl_3 is the smallest.

$$\Delta R_1 = \frac{\rho \Delta l_1}{A_{pani}} \tag{6}$$

$$\Delta R_2 = \frac{\rho \Delta l_2}{A_{pani}} \tag{7}$$

$$\Delta R_3 = \frac{\rho \Delta l_3}{A_{pani}} \tag{8}$$

$$\Delta l_2 = l_2 S_2 = l_2 S_1 \frac{A_1}{A_2} \tag{9}$$

$$\Delta l_3 = l_3 S_3 = l_3 S_1 \frac{A_1}{A_3} \tag{10}$$

We compared the Ag/H-PDMS and Ag-only samples to verify the efficiency of the buffer layer and the dumbbell shape. We also observed the surface and the resistance according to strain. The strain-dependent surface changes of the Ag thin film in the Ag/H-PDMS and the Ag samples are shown in Figure 6a–f. At the initial state (0%), the Ag/H-PDMS samples completely protected the surface (Figure 6a,c). However, the Ag samples immediately adhered to the S-PDMS and damaged the surface even in the absence of applying artificial stress (Figure 6b). In particular, the Ag/H-PDMS samples' initial condition was maintained even at 20% strain (Figure 6d). On the contrary, the Ag samples showed cracks both coincident with and perpendicular to the strain axis (Figure 6e).

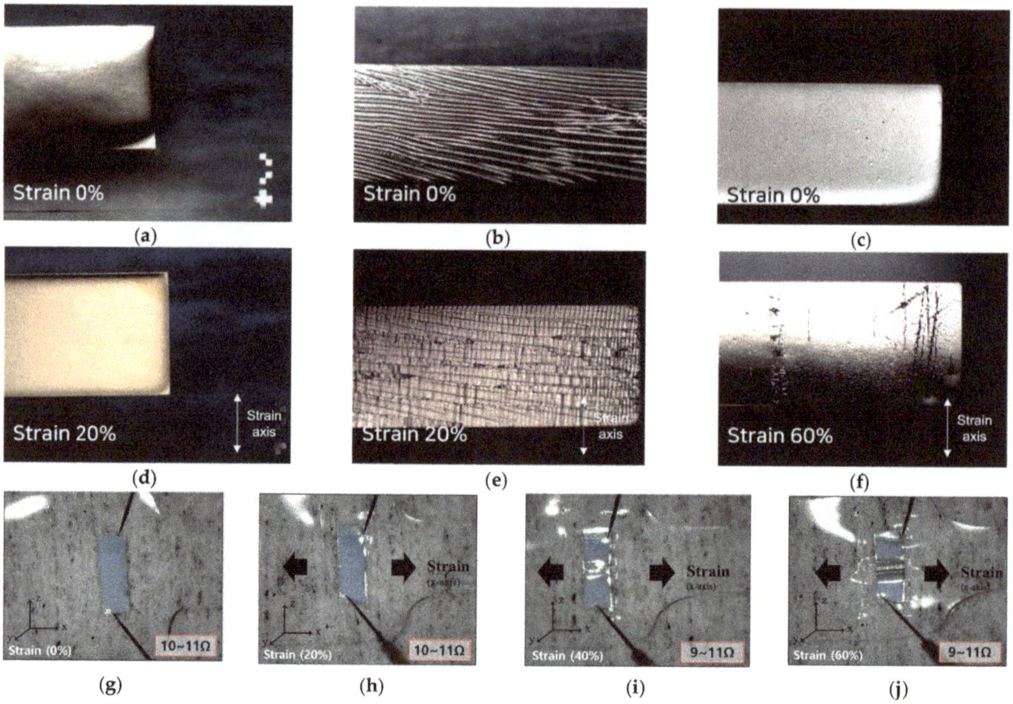

Figure 6. Ag/H-PDMS and Ag-only samples: (**a**) 0% of the Ag/H-PDMS sample; (**b**) 0% of the Ag sample; (**c**) 0% of the Ag/H-PDMS; (**d**) 20% of the Ag/H-PDMS sample; (**e**) 20% of the Ag sample; (**f**) 60% of the Ag/H-PDMS sample; (**g**–**j**) overall image of the Ag/H-PDMS sample and the measured resistance at (**g**) 0%, (**h**) 20%, (**i**) 40%, and (**j**) 60%.

Figure 6g–j shows the images of one side of the dumbbell shape, its Ag thin film, and the H-PDMS in the Ag/H-PDMS samples. Figure 6g,h shows that Ag thin film keeps its outline under deformation, under strain of up to 20%. However, it also shows distortion in the z-axis at 40% and 60% strains, causing destruction in the same direction as the strain axis. As shown in Figure 6f, the Ag/H-PDMS samples experienced damage that was parallel to the strain axis under high tension.

Although the H-PDMS had sufficient rigidity to protect the Ag films from strain, the difference in elasticity between the S-PDMS and the H-PDMS, and their high flexibility, caused these cracks (Figure 6i,j). When stress was applied in the x-axis, the S-PDMS, having high elasticity, stretched in the same axis as the stress applied, and decreased in width in the y-axis at the same time. However, the S-PDMS beneath the H-PDMS could not be stretched due to the H-PDMS being at the top, with its low stretchability. Therefore, the H-PDMS-formed regions were bent in the z-axis to compensate for the width. This implies that in the case of samples with the buffer layer, the parallel detriments appeared due to flexibility and not due to stretching.

The resistance measurement was carried out to evaluate the performance. The probes were positioned from edge to edge of the Ag thin film to maximize the effect of its breakage caused by deformation. Despite this damage, all the results measured at strains from 0% to 60%, with a 20% interval, had 9–11 Ω, indicating stability.

3.3. I−V Measurement

We measured the I−V characteristics of Samples A, B, and C to evaluate the stability and elasticity of the MCL-adapted and the buffer layer-adapted strain sensor. Figure 7 shows the I−V measurements of each sample at strains of 0%, 10%, 20%, 30%, 40%, and 50%. Through the I−V curves, the linearity and reproducibility of each sample were exhibited. GF, a parameter indicating sensitivity, verified the ohmic contact of PANI and the MCL, and their stability. GF was calculated as the change in relative resistance (ΔR) with strain (ε) (Equation (11)). The relative resistance change (ΔR) is described in Equation (12), where R_0 denotes the resistance in the initial state (0%) and R represents the resistance in the strain state. For the estimation, each resistance was described in the following colors: red for 0%; orange for 10%; yellow for 20%; green for 30%; blue for 40%; and purple for 50%. Figure 7a shows an exponential aspect and randomly overlapped values of Sample A. At the 30%, 40%, and 50% strains, the currents were decreased, with increasing strain. However, the result was almost the same at 0%, 10%, and 20%, regardless of the strain. Instability was indicated with disparate results, despite repeated measurements under the same conditions. At 30%, the first measurements had a value of approximately 0.015 mA (biased at 10 V). However, the second measurement had a value of 0.005 mA, showing an error of 67% even with the same estimation conditions. Additionally, the current at 10% strain was similar to that at 0% strain at a positive voltage bias, but the current was similar at 30% and 40% in the case of a negative voltage, rather than 0%. This asymmetry implied that PANI in Sample A, with copper tape, had unstable contact. The I−V curves of Samples B and C showed more linear curves than did Sample A, as shown in Figure 7, because the MCL formed ohmic contact with PANI. However, Sample B was insulated at 40% strain and also emitted some noises. Figure 8 is the plot of resistance vs. strain when 5 V is applied. The resistance was measured consecutively three times at each strain. The elasticity of the strain sensor damaged the MCL, which caused noises under the strain during the measurements and had a fatal effect on strain determination. The linearity or tendency of resistance according to strain was not observed in Samples A and B (Figure 8a,b). In contrast, Sample C showed clear I−V curves and resistance according to each strain (Figures 7c and 8c). By employing the MCL and the H-PDMS, noises could be suppressed and could improve the sensor's reliability and stability.

$$GF = \frac{\Delta R}{\varepsilon} \tag{11}$$

$$\Delta R = \frac{R - R_0}{R_0} \tag{12}$$

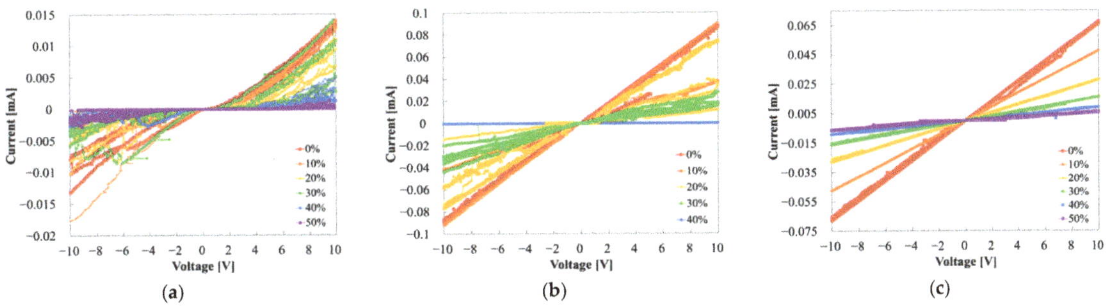

Figure 7. I−V measurement of Samples A–C: (**a**) Sample A (PANI strain sensor), (**b**) Sample B (PANI/Au strain sensor), and (**c**) Sample C (PANI/Au/H-PDMS strain sensor).

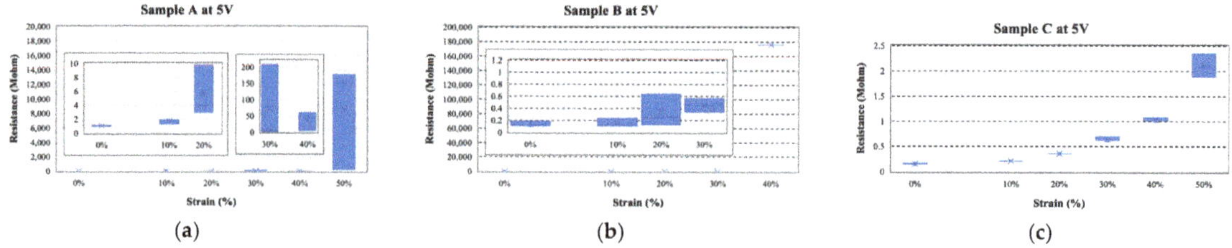

Figure 8. Resistance vs. strain of Sample A–C: (**a**) Sample A (Insets show magnified view due to large resistance differences), (**b**) Sample B (the inset shows a magnified view), and (**c**) Sample C showing distinguishable classification of the strain.

Figure 9 shows the individual graphs in each strain of Sample A. Figure 9a–f show that Sample A has strain-dependent threshold voltages: 2 V at 0% and 10% strain; 2–4 V at 20% and 30% strain; and 4–6 V at 40% and 50% strain. These values indicate that the threshold voltage increases with increasing strain. Figure 9 also shows more sporadic data distribution and hysteresis with increasing strain. Hence, we observed that the current at 10 V of each strain generally decreased with rises in strain. However, currents at 20% strain had exceptional results, having larger values than 10%. This irregularity can be interpreted according to the work function and contact resistance of copper tape and PANI, as mentioned above.

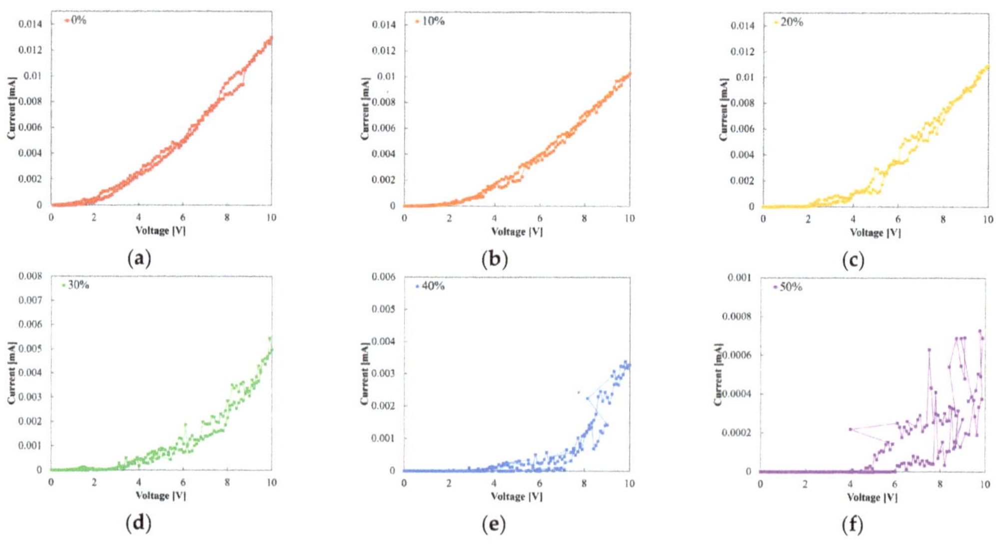

Figure 9. Individual I−V curves of Sample A based on strain: (**a**) 0% strain, (**b**) 10% strain, (**c**) 20% strain, (**d**) 30% strain, (**e**) 40% strain, and (**f**) 50% strain.

GFs of Sample A are shown in Figure 10. Considering GFs at 10%, they were presented between 0 and 0.5 when positive voltage-biased, whereas they were around 0.5 at the negative voltages, showing an asymmetric distribution. Similarly, GFs were distributed between 0 and 0.5 at the positive voltages at 30%; however, GFs were between 0.5 and 1 at the negative voltages. Notably, it was impossible to discriminate between the strain based on the GFs in these two cases because the GFs at the positive voltages were overlapped in the same range of 0–0.5. At 20% and 40%, GFs of both sides of the voltages showed a relatively symmetrical tendency. In contrast, GFs of 50% had asymmetry, with GFs of 10 or more at the positive side and 0–5 at the negative side. The unstable behavior in Sample A was illustrated through the GFs' distribution. The asymmetry of the GFs is the result of the Schottky junction of PANI and the copper tape.

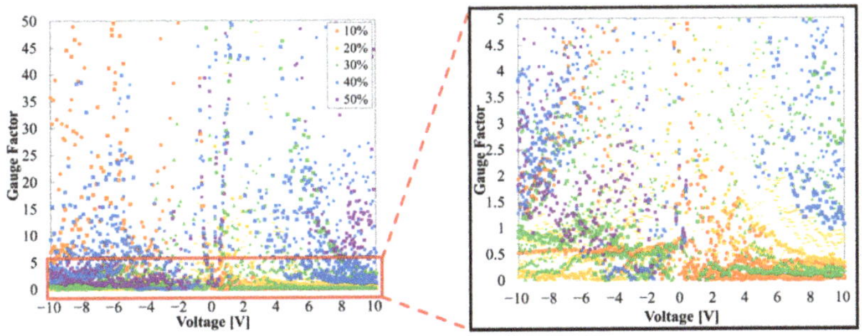

Figure 10. Overall GF of Sample A (**left**) and enlarged GF of Sample A (**right**).

The individual I−V graphs for each strain of Sample B are shown in Figure 11b–d. Each graph is the result of three consecutive measurements under the same conditions. Sample B used Au as the MCL, which has a larger work function than PANI, and a stable ohmic contact can be formed. Thus, Sample B was significantly more linear than Sample A at 0% strain. However, Sample B showed undesirable behaviors, such as large fluctuations in the I−V curves through multiple measurements when the strain was applied as shown in Figure 11b,c. The I−V curves for 10% strain shown in Figure 11b indicate that the first measurement has the lowest value, and the second and third measurements have increased. The MCL was damaged as the strain applied, and the resistance rose rapidly and exhibited its lowest value in the first trial. It partially recovered and stabilized due to the elasticity of the S-PDMS. The resistance caused the elevated currents in the second and third trials. Similarly, the graph for 20% in Figure 11c depicts that the first trial had the lowest value and current increased in the second trial and decreased in the third. As with Figure 11b, Figure 11c shows a drastic and immediate rise in the resistance at the moment when stress was applied. The damaged MCL recovered with time, leading to the current increase at the second measurement. The third measurement was lower than the second one in contrast to the interpretation described at 10% strain. This might be explained by the fact that the MCL was damaged when the strain was applied, even though the dumbbell structure and the elasticity of the S-PDMS protected the MCL. Therefore, the resistance increased, negatively affecting the sensor's behavior under the strain. At 30%, the I−V curves had the lowest value at the first trial, increasing at the second and third trials (Figure 11d). The range of current fluctuations according to the number of trials was small, unlike the previous range at 10% and 20%, which was different from the other two cases where the damage to the MCL was noticeably recovered by the stretchability of the substrate. At 30%, the MCL could not revive its properties, or it may have needed sufficient time to recuperate because of the large stress and deformation. Therefore, the MCL with the H-PDMS would be truly necessary.

The GFs of Sample B are shown in Figure 12. GFs at 40% were over 1,000,000 because of the MCL's destruction. GFs at 10%, 20%, and 30% were spread between 0 and 10 and made it difficult to discriminate among the strains, owing to their wide distributions. Additionally, fluctuation in hysteresis and the unsettled results following the damaged MCL negatively affected the sensor performance.

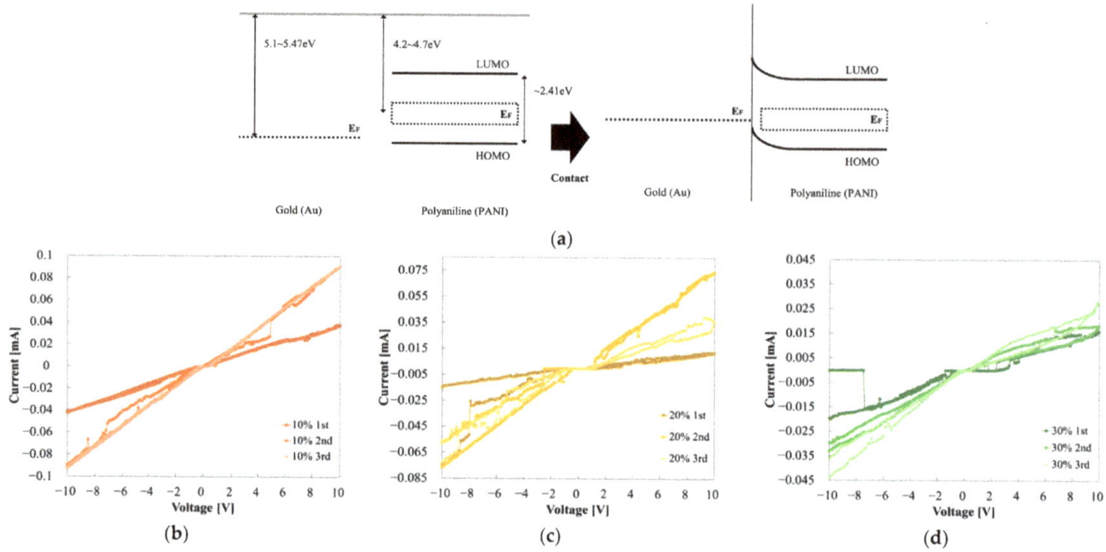

Figure 11. (**a**) Energy band diagram of PANI and the MCL before contact (**left**) and after contact (**right**); the individual I−V curves according to strain in Sample B: (**b**) 10% of strain, (**c**) 20% of strain, and (**d**) 30% of strain.

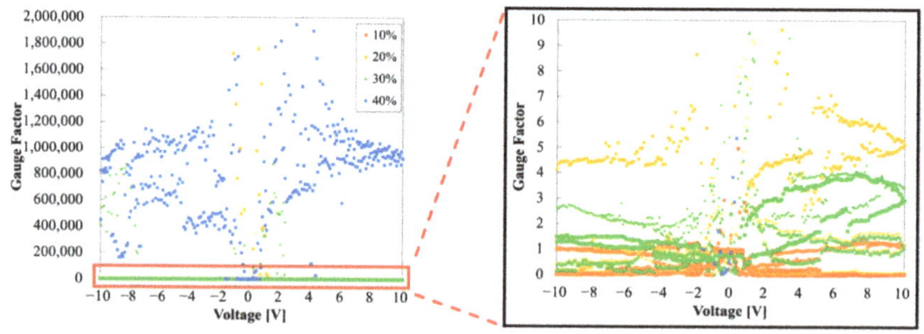

Figure 12. Overall GF of Sample B (**left**) and enlarged GF of Sample B (**right**).

As shown in Figure 7c, Sample C has clearly differentiated I−V measurements for each strain, as the most linear, stable, and repeatable. Specifically, the strain was linear and distinguishable even at the low voltages of −2 V–2 V, as shown in Figure 13a. Furthermore, when comparing the currents at 2 V in the initial state of Samples A and C, the current in the latter was 48 times larger than in the former. The MCL and PANI connection had smaller contact resistance than the copper tape and PANI connection. The GFs of Sample C are shown in Figure 13b: GFs of 0.5 at 10%, GFs of 1 at 20%, GFs of 2.5 at 30%, GFs of 4–5 at 40%, and GFs of more than 6 at 50%. The GFs were completely differentiated and characterized according to each strain. Moreover, the ohmic contact between PANI and the MCL was verified via the symmetric GFs distribution through the positive and negative sides. In general, a high GF indicates high sensitivity of the strain sensor, but at the same time, high-precision and expensive equipment is required to detect large resistance (>10^9 Ω) [7]. Table 1 shows the gauge factors, depending on the measurement contact method. Therefore, Sample C, with a relatively small but highly distinguishable GF will be promising for further body-mounted wearable device applications.

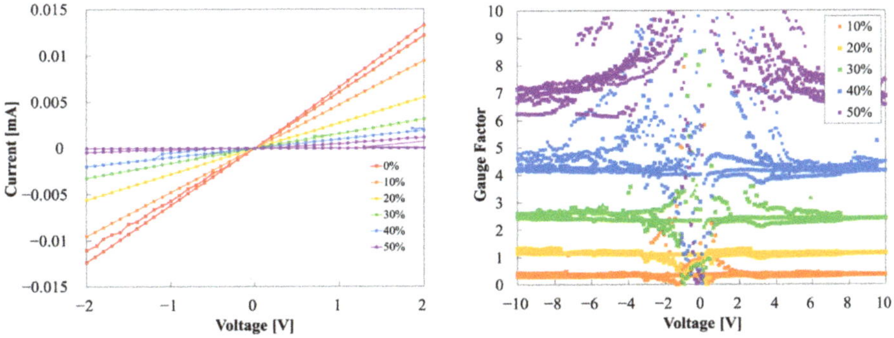

(a) (b)

Figure 13. (**a**) I−V characteristics of Sample C with a voltage range -2 V–2 V; (**b**) GFs of Sample C.

Table 1. The comparison of the measurement contact method of PANI-based strain sensors.

Measurement Contact Method	Gauge Factor	Materials	Ref.
Copper tape and silver glue	54 at 50% strain	PANI	[7]
Copper tape and silver paste	140 at 100% strain	PANI	[9]
Silver fabric	74 at 1% strain	PANI	[31]
Copper wire and silver paste	6.725 at 0~120% strain	PANI/TPU	[32]
MCL	6 at 50%	PANI	This work

Figure 14 shows the I−V curves of PANI/Au/H-PDMS strain sensors with three different PANI lengths. Each inset shows a sensor with different PANI lengths: 5020 µm, 7020 µm, and 8270 µm, respectively. The I−V measurement of the strain sensor with the PANI length of 5020 µm is shown in Figure 14a. We observed the insulation due to the rapid rise of resistance under stress with increasing strain from 20% to 30%. When the length of PANI shortened, the MCL and H-PDMS were positioned on the curved area of the dumbbell, no longer protected by the structure. The I−V measurements of the strain sensors with PANI lengths of 7020 µm and 8270 µm are shown in Figure 14b,c, respectively. They had distinctly isolated values according to deformation and sustained conductivity, even at 50% strain. Both samples, the MCL and the H-PDMS, placed on a wide area of the dumbbell, were structurally sufficient to protect it from the strain, resulting in stable outcomes.

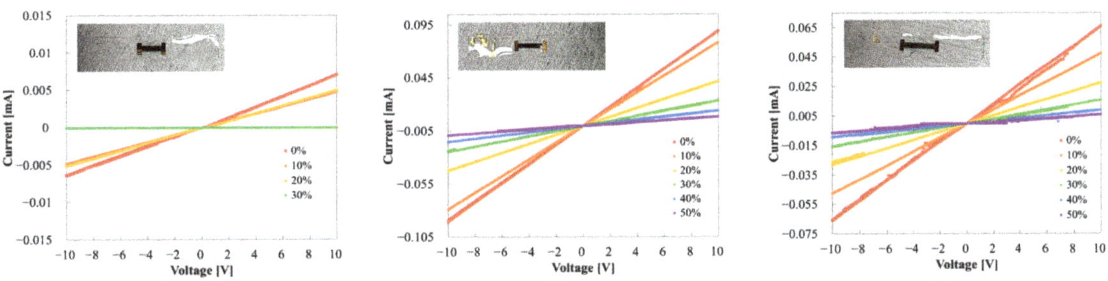

(a) (b) (c)

Figure 14. I−V curves of the PANI/Au/H-PDMS strain sensor according to the different lengths of PANI: (**a**) 5020 µm, (**b**) 7020 µm, and (**c**) 8270 µm.

3.4. Transient Measurement

Figure 15a shows the transient measurement of Sample C, which was pulled and recovered at the speed of 50 mm/min and 10 s of waiting time. It shows the relative resistance changes with time, at 10%, 20%, and 30% strain. The relative resistance change at 20% strain is two times greater than that of 10%, and that at 30% is four times greater than that of 20%. This indicates that the relative resistance change can be a parameter for classifying deformation. In particular, intense overshooting occurred with increasing strain due to the rapid and instantaneous changes in the relative resistance caused by the sudden increase in resistance at the moment of pulling. Figure 15b, which was pulled and recovered at the speed of 6 mm/min, indicates a gradual relative resistance change without the overshooting phenomenon. Consequently, it demonstrates that relative resistance change can be a parameter for classifying deformation but can be affected by tensile speed.

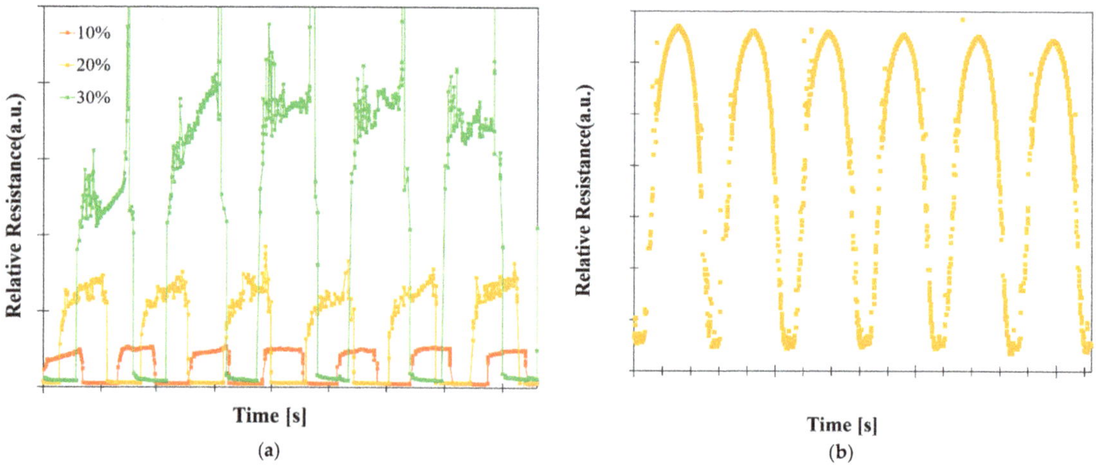

Figure 15. Transient measurement of Sample C: (**a**) relative resistance change as a function of time at 10%, 20%, and 30% strain; (**b**) transient curve at the tensile speed of 6 mm/min and a waiting time of 0.1 s.

4. Conclusions

We introduced the strain sensor to the adapting MCL to ensure ohmic contact with PANI and formed a buffer layer to protect it from the strain. The I−V curves according to each strain showed that the sample containing only PANI had the most unstable behavior and suggested that PANI and copper tape had Schottky contact, prolonging performance degradation. The strain sensor with the MCL, but without the H-PDMS, had deterioration due to damage to the MCL under deformation. The PANI/Au/H-PDMS strain sensor had stable behavior that helped in distinguishing the characterization and classification. As a result, we demonstrated the improvement of operational stability by adapting the MCL and the buffer layer. Moreover, this arrangement is more suitable for body-mounted wearable devices because it is stable even at a low voltage. The fabrication of the substrate with different stretchability regions on a single plane provided a research foundation for strain sensors' integration. We anticipate various future research into structural enhancement, such as the measurement or packaging of a strain sensor, rather than the development of materials, which can improve a device's performance and sustainability, to contribute to the commercialization of stretchable devices.

Author Contributions: Methodology, conceptualization, formal analysis, and writing—original draft, J.-Y.C. and E.-B.J.; formal analysis and data curation, H.-K.Y.; software, J.-H.H.; methodology and visualization, C.-J.Y.; software and validation, J.-H.L.; resources, funding acquisition, supervision, and writing—review and editing, H.-S.K. All authors have read and agreed to the published version of the manuscript.

Funding: This work was supported by the Mid-career Researcher Program through the National Research Foundation of Korea (NRF), funded by the Ministry of Science and ICT (No. 2019R1A2C2086747) and the MOTIE (Ministry of Trade, Industry, and Energy) in Korea, under the Fostering Global Talents for Innovative Growth Program supervised by the Korea Institute for Advancement of Technology (No. P0017307).

Institutional Review Board Statement: Not applicable.

Conflicts of Interest: The authors declare no conflict of interest.

eferences

Tan, C.; Dong, Z.; Li, Y.; Zhao, H.; Huang, X.; Zhou, Z.; Jiang, J.W.; Long, Y.Z.; Jiang, P.; Zhang, T.Y.; et al. A high performance wearable strain sensor with advanced thermal management for motion monitoring. *Nat. Commun.* **2020**, *11*, 3530. [CrossRef] [PubMed]

Amjadi, M.; Pichitpajongkit, A.; Lee, S.; Ryu, S.; Park, I. Highly stretchable and sensitive strain sensor based on silver nanowire–elastomer nanocomposite. *ACS Nano* **2014**, *8*, 5154–5163. [CrossRef]

Horev, Y.D.; Maity, A.; Zheng, Y.; Milyutin, Y.; Khatib, M.; Yuan, M.; Suckeveriene, R.Y.; Tang, N.; Wu, W.; Haick, H. Stretchable and Highly Permeable Nanofibrous Sensors for Detecting Complex Human Body Motion. *Adv. Mater.* **2021**, *33*, 2102488. [CrossRef] [PubMed]

Yu, G.F.; Yan, X.; Yu, M.; Jia, M.Y.; Pan, W.; He, X.X.; Han, W.P.; Zhang, Z.M.; Yu, L.M.; Long, Y.Z. Patterned, highly stretchable and conductive nanofibrous PANI/PVDF strain sensors based on electrospinning and in situ polymerization. *Nanoscale* **2016**, *8*, 2944–2950. [CrossRef]

Chen, J.; Zhu, Y.; Jiang, W. A stretchable and transparent strain sensor based on sandwich-like PDMS/CNTs/PDMS composite containing an ultrathin conductive CNT layer. *Compos. Sci. Technol.* **2020**, *186*, 107938. [CrossRef]

Chen, S.; Wei, Y.; Yuan, X.; Lin, Y.; Liu, L. A highly stretchable strain sensor based on a graphene/silver nanoparticle synergic conductive network and a sandwich structure. *J. Mater. Chem. C* **2016**, *4*, 4304–4311. [CrossRef]

Gong, X.X.; Fei, G.T.; Fu, W.B.; Fang, M.; Gao, X.D.; Zhong, B.N.; Zhang, L.D. Flexible strain sensor with high performance based on PANI/PDMS films. *Org. Electron.* **2017**, *47*, 51–56. [CrossRef]

Song, E.; Choi, J.W. Conducting polyaniline nanowire and its applications in chemiresistive sensing. *Nanomaterials* **2013**, *3*, 498–523. [CrossRef]

Cai, G.; Wang, J.; Lin, M.F.; Chen, J.; Cui, M.; Qian, K.; Li, S.; Cui, P.; Lee, P.S. A semitransparent snake-like tactile and olfactory bionic sensor with reversibly stretchable properties. *NPG Asia Mater.* **2017**, *9*, e437. [CrossRef]

). Amjadi, M.; Kyung, K.U.; Park, I.; Sitti, M. Stretchable, skin-mountable, and wearable strain sensors and their potential applications: A review. *Adv. Funct. Mater.* **2016**, *26*, 1678–1698. [CrossRef]

1. Tolvanen, J.; Hannu, J.; Jantunen, H. Stretchable and washable strain sensor based on cracking structure for human motion monitoring. *Sci. Rep.* **2018**, *8*, 13241. [CrossRef] [PubMed]

2. Abdulrazzaq, O.; Bourdo, S.E.; Saini, V.; Watanabe, F.; Barnes, B.; Ghosh, A.; Biris, A.S. Tuning the work function of polyaniline via camphorsulfonic acid: An X-ray photoelectron spectroscopy investigation. *RSC Adv.* **2015**, *5*, 33–40. [CrossRef]

3. Hajibadali, A.; Nejhad, M.B.; Farzi, G.; Rad, H.H. Fabrication and characterization of polyaniline based Schottky diode. In Proceedings of the 2013 21st Iranian Conference on Electrical Engineering (ICEE), Mashhad, Iran, 14–16 May 2013.

4. Ghorbani, L.; Nasirian, S. Zinc oxide nanorods assisted by polyaniline network as a flexible self-powered Ultraviolet photodetector: A comprehensive study. *Appl. Surf. Sci.* **2020**, *527*, 146786. [CrossRef]

5. Li, Q.; Li, J.; Tran, D.; Luo, C.; Gao, Y.; Yu, C.; Xuan, F. Engineering of carbon nanotube/polydimethylsiloxane nanocomposites with enhanced sensitivity for wearable motion sensor. *J. Mater. Chem. C* **2017**, *5*, 11092–11099. [CrossRef]

6. Huang, L.; Wang, H.; Wu, P.; Haung, W.; Gau, W.; Fang, F.; Cai, N.; Chen, R.; Zhu, Z. Wearable flexible strain sensor based on three-dimensional wavy laser-induced graphene and silicone rubber. *Sensors* **2020**, *20*, 4266. [CrossRef] [PubMed]

7. Zheng, Y.; Li, Y.; Dai, K.; Wang, Y.; Zheng, G.; Liu, C.; Shen, C. A highly stretchable and stable train sensor based on hybrid carbon nanofillers/polydimethylsiloxane conductive composites for large human motions monitoring. *Compos. Sci. Technol.* **2018**, *156*, 276–286. [CrossRef]

8. Mata, A.; Fleischman, A.J.; Roy, S. Characterization of polydimethylsiloxane (PDMS) properties for biomedical micro/nanosystems. *Biomed. Microdevices* **2005**, *7*, 281–293. [CrossRef] [PubMed]

9. Johnston, I.D.; McCluskey, D.K.; Tan, C.K.L.; Tracey, M.C. Mechanical characterization of bulk Sylgard 184 for microfluidics and microengineering. *J. Micromech. Microeng.* **2014**, *24*, 035017. [CrossRef]

). Jin, C.; Ma, C.; Yang, Z.; Lin, H. A force measurement method based on flexible PDMS grating. *Appl. Sci.* **2020**, *10*, 2296. [CrossRef]

1. Chen, W.; Lam, R.H.W.; Fu, J. Photolithographic surface micromachining of polydimethylsiloxane (PDMS). *Lab Chip* **2012**, *12*, 391–395. [CrossRef]

22. Guo, L.; Deweerth, S.P. An Effective Lift-Off Method for Patterning High-Density Gold Interconnects on an Elastomeric Substrat
Small **2010**, *6*, 2847–2852. [CrossRef]
23. Van Dam, R.M. *Solvent-Resistant Elastomeric Microfluidic Devices and Applications*; California Institute of Technology: Pasadena, CA USA, 2006.
24. *ASTM D412-16*; Standard Test Methods for Vulcanized Rubber and Thermoplastic Elastomer. ASTM: West Conshohocken, PA USA, 2009; 1–14.
25. Mazurek, P.; Hvilsted, S.; Skov, A.L. Green silicone elastomer obtained from a counterintuitively stable mixture of glycerol an PDMS. *Polymer* **2016**, *86*, 1–7. [CrossRef]
26. Jeronimo, K.; Koutsos, V.; Cheung, R.; Mastropaolo, E. PDMS-ZnO Piezoelectric Nanocomposites for pressure sensors. *Sensor* **2021**, *17*, 5873. [CrossRef] [PubMed]
27. Sunkara, V.; Park, D.K.; Cho, Y.K. Versatile method for bonding hard and soft materials. *RSC Adv.* **2012**, *2*, 9066–9070. [CrossRef
28. Dumond, J.J.; Low, H.Y.; Lee, H.P.; Fuh, J.Y. Multi-functional silicone stamps for reactive release agent transfer in UV roll-to-ro nanoimprinting. *Mater. Horiz.* **2016**, *3*, 152–160. [CrossRef]
29. Schmid, H.; Michel, B. Siloxane polymers for high-resolution, high-accuracy soft lithography. *Macromolecules* **2000**, *33*, 3042–304 [CrossRef]
30. Hill, S.; Qian, W.; Chen, W.; Fu, J. Surface micromachining of polydimethylsiloxane for microfluidics applications. *Biomicrofluidic* **2016**, *10*, 054114. [CrossRef] [PubMed]
31. Kang, S.; Rachim, V.P.; Baek, J.H.; Lee, S.Y.; Park, S.M. A flexible patch-type strain sensor based on polyaniline for continuou monitoring of pulse wave. *IEEE Access* **2020**, *8*, 152105–152115. [CrossRef]
32. Tong, L.; Wang, X.X.; He, X.X.; Nie, G.D.; Zhang, J.; Zhang, B.; Guo, W.Z.; Long, Y.Z. Electrically conductive TPU nanofibrou composite with high stretchable for flexible strain sensor. *Nanoscale Res. Lett.* **2018**, *13*, 86. [CrossRef] [PubMed]

sensors

rticle

Classification of Chaotic Squeak and Rattle Vibrations by CNN Using Recurrence Pattern

aehyeon Nam [1] and Jaeyoung Kang [2,*]

[1] Future Automotive Intelligent Electronics Core Technology Center, Kongju National University, Cheonan 31080, Korea; jaehyeon86@kongju.ac.kr
[2] Department of Mechanical Engineering, Inha University, Incheon 22212, Korea
* Correspondence: kangj@inha.ac.kr

Abstract: The chaotic squeak and rattle (S&R) vibrations in mechanical systems were classified by deep learning. The rattle, single-mode, and multi-mode squeak models were constructed to generate chaotic S&R signals. The repetition of nonlinear signals generated by them was visualized using an unthresholded recurrence plot and learned using a convolutional neural network (CNN). The results showed that even if the signal of the S&R model is chaos, it could be classified. The accuracy of the classification was verified by calculating the Lyapunov exponent of the vibration signal. The numerical experiment confirmed that the CNN classification using nonlinear vibration images as the proposed procedure has more than 90% accuracy. The chaotic status and each model can be classified into six classes.

Keywords: squeak; rattle; convolutional neural network; Lyapunov exponent; chaos; recurrence patterns

1. Introduction

Chaotic squeak and rattle (S&R) vibrations are a significant factor for evaluating the quality of automotive parts. Early S&R was detected with a find-and-fix approach by a subjective evaluation from engineers. Therefore, highly skilled experts are needed to detect S&R. In addition, the evaluation was made subjectively because of the differences in training and expertise and the use of different measurement tools. For this reason, an objective tool is needed for quantitative measurements. On the other hand, the S&R indices developed for such an evaluation must define the threshold value of the index, and the threshold value must be based on a subjective evaluation [1–4]. In addition, S&R problems occurring in mechanical systems are challenging to analyze because they include extreme nonlinearities, such as impact and friction [5–9].

Squeak is a self-excited vibration caused by friction that frequently occurs in automobile brakes, artificial hip joints, and gear systems [8,10,11]. Many studies have examined the vibration instability caused by friction based on an analysis of brake squill noise. The method to solve these problems was studied mainly by analyzing the instability using vibration equations, including nonlinearity of friction and linear stability through the linearization of nonlinear terms. On the other hand, the linearized approach can only be investigated near equilibrium. Kang [8] described complex models, such as instability caused by friction curves, modal coupling instability, gyroscopic, and friction damping occurring in automobile disc brake systems. They also analyzed the influence of squeal. Nam et al. [10] investigated the vibration instability in the lead screw system experimentally and analyzed the instability mechanism using the finite element method (FEM). Ouenzerfi et al. [11] examined the frictional instability occurring in an artificial hip joint and investigated the instability through a detailed FEM. In addition, the friction force is expressed as a function of the velocity in the dynamic instability of the friction-induced model. Higher-order nonlinear problems, such as chaos, were described because the model includes extreme nonlinearity in the creep section [12–16]. Kang [15] used a two-degree-of-freedom friction model to show that the chaotic phenomenon is generated by self-excited vibrations and investigated the parameters that create chaos.

Rattle can cause chaos due to extreme discontinuity caused by the impact force, including vibrations due to the impact vibrations induced by the excitation. For this reason, the dynamics of impact motion have been studied extensively for a study of chaos. Serweta et al. [17,18] examined the chaotic characteristics by calculating the Lyapunov exponent of an impact oscillator with symmetrical soft stops and rigid stop. Kang [19] analyzed the chaotic factors by calculating the Lyapunov exponent for the truncated number of modes of the impact beam under a distributed contact using the continuum beam model.

In addition to the theoretical approach, the analysis of such a nonlinear vibration signal has been performed using a visualization method and a quantified index. A general signal analysis method takes an FFT in the time series and analyzes the dynamic characteristics in the frequency domain. Furthermore, dynamical characteristics were examined through the trajectory of the attractor in the phase space. As many studies on signal analysis have been carried out, signal visualization methods, such as Gauss wavelets [20,21] and a recurrence plot (RP), have been developed. Marwan et al. [22] introduced various RP methods to visualize the dynamic characteristics in a complex system. RP is a power tool that visualizes and analyzes the recurrence characteristics of dynamic systems. In addition, recurrence can be visualized efficiently and developed formally using a matrix. The reciprocal of the longest diagonal of an RP is proportional to the largest Lyapunov exponent. This shows that RP can express both the recurrence and chaos characteristics well. An RP is represented on the reconstructed phase space that is determined using the time delay method [23–26]. Recurrence quantification analysis (RQA) can quantify the repetition characteristics through indices expressed as found in the recurrence rate (RR), the determinism (DET), and the average diagonal line length based on the RP. RQA is a good technique for quantifying recurrence properties, but the results are presented only in indices [22]. In addition, higher-order spectrum analysis (HOSA) and clustering techniques are used to analyze various methods, including high-dimensional nonlinearity [27].

Dynamics problems involving extreme nonlinearities, such as S&R, can be accompanied by chaos. The most accurate way to determine chaos is the Lyapunov exponent. Wolf et al. described a method called the spectrum of the largest Lyapunov exponent. A calculation algorithm was also developed [28], and chaos could be determined by parameter analysis. On the other hand, this algorithm cannot be applied in nonlinear dynamical systems, including discontinuities, and can only be used in smooth dynamical systems. In contrast, Muller's algorithm can be applied to a non-smooth dynamical system through an indicator function and transition condition [16]. Determining the chaos through the Lyapunov exponents is advantageous if the governing equation for the system is known or the available observations are very long [29].

Recently, with the rapid development of artificial intelligence, many algorithms using machine learning have been developed. In particular, for image classification, numerous CNN models based on a convolutional neural network (CNN) have been established, and ResNet, which was released in 2015, transcends human cognitive ability [30]. Hsueh et al. [31] showed that the fault signal of a motor through the experiment could be classified in binary by a CNN. Nam et al. [32] reported that, even if the vibration signal includes discontinuous nonlinearities, such as impact, the chaotic signal can be classified by a CNN using the image visualized with an unthresholded RP. On the other hand, it only performed a dichotomous classification for chaos and non-chaos. Therefore, the binary classification study was extended to perform a multi-class classification of chaotic S&R vibration signals.

This study examined whether the rattle and squeak signals can be classified through a CNN, even if they are chaotic, by applying a signal visualization technique. Because

CNN is an image-based classification technique, an RP-based dataset was constructed to express the repetition of a dynamic system quantitatively. A single-mode squeak, multi-mode squeak, and rattle model were built. A methodology that classifies six classes for the chaotic S&R model with high accuracy through a CNN is proposed.

2. Methods

In this study, a theoretical model of S&R vibration, a representative nonlinear vibration that can occur in a mechanical system, is used. Figure 1 presents the rattle model considering the mass, linear spring, nonlinear elastic contact, and damping. k is a linear spring coefficient. This system is excited with amplitude f_0 and excitation frequency ω_{ex}. As shown in Figure 1a, the distance from the impact surface at the static equilibrium position of the system is L. In addition, the nonlinear elastic model of the impact force was defined as Hertz's contact model [17,18]. The coordinate x^r describes the vibration motion of the rattle model at the static equilibrium position. Figure 1b shows the impact force of Hertz's contact model for a relative displacement. All systems consider a mass m attached to a spring with a stiffness coefficient k and coefficient of viscous damping c.

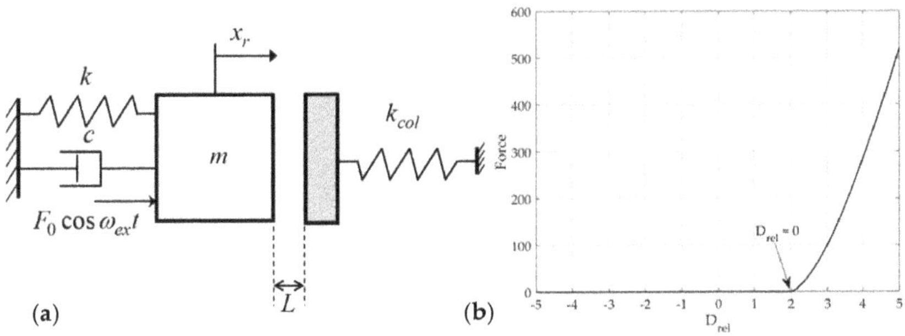

Figure 1. Rattle model: (**a**) 1-D model; (**b**) impact force.

The single-mode squeak model is excited with the same amplitude f_0 and excitation frequency ω_{ex} as the rattle model shown in Figure 2a. On the contact surface, a frictional force is generated by the normal force and the relative velocity. The friction force includes the creep region and negative slope for the sliding speed from Coulomb's law of friction, as shown in Figure 2b. The multi-mode squeak model has an added mass m_1, spring stiffness k_1, and coefficient of viscous damping c_1, and the frictional force generated by each mass is the same as that of the single-mode squeak model, as shown in Figure 3.

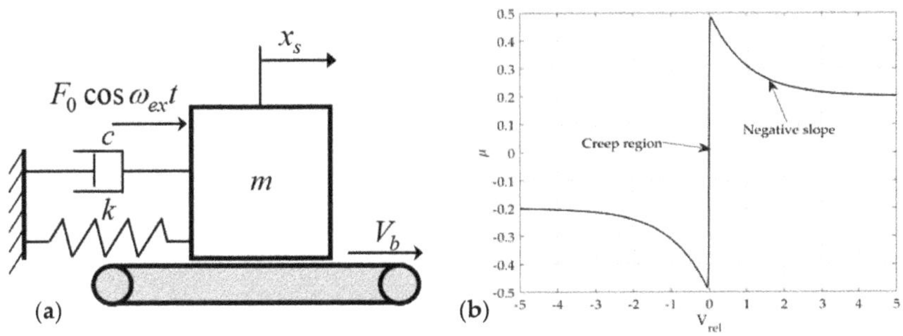

Figure 2. Single-mode squeak model: (**a**) 1-D model (**b**) friction–velocity curve.

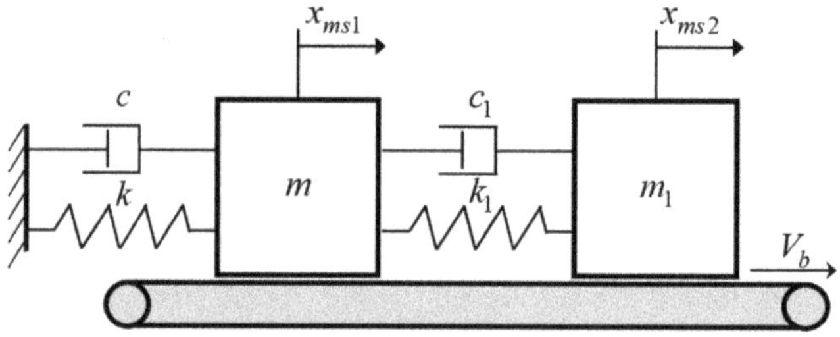

Figure 3. Multi-mode squeak model.

For the rattle model shown in Figure 1a

$$m\ddot{x}_r + c\dot{x}_r + kx_r = F_0 \cos \omega_{ex} t + F_c \tag{1}$$

where the contact model is the impact force by Hertz's nonlinear elastic model as follows

$$\begin{array}{ll} F_c = 0 & \text{if } x_r < L \\ F_c = k_{col}(x_r - L)^{3/2} & \text{if } x_r \geq L \end{array} \tag{2}$$

Using the dimensionless time $\tau = t\sqrt{k/m}$ and the coordinate transformation $x_r(t) = X_r(\tau)$, the dimensionless equation of motion can be written as

$$X''_r + \beta X'_r + X_r = f_0 \cos \eta\tau + F_c \frac{1}{k} \tag{3}$$

where prime is differentiation for $\tau(\geq 0)$ and the dimensionless parameter is defined as $\Omega = \sqrt{k/m}$, $\eta = \omega_{ex}/\Omega$, $f_0 = F_0/k$, $\beta = \sqrt{c^2/mk}$, $k_c = k_{col}\sqrt{f_0}/k$ and $u = X_r/f_0$. Therefore, the dimensionless equation of motion for the rattle model can be written as

$$u'' + \beta u' + u = \cos \eta\tau + Hf_c \tag{4}$$

and the dimensionless form of the impact force is rewritten as follows

$$\begin{array}{ll} f_c = 0 & \text{if } u < r \\ f_c = k_c(u - r)^{3/2} & \text{if } u \geq r \end{array} \tag{5}$$

Equation of motion for the rattle model is expressed in vector form as follows

$$\mathbf{u} = \begin{bmatrix} u & u' \end{bmatrix}^{\mathrm{T}} \tag{6}$$

$$\mathbf{u}' = \mathbf{f}_r(\mathbf{u}),\ \mathbf{u}(0) = \mathbf{u}_0 \tag{7}$$

$$\mathbf{f}(\mathbf{u}) = \begin{bmatrix} u_2 \\ -\beta u_2 - u_1 + \cos(u_3) + Hf_c \\ \eta \end{bmatrix} \tag{8}$$

where H is the Heaviside function and $\mathbf{u}_0$ is the initial condition of the rattle model.

Equation (7) is a dynamic system with discontinuities involving the discontinuous impact effects in the rattle model. Therefore, it can be rewritten as follows from Muller's method that includes the instantaneous discontinuity of impact. Here, $\tau = \tau_i$ is a discontinuous moment. Let $\mathbf{z}$ be a state variable $\mathbf{u}$ of the rattle model.

$$\tau_{i-1} < \tau < \tau_i\text{: } \mathbf{z}' = \mathbf{f}_i(\mathbf{z}),\ \mathbf{z}(\tau_{i-1}) = \mathbf{z}(\tau_{i-1}^{+}) \tag{9}$$

$$\tau = \tau_i\text{: } 0 = \mathbf{h}(\mathbf{z}(\tau_i^{-})) \tag{10}$$

$$\mathbf{z}(\tau_i^+) = \mathbf{g}(\mathbf{z}(\tau_i^-)) \tag{11}$$

$$\tau_i < \tau < \tau_{i+1}: \mathbf{z}' = \mathbf{f}_{i+1}(\mathbf{z}), \mathbf{z}(\tau_i) = \mathbf{z}(\tau_i^+) \tag{12}$$

where τ^- and τ^+ denote before and after the discontinuously condition. The perturbed trajectory is given by

$$\tilde{\mathbf{z}}(\tau) = \mathbf{z}(\tau) + \delta\mathbf{z}(\tau) \tag{13}$$

$$\tilde{\tau}_i = \tau_i + \delta\tau_i \tag{14}$$

and the perturbed trajectory satisfies the following equation

$$\tilde{\tau}_{i-1} < \tau < \tilde{\tau}_i: \tilde{\mathbf{z}}' = \mathbf{f}_i(\tilde{\mathbf{z}}), \tilde{\mathbf{z}}(\tilde{\tau}_{i-1}) = \tilde{\mathbf{z}}(\tilde{\tau}_{i-1}^+) \tag{15}$$

$$\tau = \tilde{\tau}_i: 0 = \mathbf{h}(\tilde{\mathbf{z}}(\tilde{\tau}_i^-)) \tag{16}$$

$$\tilde{\mathbf{z}}(\tilde{\tau}_i^+) = \mathbf{g}(\tilde{\mathbf{z}}(\tilde{\tau}_i^-)) \tag{17}$$

$$\tilde{\tau}_i < \tau < \tilde{\tau}_{i+1}: \tilde{\mathbf{z}}' = \mathbf{f}_{i+1}(\tilde{\mathbf{z}}), \tilde{\mathbf{z}}(\tilde{\tau}_i) = \tilde{\mathbf{z}}(\tilde{\tau}_i^+) \tag{18}$$

where each interval of discontinuities is smooth. $\mathbf{h}(\mathbf{z})$ and $\mathbf{g}(\mathbf{z})$ are the indicator function and the transition condition, respectively. The plus and minus signs denote the right- and left-sided limits, and

$$\delta\tau_i = \tilde{\tau}_i - \tau_i = -\frac{D\mathbf{h}(\mathbf{z}_i^-)\delta\mathbf{z}_i^-}{D\mathbf{h}(\mathbf{z}_i^-)\mathbf{f}_i(\mathbf{z}_i^-)} \tag{19}$$

$$\delta\mathbf{z}_i^+ = D\mathbf{g}(\mathbf{z}_i^-)\delta\mathbf{z}_i^- + [D\mathbf{g}(\mathbf{z}_i^-)\mathbf{f}_i(\mathbf{z}_i^-) - \mathbf{f}_{i+1}(\mathbf{z}_i^+)]\delta\tau_i \tag{20}$$

in which

$$D\mathbf{h}(\mathbf{z}_i^-) = \left.\frac{\partial\mathbf{h}(\mathbf{z})}{\partial\mathbf{z}}\right|_{\mathbf{z}=\mathbf{z}_i^-}, D\mathbf{g}(\mathbf{z}_i^-) = \left.\frac{\partial\mathbf{g}(\mathbf{z})}{\partial\mathbf{z}}\right|_{\mathbf{z}=\mathbf{z}_i^-} \tag{21}$$

are the Jacobian matrix of indicator function and transition condition at point $\mathbf{z}_i^-$, respectively, where $\mathbf{z}_i^- = \mathbf{z}(\tau_i^-)$ and $\mathbf{z}_i^+ = \mathbf{z}(\tau_i^+)$. For an impact oscillator with Hertz's model of contact, the Jacobian matrix of the transition condition and indicator function becomes the following matrix [17]

$$D\mathbf{h}(\mathbf{z}_i^-) = \begin{bmatrix} 1 & 0 & 0 \end{bmatrix}^{\mathrm{T}}, D\mathbf{g}(\mathbf{z}_i^-) = \mathbf{I} \tag{22}$$

Therefore, the deviated trajectory can be written as

$$\delta\mathbf{z}' = \left.\frac{\partial\mathbf{f}}{\partial\mathbf{z}}\right|_{\mathbf{z}=\mathbf{z}_i} \cdot \delta\mathbf{z} + O_{(2)}, \delta\mathbf{z}(\tau_0) = \delta\mathbf{z}_0 \tag{23}$$

By letting $\delta\mathbf{z} = [\mathbf{\Phi}_\tau(\mathbf{z}_0)]\delta\mathbf{z}_0$ substitute into the perturbation Equation (23), $\delta\mathbf{z}_i^+$ at the discontinuous region is estimated using Equation (20). The variation equation is also calculated at the same time as

$$\left[\mathbf{\Phi}'_\tau(\mathbf{z}_0)\right] = [D_z\mathbf{f}][\mathbf{\Phi}_\tau(\mathbf{z}_0)], [\mathbf{\Phi}_{\tau_0}(\mathbf{z}_0)] = [\mathbf{I}] \tag{24}$$

where $[D_z\mathbf{f}]$, $[\mathbf{I}]$, and $[\mathbf{\Phi}_\tau(\mathbf{z}_0)]$ denote the Jacobian matrix, identity matrix, and solution of the variational equation, respectively.

For the single-mode squeak model shown in Figure 2a

$$m\ddot{x}_s + c\dot{x}_s + kx_s = f_0 \cos\omega_{ex}t + F_\mu^{x_s} \tag{25}$$

where $F_\mu^{x_s}$ is the friction force. The friction force of the single-mode squeak model is expressed as α, and h are the control parameters that determine the negative slope. μ_s and μ_k are the static and dynamic friction coefficients, respectively.

$$F_\mu^{x_s} = \tanh(\alpha(V_b - \dot{x}_s))\{\mu_k + (\mu_s - \mu_k)\exp(-h|V_b - \dot{x}_s|)\}N \tag{26}$$

Using the dimensionless time $\tau = t\sqrt{k/m}$ and the coordinate transformation $x_s(t) = X_s(\tau)$, the dimensionless equation of motion can be written as

$$X_s'' + \beta X_s' + X_s = f_0 \cos \eta\tau + \frac{1}{k}F_\mu^{x_s} \tag{27}$$

where prime is the differentiation for $\tau(\geq 0)$, and the dimensionless parameters are defined as $\Omega = \sqrt{k/m}$, $\eta = \omega_{ex}/\Omega$, $f_0 = F_0/k$, $\beta = \sqrt{c^2/mk}$, and $u = X_r/f_0$. Therefore, the dimensionless friction force of the single-mode squeak model can be expressed as

$$f_\mu^v = \tanh(\alpha(V_b - v'))\{\mu_k + (\mu_s - \mu_k)\exp(-h|V_b - v'|)\}n_0 \tag{28}$$

where $\alpha\Omega f_0 \to \alpha$, $V_b/\Omega f_0 \to V_b$, $h\Omega f_0 \to h$, and $N/kf_0 = n_0$ are dimensionless parameters. Therefore, the dimensionless equation of motion for the single-mode squeak model can be written as

$$v'' + \beta v' + v = \cos \eta\tau + f_\mu^v \tag{29}$$

The equation of motion for the single-mode squeak model is expressed in vector form

$$\mathbf{v} = \begin{bmatrix} v & v' \end{bmatrix}^{\mathbf{T}} \tag{30}$$

$$\mathbf{v}' = \mathbf{f}_s(\mathbf{v}),\ \mathbf{v}(0) = \mathbf{v}_0 \tag{31}$$

$$\mathbf{f}_s(\mathbf{v}) = \begin{bmatrix} v_2 \\ -\beta v_2 - v_1 + \cos(v_3) + f_\mu^v \\ \eta \end{bmatrix} \tag{32}$$

where $\mathbf{v}_0$ is the initial condition of the single-mode squeak model.

For the multi-mode squeak model shown in Figure 3

$$\begin{aligned} m\ddot{x}_{ms1} + (c + c_1)\dot{x}_{ms1} - c_1\dot{x}_{ms2} + (k + k_1)x_{ms1} - k_1x_{ms2} = F_\mu^{x_{ms1}} \\ m_1\ddot{x}_{ms2} - c_1\dot{x}_{ms1} + c_1\dot{x}_{ms2} - k_1x_{ms1} + k_1x_{ms2} = F_\mu^{x_{ms2}} \end{aligned} \tag{33}$$

where F_μ^{ms1} and F_μ^{ms2} are the friction forces acting on each mass. The friction force of the multi-mode squeak model is expressed as

$$F_\mu^{x_{ms1}} = \tanh(\alpha(V_b - \dot{x}_{ms1}))\{\mu_k + (\mu_s - \mu_k)\exp(-h|V_b - \dot{x}_{ms1}|)\}N \tag{34}$$

$$F_\mu^{ms2} = \tanh(\alpha(V_b - \dot{x}_{ms2}))\{\mu_k + (\mu_s - \mu_k)\exp(-h|V_b - \dot{x}_{ms2}|)\}N \tag{35}$$

Using the dimensionless time $\tau = t\sqrt{k/m}$ and the coordinate transformation $x_{ms1}(t) = X_{ms1}(\tau)$ and $x_{ms2}(t) = X_{ms2}(\tau)$, the dimensionless equation of motion can be expressed as

$$\begin{aligned} X_{ms1}'' + (\beta + \beta_1)X_{ms1}' - \beta_1X_{ms2}' + (1 + \gamma)X_{ms1} - \gamma X_{ms2} = \tfrac{1}{k}F_\mu^{x_{ms1}} \\ X_{ms2}'' - \sigma\beta_1X_{ms1}' + \sigma\beta_1X_{ms2}' - \sigma\gamma X_{ms1} + \sigma\gamma X_{ms2} = \sigma\tfrac{1}{k}F_\mu^{x_{ms2}} \end{aligned} \tag{36}$$

where prime is the differentiation with respect to $\tau(\geq 0)$, and the dimensionless parameter is defined as $\Omega = \sqrt{k/m}$, $\eta = \omega_{ex}/\Omega$, $f_0 = F_0/k$, $\beta = \sqrt{c^2/mk}$, $\beta_1 = \sqrt{c_1^2/mk}$, $\gamma = k_1/k$, $\sigma = m/m_1$, $w = X_{ms1}/f_0$, and $s = X_{ms2}/f_0$. Therefore, the dimensionless friction force of the multi-mode squeak model can be rewritten as

$$f_\mu^w = \tanh(\alpha(V_b - w'))\{\mu_k + (\mu_s - \mu_k)\exp(-h|V_b - w'|)\}n_0 \tag{37}$$

$$f_\mu^s = \tanh(\alpha(V_b - s'))\{\mu_k + (\mu_s - \mu_k)\exp(-h|V_b - s'|)\}n_0 \tag{38}$$

where $\alpha\Omega f_0 \to \alpha$, $V_b/\Omega f_0 \to V_b$, $h\Omega f_0 \to h$, and $N/kf_0 = n_0$ are dimensionless parameters. Therefore, the dimensionless equation of motion for the simple model can be expressed as

$$\begin{array}{l} w'' + (\beta+\beta_1)w' - \beta_1 s' + (1+\gamma)w - \gamma s = f_\mu^w \\ s'' - \sigma\beta_1 w' + \sigma\beta_1 s' - \sigma\gamma w + \sigma\gamma s = f_\mu^s \end{array} \quad (39)$$

The equation of motion for the multi-mode squeak model is expressed in vector form as follows

$$\mathbf{w} = \begin{bmatrix} w & w' & s & s' \end{bmatrix}^{\mathbf{T}} \quad (40)$$

$$\mathbf{w}' = \mathbf{f}_{ms}(\mathbf{w}),\ \mathbf{w}(0) = \mathbf{w}_0 \quad (41)$$

$$\mathbf{f}_{ms}(\mathbf{w}) = \begin{bmatrix} w_2 \\ -(\beta+\beta_1)w_2 + \beta_1 s_2 - (1+\gamma)w_1 + \gamma s_1 + f_\mu^w \\ s_2 \\ \sigma\left(\beta_1 w_2 - \beta_1 s_2 + \gamma w_1 - \gamma s_1 + f_\mu^s\right) \end{bmatrix} \quad (42)$$

where $\mathbf{w}_0$ is the initial condition of the multi-mode squeak model.

Because the squeak models are a dynamic system without discontinuities, Equations (9)–(22) are unnecessary. Therefore, the Lyapunov exponent of the squeak model can be obtained directly from the eigenvalue of the variation Equation (24).

The Lyapunov exponents can be defined as

$$\lambda_i = \lim_{\tau\to\infty} \frac{1}{\tau} \ln|m_i(\tau)| \quad (43)$$

where $m_i(t)$ are the eigenvalues of Equation (24). On the other hand, the definition cannot be used directly in the numerical calculation. If there is a considerable time, the variation equations tend to be the ill-condition [17]. Therefore, the spectrum of the Lyapunov exponent for the linearized equation was estimated using Wolf's algorithm via the QR-factorization orthonormalization [33].

As mentioned earlier, in this study, an image of a dynamic signal was constructed based on the RP that visualized the dynamic characteristics most effectively. The recurrence is a fundamental property in a dynamic systems, and RP is a tool that visualizes the iteration of the state of the system. The corresponding RP is based on the following recurrence matrix as follows

$$\mathbf{RP}_{i,j} = H(\varepsilon - \|\mathbf{z}_i - \mathbf{z}_j\|), i, j = 1, \ldots, N \quad (44)$$

where $\mathbf{RP}_{i,j}$ is called the RP or threshold RP. $\|\cdot\|$ is an L-2 norm; N is the measured points, and $\{\mathbf{z}_i\}_{i=1}^N$ is trajectories of a system in its phase space. ε is the threshold. The threshold is a critical parameter that can be obtained differently depending on the system, but it was quantified probabilistically ($\varepsilon > 5\sigma$) by Thiel et al. [34]. Therefore, an unthresholded RP without the influence of the threshold can be expressed as

$$\mathbf{RP}_{i,j}^{un} = \|\mathbf{z}_i - \mathbf{z}_j\|, i, j = 1, \ldots, N \quad (45)$$

Here the element of phase space indicates the possible state of the system for the time evolution law. In such a case, the phase space needs to be reconstructed. The method for reconstruction is generally conducted using the time delay method. Thus, the reconstructed state variable can be expressed as

$$\mathbf{z}_i \to \hat{\mathbf{z}}_i = \sum_{j=1}^{m} q_{i+(j-1)\nu} \mathbf{e}_j \quad (46)$$

where $q_i = q(i\Delta\tau)$, $\Delta\tau$, m, ν, and $\mathbf{e}_j$ are the discrete time series, sampling rate, embedding dimension, time delay, and unit vectors, respectively. The reconstruction does not change the dynamic properties, and the reconstructed phase space can be expressed through an appropriately selected embedding dimension and time delay. In general, the time delay can be selected appropriately using the mutual information method.

During time delay reconstruction, all self-crossing trajectories in the dimension D_A of the attractor can disappear when the embedding dimension $D > 2D_A$ is set. On the other hand, it is imperative to determine the minimum embedding dimension to minimize the Lyapunov exponents and computational calculations from a physical perspective. From Equation (46), in dimension d, $\mathbf{z}_r$ is the rth nearest neighbor of $\mathbf{z}$, and the square of the Euclidean distance between the two vectors is

$$R_d^2(i,r) = \sum_{e=0}^{v-1} \left[q_{(i+ev)} - q_{r,(i+ev)} \right]^2 \tag{47}$$

Here, as the time delay embedding extends from dimension d to dimension $d+1$, the Euclidean distance between the rth neighbors for dimension rth can be written as follows

$$R_{d+1}^2(i,r) = R_d^2(i,r) + \left[q_{(i+ev)} - q_{r,(i+ev)} \right]^2 \tag{48}$$

where the error for the minimum embedding dimension can be determined from the rate of change in the Euclidean distance.

$$\sqrt{\frac{R_{d+1}^2(i,r) - R_d^2(i,r)}{R_{d+1}^2(i,r)}} > R_{tol} \tag{49}$$

where R_{tol} is the threshold. Kennel et al. [24] reported that false neighbors could be identified clearly in $R_{tol} \geq 10$. Another condition for determining false neighbors defined based on the actual value of $R_d(i) \equiv R_d(i, r = 1)$ is similar to the standard variation R_A of the attractor using finite data of the noisy signal. Thus, the Euclidean distance for the dimension $d+1$ becomes $R_{d+1}(i) \approx 2R_A$, and the second criterion for determining false neighbors can be written as

$$\frac{R_{d+1}(i)}{R_A} > A_{tol} \tag{50}$$

Therefore, the minimum embedding dimension can be obtained by discriminating as false nearest neighbors (FNN) under the conditions in Equations (49) and (50).

Furthermore, the classified features were visualized through Class Activation Mapping (CAM) [35]. The procedure for CAM is as follows

$$F^k = \sum_{x,y} f_k(x,y) \tag{51}$$

where $f_k(x,y)$ represents the activation of the kth unit of the last convolutional layer at the spatial location (x,y). Therefore, the value obtained by Global Average Pooling (GAP) on the kth unit becomes F^k. Accordingly, the input softmax for c classes is as follows

$$S_c = \sum_k w_k^c \sum_{x,y} f_k(x,y) = \sum_k \sum_{x,y} w_k^c f_k(x,y) \tag{52}$$

where w_k^c is the weight corresponding to class c for kth units, the learned weight represents an optimized model for class c. The output probability of softmax for class c is as follows

$$P_c = \frac{e^{S_c}}{\sum_c e^{S_c}} \tag{53}$$

Therefore, CAM for the class is defined, and the elements on each space are given as follows

$$M_c(x,y) = \sum_k w_k^c f_k(x,y) \quad (54)$$

Finally, the features of the learning result using CNN can be visualized as a heat map using Equation (54).

3. Results

For preliminary analysis, the Rossler model, a representative chaotic system, was used [36]. The Rossler model has already been studied extensively. It is a simple chaotic vibration system because it can produce a section that always vibrates in response to a parameter change. c is selected as the control parameter. The other parameters are $a = 0.2$ and $b = 0.2$, and the initial condition is $\mathbf{q}(0) = \begin{bmatrix} 1 & 1 & 1 \end{bmatrix}^T$.

$$\dot{\mathbf{s}} = \mathbf{f}(\dot{\mathbf{s}}) = \begin{bmatrix} -s_2 - s_3 \\ s_1 + a s_2 \\ b + s_3(s_1 - c) \end{bmatrix} \quad (55)$$

Figure 4 presents the signal s_1 for the control parameters of the Rossler model. Figure 4a is a time series analysis for $c = 3.5$ (dash line) and $c = 10$ (solid line). Figure 4b shows the corresponding phase space. $c = 3.5$ shows a clear period-2 in phase space; $c = 10$ shows the trajectory in the phase space and the aperiodic infinite trajectory in a finite boundary. On the other hand, the chaos cannot be identified clearly as a phase plot. To determine chaos, the Lyapunov exponents need to be calculated. Figure 5 presents the flow chart of the proposed methodology for applying signal classification using deep learning.

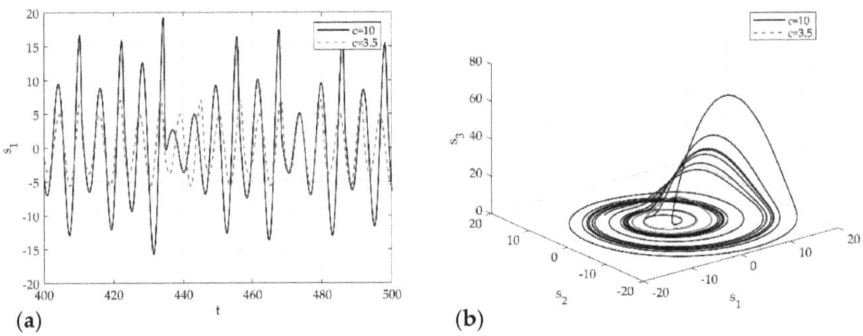

Figure 4. Dynamic solutions for the Rossler model for various c: (**a**) time analysis; (**b**) 3-D phase portrait.

As shown in the flow diagram, the proposed method classifies the characteristics of the nonlinear vibration signals not included in learning after learning a dataset of nonlinear vibration signals composed of images using the CNN architecture. In other words, the focus of this study was to learn the vibrating signal visualized based on RP by machine learning and to distinguish between the causes of vibration, such as friction or impact and chaotic characteristics. Details of the proposed method are as follows. First, the nonlinear time series data of the parametric deterministic dynamic system was obtained by numerical analysis using the Runge–Kutta method. The Lyapunov exponent was calculated for the time series data and chaos was determined. The image visualization method of the vibration signal used the FNN algorithm to determine the embedding dimension and reconstruct the phase space. The reconstructed signal was expressed as an unthresholded RP to visualize the dynamic characteristics. Finally, the dataset composed of the visualized signals was trained by the CNN model and verified using the Lyapunov exponent.

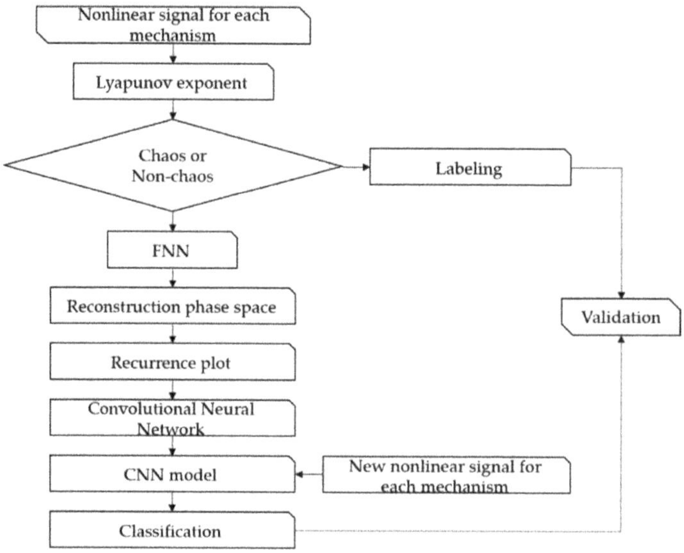

Figure 5. Flow diagram of the proposed methodology.

The architecture is structured relatively simply. However, GAP was used instead of Fully Connected (FC) to activate the CAM in the last layer. GAP is relatively less accurate than FC [35]. On the other hand, the purpose of this paper was to show that even if the S&R model is chaotic, it is possible to classify it through deep learning using imaged vibration signals. Hence, the architecture was constructed simply with the aim of approximately 90% accuracy. For preliminary analysis, Table 1 lists the layer type of the CNN model, filter size, and shape of each layer. Figure 6 presents a flow diagram of the Rossler system's CNN model.

Table 1. CNN model summary.

Layer (Type)	Output Shape	Param #
Conv2d	(None, 200, 200, 32)	896
Batch normalization	(None, 200, 200, 32)	128
Max pooling 2d	(None, 100, 100, 32)	0
Conv2d_1	(None, 100, 100, 64)	18,496
Batch normalization_1	(None, 100, 100, 64)	256
Max pooling 2d_1	(None, 50, 50, 64)	0
Conv2d_2	(None, 50, 50, 128)	73,856
Batch normalization_2	(None, 50, 50, 128)	512
Max pooling 2d_2	(None, 25, 25, 128)	0
Conv2d_3	(None, 25, 25, 256)	295,168
Batch normalization_3	(None, 25, 25, 256)	1024
Max pooling 2d_3	(None, 12, 12, 256)	0
Conv2d_4	(None, 12, 12, 512)	1,180,160
Global Average Pooling 2d	(None, 512)	0
Dense	(None, 2)	1026

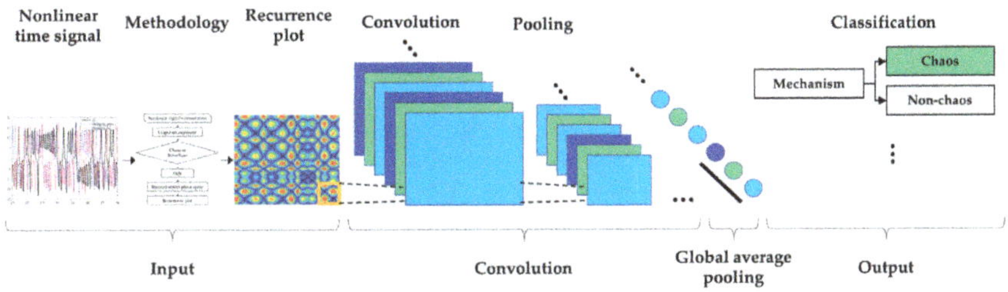

Figure 6. Flow diagram of CNN model.

Figure 7 presents the chaos analysis of the Rossler model. Figure 7a is a bifurcation diagram of the Rossler's model for parameter c change, and Figure 6b shows the corresponding Lyapunov exponents. The explanation of the critical c for the Lyapunov exponents of the Rossler model has been studied extensively. Briefly, to summarize this system, the first bifurcation appears near $c \approx 2.866$, and becomes period 2. The bifurcation appears again near $c \approx 3.86$ and becomes period 4. In other words, chaos occurs as period-doubling occurs at each point. As shown in Figure 4, if the time analysis result is $c = 3.5$, chaos is expressed as period 2, and $c = 10$. This agrees well with the time analysis results. In 3D phase space, the Lyapunov exponent has four types of attractors: stable fixed points ($\lambda_i < 0, i = 1,2,3$), stable limit cycles ($\lambda_1 < 0, \lambda_i < 0, i = 2,3$), stable two-torus ($\lambda_1 = \lambda_2 = 0, \lambda_i < 0, i = 3$), and strange attractors ($\lambda_1 > 0$). In the calculated system, however, only the classification of the S&R model and the existence of chaos were classified (Rossler system only distinguished between chaos and non-chaos). In other words, the strange attractor ($\lambda_1 > 0$) and dynamic characteristics of the deterministic dynamic system can be obtained from the flow of the proposed method, and an unthresholded recurrence plot was learned using CNN. Subsequently, an attempt was made to classify the signals and chaos generated by the S&R model that were not used for training. Figure 8 shows the visualized chaos and non-chaos signals for the randomly extracted Rossler model.

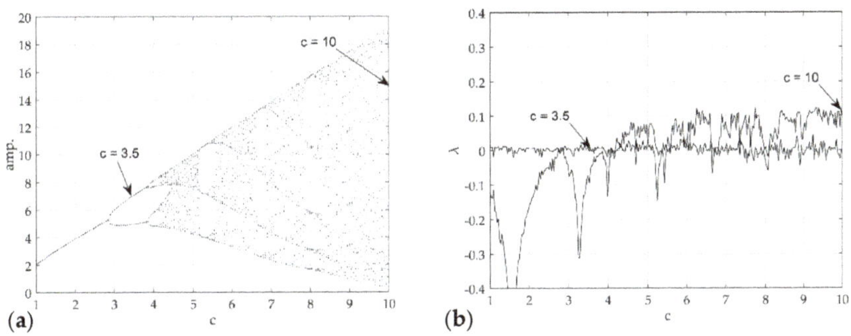

Figure 7. Chaotic analysis for Rossler model (**a**) bifurcation diagram of displacement (**b**) and largest Lyapunov exponent with respect to c.

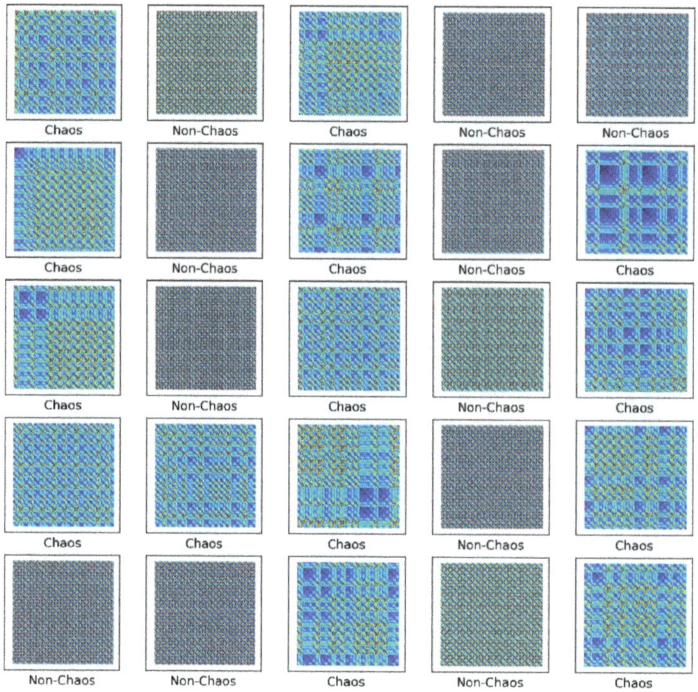

Figure 8. Unthresholded recurrence plot of the Rossler model.

Image classification using CNN has developed many sophisticated models that transcend the human cognitive abilities, but the design of a sophisticated architecture was not the goal of the present study. Therefore, the architecture is composed of a simple five-level structure, as shown in Table 1. Each step includes the convolution layer, activation function, and pooling layer. The proposed model comprises five convolution layers with a 32-3 × 3 filter, 64-3 × 3 filter, 128-3 × 3 filter, 256-3 × 3 filter, and 512-3 × 3 filter in each step. As mentioned earlier, RP is a tool to visualize the recurrence characteristics of a dynamic system. The filter size was set as small as possible because the chaos system can occur within a very short interval. In addition, three max pooling layers were used. Through five convolution layers, the feature map classifies the features of the image into six classes. To use CAM, GAP was used instead of the FC layer as the last layer. Softmax was used as the activation function of the output value. One of the gradient-based optimization methods was used. The Adam optimizer is an optimization function based on the gradient descent algorithm and was used to achieve faster convergence [37]. The weight initialization is one of the fundamental problems. Incorrect weight setting causes many problems, such as convergence problems and local minima. LeCun initialization follows a Gaussian distribution and uniform distribution of weight initialization for effective backpropagation [38]. Xavier initialization sets the initial weight depending on the number of previous and next nodes [39]. This is the most generalized method, but the output value shows inefficient results when used in the ReLU function. The He initialization was developed to compensate for this [25]. For the weight initialization in the proposed CNN model, the He initialization method following a Gaussian distribution was used.

The Rossler model consisted of 3200 datasets and 200 × 200-pixel images. The intervals of the time step for the ODE and orthonormalization for the Lyapunov exponent were 0.05 and 0.1, respectively. The dataset is usually divided into three parts. The 3200 datasets were divided into 70% for the training dataset and 30% of the test dataset. The validation dataset consisted of 30% of the training dataset. Table 2 lists the dataset samples used for training. The errors due to sequential datasets were removed by shuffling the dataset because the images are generated sequentially for parameter analysis. Deep learning requires a high-performance computer. The hardware used was a GPU machine (NFEC-2021-01-267120, Future Automotive Intelligent Electronics Core Technology Center) with NVIDIA GPU V100.

Table 2. Dataset split ratio for the Rossler model.

Data	Percentage	Number of Samples
Training	56%	2240
Validation	14%	560
Testing	30%	1200

Figure 9 shows the results of a numerical experiment for the proposed procedure. Chaos characteristics were found in the training dataset for 2240. At the same time, it was verified through 560 validation data in each epoch. After that, the tests were performed on 1200 testing datasets on the trained CNN model. The batch size was set to 10, and the learning rate of the optimization function was 0.0001. Figure 9a,b shows the accuracy and loss function of the training data and validation data for each epoch. As shown in the learning result, the accuracy showed a logarithmic function and converged to approximately 100, and the loss also showed a negative exponential function and converged close to zero. The accuracy and loss of validation data and the training data almost coincide, suggesting that the training proceeded well without overfitting. This suggests that the proposed procedure detects the chaos characteristics of the Rossler model well. In addition, 1200 testing datasets that were not used for training were also classified with 99% accuracy.

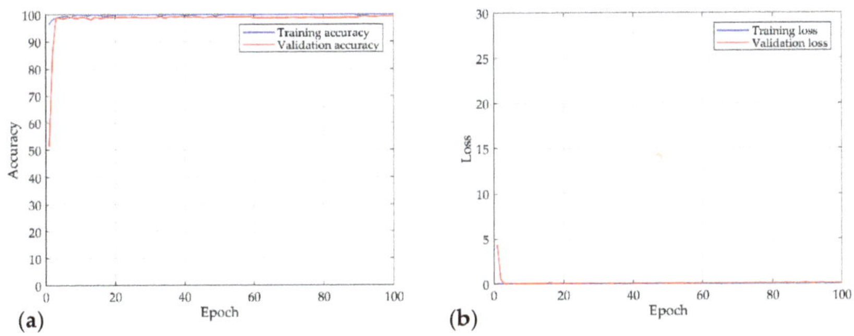

Figure 9. Results of the numerical experiment for the Rossler model: (**a**) accuracy; (**b**) and loss curves over 100 epochs.

Figure 10 shows the heat map using CAM. The heat map shows the spatial importance for each class, and red is the most important part. As shown in the heat map, when the iteration of the Rossler model is non-chaos, the characteristics of the image tend to be uniform and symmetrical, and the chaos is irregular.

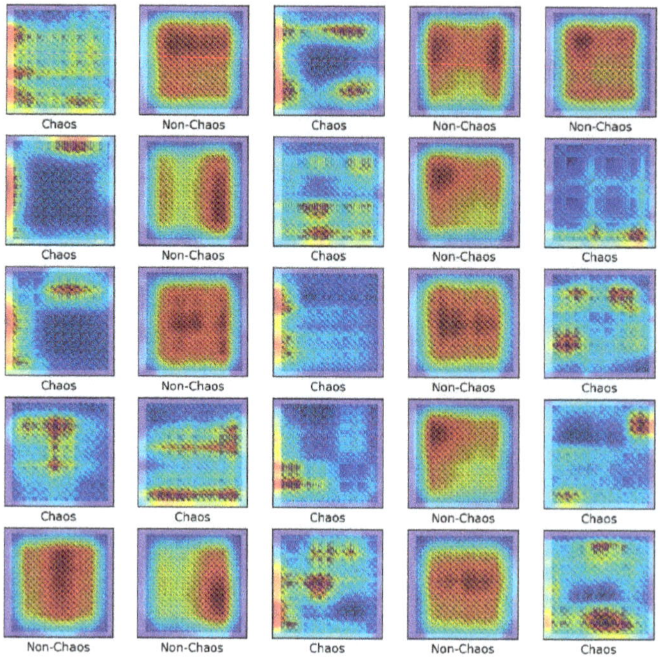

Figure 10. Class activation mapping for the Rossler model.

In the Rossler model, the cycle is long and clearly expressed for the set control parameters. Therefore, the visualized repetition was distinguished easily. On the other hand, the S&R model, which includes the friction force and impact force, contains extreme nonlinearities so that the repeatability can be very complex. Here, because the rattle problem contains discontinuities, the Lyapunov exponent was calculated by considering Muller's method. The squeak problem has a continuity, including creep, so the Lyapunov was calculated as a continuous dynamic problem.

Figure 11 shows the results of chaotic analysis for the model control parameters corresponding to each model. Figure 11a is the calculation result of the largest Lyapunov exponent of the rattle model, and Figure 11b shows the corresponding bifurcation diagram. The other parameters and initial conditions are $r = 2$, $\xi = 0.05$, $k_h = 100$, and $\mathbf{u}_0 = \begin{bmatrix} 0.4 & -1.1 & 0 \end{bmatrix}^{\mathrm{T}}$. Figure 11c,d shows the Lyapunov exponent calculation result for the control parameter of the single-mode squeak model and the corresponding bifurcation diagram. The other parameters and initial conditions are $V = 2$, $n_0 = 2.5$, $\beta = 0.002$, $\mu_s = 0.5$, $\mu_k = 0.1$, $h = 1$, $\alpha = 65$, and $\mathbf{v}_0 = \begin{bmatrix} 0.1 & 1.2 & 0 \end{bmatrix}^{\mathrm{T}}$. Figure 11e,f presents the Lyapunov exponent calculation result and bifurcation diagram for the control parameter of the multi-modes squeak model. Other parameters and initial conditions were $V = 1$, $n_0 = 2.5$, $\beta = \beta_1 = 0$, $\mu_s = 0.5$, $\mu_k = 0.3$, $h = 1$, $\alpha = 65$, $\gamma = 0.1$, and $\mathbf{w}_0 = \begin{bmatrix} 0.1 & 0.1 & 0.1 & 1.1 \end{bmatrix}^{\mathrm{T}}$. In this study, only each model and chaotic characteristics were distinguished, so other detailed types of attractors were not considered. The analysis results show that the S&R model changes with extreme nonlinearity in the largest Lyapunov exponent for the change in the control parameter.

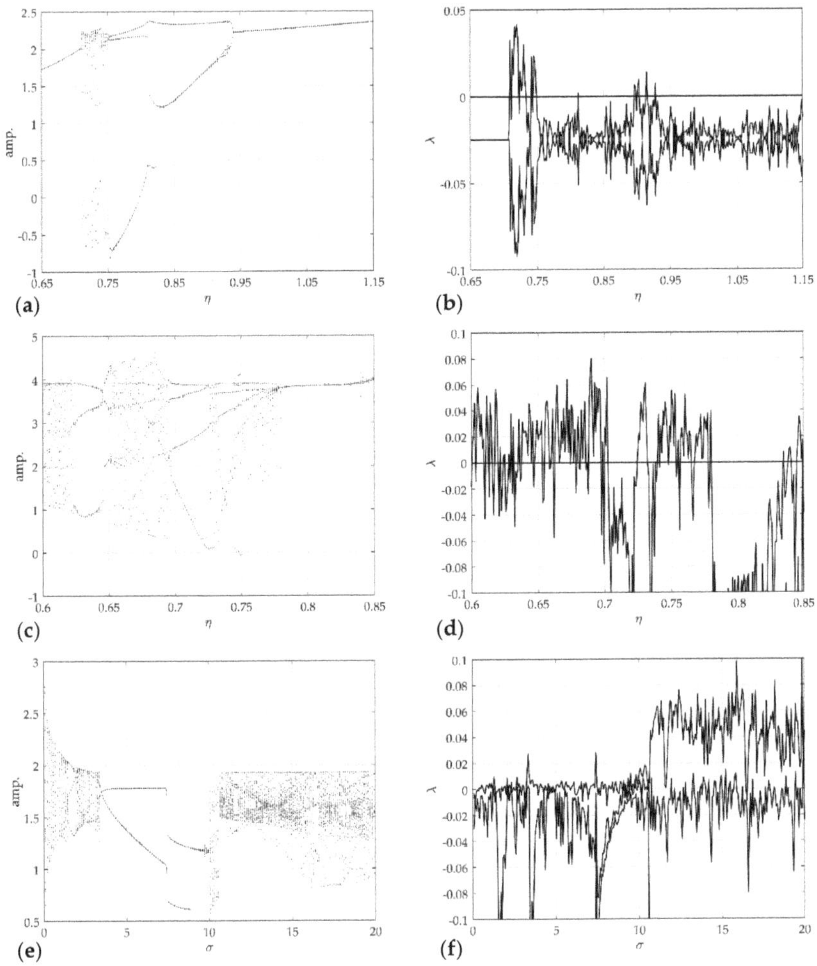

Figure 11. Chaotic analysis for the S&R model: (**a**) bifurcation diagram of displacements; (**b**) largest Lyapunov exponent, for rattle model with respect to η; (**c**) bifurcation diagram of displacements; (**d**) largest Lyapunov exponent, for single-mode squeak model with respect to η; (**e**) bifurcation diagram of displacements; (**f**) largest Lyapunov exponent, for multi-modes squeak model with respect to σ.

As mentioned earlier, Figures 12–14 presents the representative attractors of each system divided into chaos and non-chaos, and show the rattle, single-mode squeak, and multi-mode squeak models, respectively. Figure 12a,b shows the time series plot and phase portrait of displacement for the rattle model in $\eta = 0.7202$ and $\eta = 0.6801$. The dotted line oscillates constantly, and the solid line vibrates with an irregular amplitude. In phase space at $\eta = 0.6801$, it produces one stable limit cycle without impact and oscillates stably. On the other hand, in $\eta = 0.7202$, the system includes impact, and the trajectory appears without a specific period. In other words, it expresses chaos.

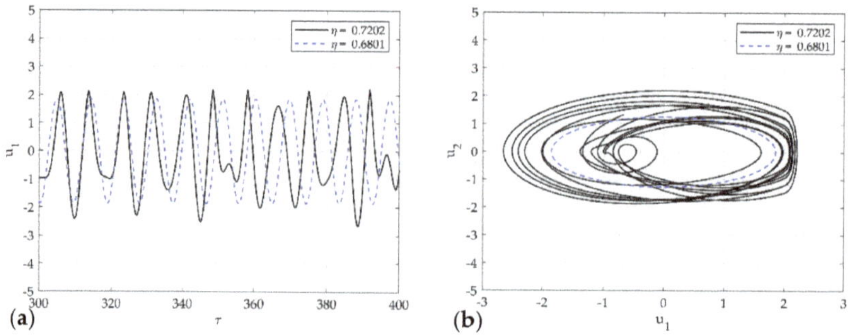

Figure 12. Dynamic solutions for the rattle model for various η: (**a**) time analysis; (**b**) phase portrait corresponding to (**a**).

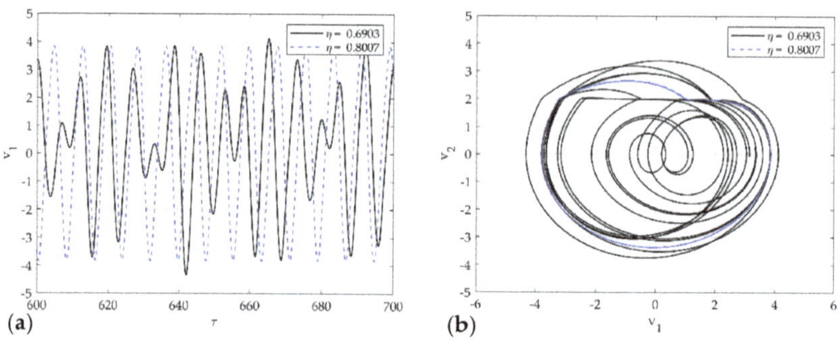

Figure 13. Dynamic solutions for single-mode squeak model for various η: (**a**) time analysis; (**b**) phase portrait corresponding to (**a**).

Figure 13a,b shows the time series plot and phase portrait of displacement for the single-mode squeak model in $\eta = 0.8007$ and $\eta = 0.6903$. The dotted line oscillates constantly, and the solid line vibrates with an irregular amplitude. In the phase plot, the flat phase means the stick phase in the stick-slip. When $\eta = 0.8007$ produces an unstable limit cycle in phase space, it vibrates unstably for 1 period within the limit cycle. On the other hand, in $\eta = 0.6903$, the system produces an unstable limit cycle and generates chaos without a constant cycle.

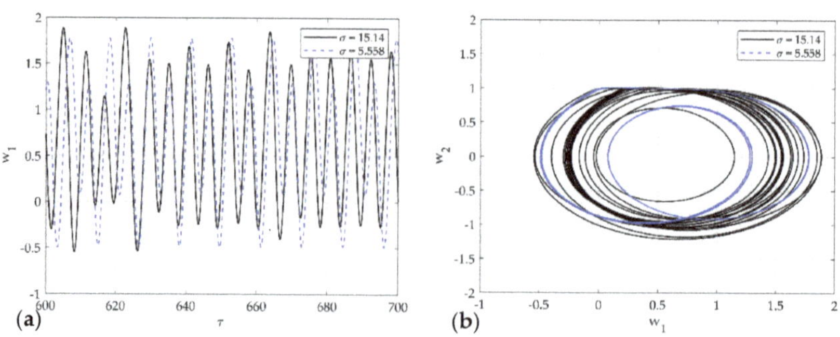

Figure 14. Dynamic solutions for single-mode squeak model for various σ: (**a**) time analysis; (**b**) phase portrait corresponding to (**a**).

Figure 14a,b shows the time series plot and phase portrait of displacement for the multi-mode squeak model in $\sigma = 15.14$ and $\sigma = 5.558$. The dotted line constantly oscillates with two amplitudes, and the solid line oscillates with an irregular amplitude. In phase space at $\sigma = 15.14$, it vibrates unstably for two periods within the limit cycle. On the other hand, at $\sigma = 5.558$, the system generates an unstable limit cycle and generates chaos.

The S&R model was classified into six classes to distinguish between chaos and non-chaos, and an unthresholded RP was shown. Here, the six classes were divided into the rattle, single-mode squeak and multi-modes squeak, and chaos and non-chaos for each model. Table 3 lists the dataset sample, and Figure 15 presents a part of the training dataset.

Table 3. Dataset split ratio for the S&R model.

Data	Percentage	Number of Samples
Training	56%	4160
Validation	14%	1040
Testing	30%	1800

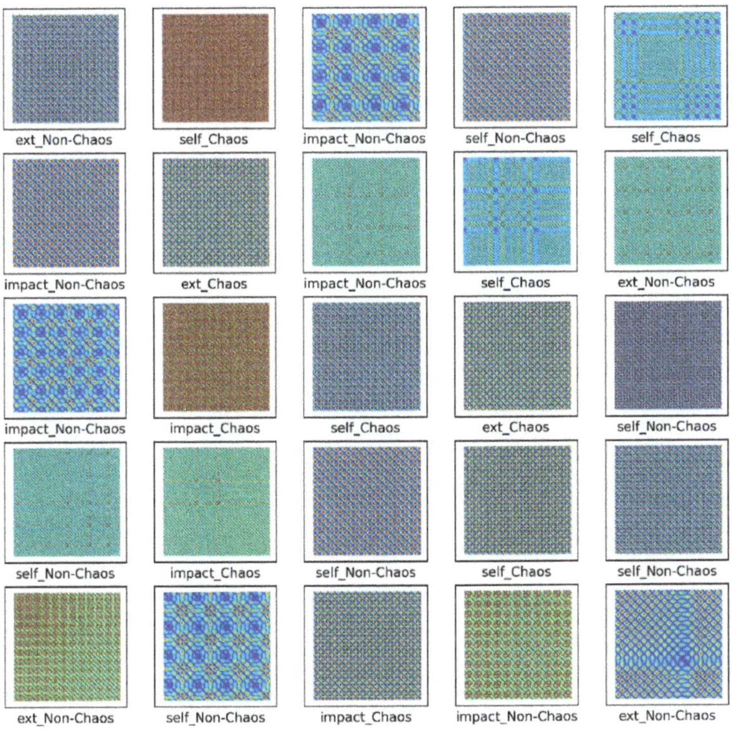

Figure 15. Unthresholded RP for the S&R model.

As shown in Figure 15, images by chaos and images by non-chaos are always vibrating, so repetition appears in a complex form. Six thousand data were used in the total dataset and consisted of 200 × 200-pixel images. To escape the local minima and converge to a lower loss, and prevent overfitting, the learning rate was adjusted for each specific step using a callback function. Each step consisted of five convolution layers with 32-2 × 2 filter, 64-2 × 2 filter, 128-2 × 2 filter, 256-2 × 2 filter, and 512-2 × 2 filter.

Figure 16 shows the results of the numerical experiment. The validation loss and accuracy do not decrease until 20 epochs but escape from the local minima by adjusting the learning rate. This shows that the learning rate is adjusted again at 70 epochs and converges with a certain accuracy and loss. In the final test, the model of each system and the chaos problem were classified with approximately 90% accuracy. Figure 17 shows the characteristics of the nonlinear vibration signal of each system for data extracted randomly by CAM. The most important characteristic of the recurrence plot is on the main diagonal line, meaning $\mathbf{RP}^{un}_{i,j} = \mathbf{RP}^{un}_{j,i}$. If $\mathbf{z}_i \neq \mathbf{z}_j$, an aperiodic pattern representing chaos may appear. As shown in Figure 17, macroscopically, the pattern is almost similar, but if the measurement length is long or the period is fast, a very small aperiodic pattern can be indicated. In other words, the CAM results in the first row and second column appear macroscopically similar, but physically show completely different characteristics.

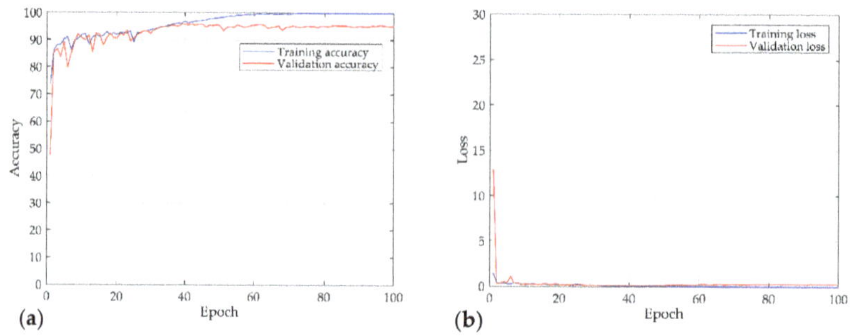

Figure 16. Results of the numerical experiment for the S&R model: (**a**) accuracy; (**b**) and loss curves over 100 epochs.

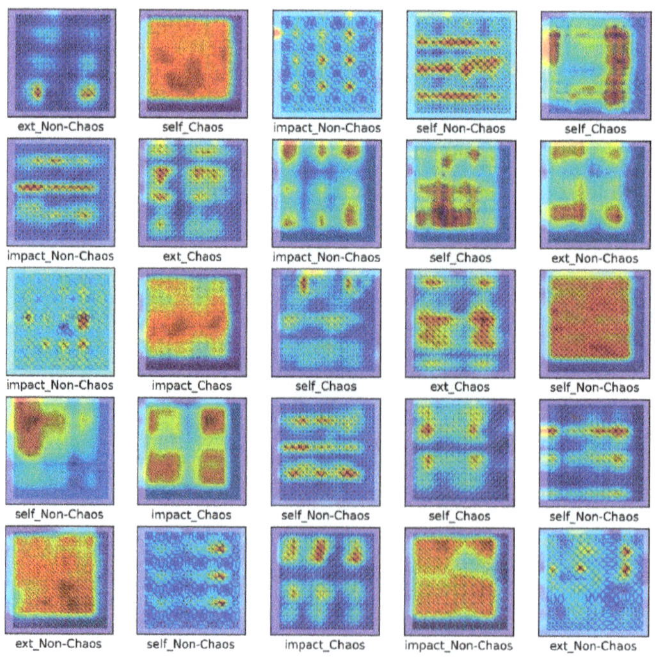

Figure 17. Class activation map for the S&R model.

The areas of difference for the recurrence characteristics for each model were detected successfully using deep learning. The proposed procedure can classify the nonlinear vibration characteristics generated in the mechanical system with high accuracy. In other words, the causes of complex signals can be classified due to nonlinear vibrations generated in mechanical systems with high accuracy. This shows that the characteristics of the complex signals, such as BSR noise, that humans cannot recognize can be classified with high accuracy by a CNN.

4. Discussion and Future Work

Visualization was performed with the proposed method for the vibration signals with extreme nonlinearity occurring in different models. The CNN was used to classify the S&R model and its chaotic characteristics. This result was verified by calculating the Lyapunov exponent for each model. The chaotic characteristics can distinguish the signals generated in a deterministic system by calculating the Lyapunov exponent, but calculating this is very complex. The signal visualization method analyzes the dynamic signals, but signals containing nonlinearity are complicated for engineers to analyze. In other words, complex signals are difficult to classify by human cognitive ability. Therefore, a procedure for classifying S&R models, including chaos and non-chaos, was proposed and verified with approximately 91% accuracy using a simple CNN model. In future work, we will study more complex models to analyze signals with added noise. It will also conduct detailed experimental studies.

Author Contributions: Conceptualization, J.K.; methodology, J.K.; software, J.K.; validation, J.K. and J.N.; formal analysis, J.K. and J.N.; investigation, J.K. and J.N.; resources, J.K.; data curation, J.K.; writing, original draft preparation, J.K. and J.N.; writing, review and editing, J.K. and J.N.; visualization, J.K. and J.N.; project administration, J.K.; funding acquisition, J.K. All authors have read and agreed to the published version of the manuscript.

Funding: This work was supported by Inha University Research Grant (INHA-65389-01).

Institutional Review Board Statement: Not applicable.

Informed Consent Statement: Not applicable.

Conflicts of Interest: The authors declare no conflict of interest.

eferences

Lee, G.; Kim, K.; Kim, J. Development of an algorithm to automatically detect and distinguish squeak and rattle noises. *SAE Tech. Pap.* **2015**, *1*, 2258. [CrossRef]

Cerrato-Jay, G.; Gabiniewicz, J.; Gatt, J.; Pickering, D.J. Automatic Detection of Buzz, Squeak and Rattle Events. *SAE Trans.* **2001**, *110*, 1763–1770.

Wang, Y.S.; Lee, C.; Kim, D.; Xu, Y. Sound-Quality Prediction for Nonstationary Vehicle Interior Noise Based on Wavelet Pre-Processing Neural Network Model. *J. Sound Vib.* **2007**, *299*, 933–947. [CrossRef]

Kavarana, F.; Rediers, B. *Squeak and Rattle-State of the Art and Beyond*; SAE Technical Paper 1999-01-1728; SAE International: Warrendale, PA, USA, 1999. [CrossRef]

Zheng, X.; Agarwal, V.; Liu, X.; Balachandran, B. Nonlinear Instabilities and Control of Drill-String Stick-Slip Vibrations with Consideration of State-Dependent Delay. *J. Sound Vibrat.* **2020**, *473*, 115235. [CrossRef]

Arvin, H.; Arena, A.; Lacarbonara, W. Nonlinear Vibration Analysis of Rotating Beams Undergoing Parametric Instability: Lagging-Axial Motion. *Mech. Syst. Signal Process.* **2020**, *144*, 106892. [CrossRef]

Dong, C.; Mo, J.; Yuan, C.; Bai, X.; Tian, Y. Vibration and Noise Behaviors during Stick–slip Friction. *Tribol. Lett.* **2019**, *67*, 1–12. [CrossRef]

Kang, J.; Krousgrill, C.M.; Sadeghi, F. Comprehensive Stability Analysis of Disc Brake Vibrations Including Gyroscopic, Negative Friction Slope and Mode-Coupling Mechanisms. *J. Sound Vib.* **2009**, *324*, 387–407. [CrossRef]

Niknam, A.; Farhang, K. Vibration Instability in a Large Motion Bistable Compliant Mechanism due to Stribeck Friction. *J. Vib. Acoust.* **2018**, *140*, 061017. [CrossRef]

). Nam, J.; Do, H.; Kang, J. Investigation of Friction Induced Vibration in Lead Screw System using FE Model and its Experimental Validation. *Appl. Acoust.* **2017**, *122*, 98–106. [CrossRef]

1. Ouenzerfi, G.; Massi, F.; Renault, E.; Berthier, Y. Squeaking Friction Phenomena in Ceramic Hip Endoprosthesis: Modeling and Experimental Validation. *Mech. Syst. Signal Process.* **2015**, *58*, 87–100. [CrossRef]

12. Oberst, S.; Lai, J. Nonlinear Transient and Chaotic Interactions in Disc Brake Squeal. *J. Sound Vib.* **2015**, *342*, 272–289. [CrossRef
13. Li, Z.; Cao, Q.; Nie, Z. Stick-Slip Vibrations of a Self-Excited SD Oscillator with Coulomb Friction. *Nonlinear Dyn.* **2020**, *10* 1419–1435. [CrossRef]
14. Wei, D.; Ruan, J.; Zhu, W.; Kang, Z. Properties of Stability, Bifurcation, and Chaos of the Tangential Motion Disk Brake. *J. Soun Vib.* **2016**, *375*, 353–365. [CrossRef]
15. Kang, J. Lyapunov Exponent of Friction-Induced Vibration under Smooth Friction Curve. *J. Mech. Sci. Technol.* **2018**, *32*, 3563–356 [CrossRef]
16. Müller, P.C. Calculation of Lyapunov Exponents for Dynamic Systems with Discontinuities. *Chaos Solitons Fractals* **1995**, 1671–1681. [CrossRef]
17. Serweta, W.; Okolewski, A.; Blazejczyk-Okolewska, B.; Czolczynski, K.; Kapitaniak, T. Lyapunov Exponents of Impact Oscillator with Hertz's and Newton's Contact Models. *Int. J. Mech. Sci.* **2014**, *89*, 194–206. [CrossRef]
18. Serweta, W.; Okolewski, A.; Blazejczyk-Okolewska, B.; Czolczynski, K.; Kapitaniak, T. Mirror Hysteresis and Lyapunov Exponent of Impact Oscillator with Symmetrical Soft Stops. *Int. J. Mech. Sci.* **2015**, *101*, 89–98. [CrossRef]
19. Kang, J. Calculation of Lyapunov Exponents in Impacted Beam on Distributed Contact. *J. Sound Vib.* **2018**, *431*, 295–303. [CrossRef
20. Awrejcewicz, J.; Krysko, A.V.; Erofeev, N.P.; Dobriyan, V.; Barulina, M.A.; Krysko, V.A. Quantifying Chaos by various Computa tional Methods. Part 1: Simple Systems. *Entropy* **2018**, *20*, 175. [CrossRef]
21. Astaf'eva, N.M. Wavelet Analysis: Basic Theory and some Applications. *Phys. Usp.* **1996**, *39*, 1085. [CrossRef]
22. Marwan, N.; Romano, M.C.; Thiel, M.; Kurths, J. Recurrence Plots for the Analysis of Complex Systems. *Phys. Rep.* **2007**, *43* 237–329. [CrossRef]
23. Cao, L. Practical Method for Determining the Minimum Embedding Dimension of a Scalar Time Series. *Phys. D Nonlinear Phenom* **1997**, *110*, 43–50. [CrossRef]
24. Kennel, M.B.; Brown, R.; Abarbanel, H.D. Determining Embedding Dimension for Phase-Space Reconstruction using a Geometri cal Construction. *Phys. Rev. A* **1992**, *45*, 3403. [CrossRef]
25. March, T.K.; Chapman, S.C.; Dendy, R.O. Recurrence Plot Statistics and the Effect of Embedding. *Physica D* **2005**, *200*, 171–18 [CrossRef]
26. Wallot, S.; Mønster, D. Calculation of Average Mutual Information (Ami) and False-Nearest Neighbors (Fnn) for the Estimation o Embedding Parameters of Multidimensional Time Series in Matlab. *Front. Psychol.* **2018**, *9*, 1679. [CrossRef]
27. Butusov, D.N.; Karimov, A.I.; Pesterev, D.O.; Tutueva, A.V.; Okoli, G. Bifurcation and Recurrent Analysis of Memristive Circuit In Proceedings of the 2018 IEEE Conference of Russian Young Researchers in Electrical and Electronic Engineering (EIConRus Moscow/St. Petersburg, Russia, 29 January–1 February 2018; pp. 178–183.
28. Wolf, A.; Swift, J.B.; Swinney, H.L.; Vastano, J.A. Determining Lyapunov Exponents from a Time Series. *Phys. D Nonlinear Phenom* **1985**, *16*, 285–317. [CrossRef]
29. Zou, Y.; Donner, R.V.; Donges, J.F.; Marwan, N.; Kurths, J. Identifying Complex Periodic Windows in Continuous-Time Dynamica Systems using Recurrence-Based Methods. *Chaos Int. J. Nonlinear Sci.* **2010**, *20*, 043130. [CrossRef] [PubMed]
30. He, K.; Zhang, X.; Ren, S.; Sun, J. Deep Residual Learning for Image Recognition. In Proceedings of the IEEE Conference o Computer Vision and Pattern Recognition, Las Vegas, NV, USA, 27–30 June 2016; pp. 770–778.
31. Hsueh, Y.; Ittangihala, V.R.; Wu, W.; Chang, H.; Kuo, C. Condition Monitor System for Rotation Machine by CNN with Recurrenc Plot. *Energies* **2019**, *12*, 3221. [CrossRef]
32. Nam, J.; Kang, J. Classification of Chaotic Signals of the Recurrence Matrix using a Convolutional Neural Network and Verificatio through the Lyapunov Exponent. *Appl. Sci.* **2021**, *11*, 77. [CrossRef]
33. Parker, T.S.; Chua, L. *Practical Numerical Algorithms for Chaotic Systems*; Springer Science and Business Media: New York, N USA, 2012.
34. Thiel, M.; Romano, M.C.; Kurths, J.; Meucci, R.; Allaria, E.; Arecchi, F.T. Influence of Observational Noise on the Recurrenc Quantification Analysis. *Phys. D Nonlinear Phenom.* **2002**, *171*, 138–152. [CrossRef]
35. Zhou, B.; Khosla, A.; Lapedriza, A.; Oliva, A.; Torralba, A. Learning Deep Features for Discriminative Localization. In Proceeding of the IEEE Conference on Computer Vision and Pattern Recognition, Las Vegas, NV, USA, 27–30 June 2016; pp. 2921–2929.
36. Barrio, R.; Blesa, F.; Serrano, S. Qualitative Analysis of the Rössler Equations: Bifurcations of Limit Cycles and Chaotic Attractor *Phys. D Nonlinear Phenom.* **2009**, *238*, 1087–1100. [CrossRef]
37. Kingma, D.P.; Ba, J. Adam: A Method for Stochastic Optimization. *arXiv* **2014**, arXiv:1412.6980.
38. LeCun, Y.A.; Bottou, L.; Orr, G.B.; Müller, K. Efficient backprop. In *Neural Networks: Tricks of the Trade*; Springer: Berlin, German 2012; pp. 9–48.
39. Glorot, X.; Bengio, Y. Understanding the Difficulty of Training Deep Feedforward Neural Networks. In Proceedings of th Thirteenth International Conference on Artificial Intelligence and Statistics, Chia Laguna Resort, Sardinia, Italy, 13–15 May 201 pp. 249–256.

sensors

MDPI

rticle

Autonomous Assessment of Delamination Using Scarce Raw Structural Vibration and Transfer Learning

sif Khan [1], Salman Khalid [1], Izaz Raouf [1], Jung-Woo Sohn [2,*] and Heung-Soo Kim [1,*]

1 Department of Mechanical, Robotics and Energy Engineering, Dongguk University Seoul, 30 Pildong-ro 1 Gil, Jung-gu, Seoul 04620, Korea; asif_dgu@dongguk.edu (A.K.); salmankhalid@dongguk.edu (S.K.); izazraouf@dongguk.edu (I.R.)

2 Department of Mechanical Design Engineering, Kumoh National Institute of Technology, Gumi 39177, Korea

* Correspondence: jwsohn@kumoh.ac.kr (J.-W.S.); heungsoo@dgu.edu (H.-S.K.)

Abstract: Deep learning has helped achieve breakthroughs in a variety of applications; however, the lack of data from faulty states hinders the development of effective and robust diagnostic strategies using deep learning models. This work introduces a transfer learning framework for the autonomous detection, isolation, and quantification of delamination in laminated composites based on scarce low-frequency structural vibration data. Limited response data from an electromechanically coupled simulation model and from experimental testing of laminated composite coupons were encoded into high-resolution time-frequency images using SynchroExtracting Transforms (SETs). The simulated and experimental data were processed through different layers of pretrained deep learning models based on AlexNet, GoogleNet, SqueezeNet, ResNet-18, and VGG-16 to extract low- and high-level autonomous features. The support vector machine (SVM) machine learning algorithm was employed to assess how the identified autonomous features were able to assist in the detection, isolation, and quantification of delamination in laminated composites. The results obtained using these autonomous features were also compared with those obtained using handcrafted statistical features. The obtained results are encouraging and provide a new direction that will allow us to progress in the autonomous damage assessment of laminated composites despite being limited to using raw scarce structural vibration data.

Keywords: laminated composites; structural vibration; synchroextracting transform; scarce data; autonomous features

1. Introduction

The use of composite materials is growing continuously in a variety of industries (e.g., aerospace, automotive, wind energy) and remains at the forefront of contemporary research [1–5]. These materials have preferential properties of high specific stiffness, high specific strength, high corrosion resistance, being lightweight, and offering design flexibility. Despite their benefits, the orthotropic layered nature of laminated composites makes them prone to complex failure modes (matrix cracks, fiber breakage, delamination, etc.) during their manufacturing and in-service life [6]. Among the various defects, delamination is one of the most commonly occurring failure mode in laminated composites. The presence and propagation of delamination can result in up to a 60% loss in structural stiffness without any visible change to the surface of the material [7,8]. Owing to these severe consequences, developing a diagnostic strategy to continuously assess laminated composites for the presence of delamination is imperative. Commonly employed non-destructive evaluation methods (ultrasonic C-scan, shearography, computed tomography, etc., [9,10] are not suitable for continuous real time in-service damage assessments.

In recent years, AI-based machine learning and deep learning techniques have emerged as powerful tools that can be used for early damage detection in composite structures in various applications such as helping airplanes avoid catastrophic failures. Ai et al. [11] adopted a model based on random forest and deep learning methods to perform impact localization in aircraft composite structures using acoustic emission (AE) data. The analysis of the results showed that the random forest and deep learning model achieved better event localization performance than previous conventional artificial neural networks. Brien et al. [12] and his team used acoustic emission signals generated from healthy and faulty composite beams to test their proposed pattern recognition system that employs an artificial neural network classifier to detect and classify damage in composite beams. The proposed methodology was proven to be able to effectively classify damage into one of four different levels with 97% accuracy. Khan et al. [13] proposed a Convolutional Neural Network (CNN) approach to analyse the low-frequency structural vibration outputs generated by composite laminates that came from either healthy cases or from 12 delaminated cases. The output results showed the proposed model could classify cases as healthy or delaminated with 90.1% accuracy. Loutas et al. [14] developed an intelligent structural health monitoring system for composite structures used in aerospace applications that used multi-sensor data fusion in a feature level approach to classify thirteen different damage states. The results indicated that using SVM with a nonlinear kernel could detect faults with up to 99.3% accuracy. Further, various authors have applied machine and deep learning models in these references [15–18] for damage detection and classification of composite structures. Although deep learning models with several hidden layers are able to autonomously discover discriminative features from target images, a substantial amount of annotated data is required to optimize the network parameters in these systems. In engineering applications, acquiring sufficient data with faulty states is difficult due to the severe consequences of operating a system in the presence of any kind of fault.

From the literature survey above, it is clear AI-based models are able to detect and classify faults in composite structures with a high level of accuracy. In cases with small training datasets, the trained model will have poor generalization. The small scale of the data in these situations cannot represent the distribution to which they belong, and the training and testing datasets must have the same distribution. On the other hand, large-scale training data is expensive and challenging to collect in certain scenarios such as aerospace applications. To overcome this issue, transfer learning-based models have been introduced recently for intelligent fault detection. Gong et al. [19] proposed a novel deep transfer learning model for aeronautics composite materials (ACM) that combined deep learning with a sliding window approach. The proposed model used X-ray images of ACM but these samples were scarce. Experimental results showed that the proposed model could classify faults up to 96% classification accuracy. Outside of fault detection systems, Milad et al. [20] used a transfer learning approach with smart manufacturing processes of composite materials. The historical data generated from their autoclave curing process was successfully used to learn new cycle settings with limited data. Yang et al. [21] proposed a feature-based transfer neural network for bearing fault detection that used transferred diagnosis knowledge from laboratory bearings to detect faults in locomotive bearings. The results showed that the proposed model could successfully identify transferable features from the laboratory scale bearings to detect faults in locomotive bearings. Other literature on applying transfer learning to intelligent fault diagnosis can be found in the references [22–25].

The current work proposes aiding autonomous feature extraction from limited data using off-the-shelf pretrained deep learning models for the assessment of delamination in laminated composites. An electromechanically coupled finite element model was employed to study the effect of different sized delaminations based on a numerical framework. For experimental verification of the proposed approach, laminated composite samples were manufactured using a hot press machine and water cutting jet. Low-frequency structural vibration responses were obtained from simulations and from the experimental samples. The time series data was transformed into time-frequency images using the Synchroextracting Transform (SET) technique [26,27]. Pretrained deep learning models based on AlexNet [28], GoogleNet [29], SqueezeNet [30], ResNet-18 [31], and VGG-16 [32] were employed to extract low- and high-level features from the images based on the limited simulated and experimental data. A conventional machine learning algorithm, support vector machines (SVM), was used to assess the extracted autonomous features for their usefulness in the detection, isolation, and quantification of delamination in laminated composites. The classification performance using the autonomous features was also compared with the results from using handcrafted statistical features.

2. Proposed Methodology

This section describes the proposed methodology. The general workflow of the current work is shown in Figure 1.

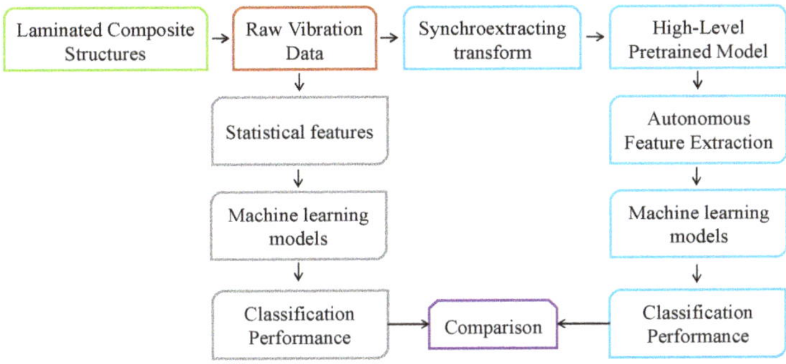

Figure 1. The general workflow of the proposed methodology.

First, structural vibration responses were obtained from the laminated composite plates in both pristine and delaminated states. The raw vibration signals were processed via the Synchroextracting Transform (SET) technique to produce high resolution time-frequency "images" that represent the signals in such a way that they can be used as inputs to deep learning-based diagnostic systems. High-level deep learning models were employed to autonomously extract discriminative feature from the limited data encoded in the time-frequency images. These autonomous features were processed using different machine learning models and the performance of each was evaluated in terms of training, validation, testing, and confusion matrices. To show the effectiveness of the proposed approach to diagnose laminated composites, handcrafted statistical features from the time and frequency domains were extracted from the raw vibration signals and processed with machine learning models. The results achieved using the handcrafted statistical features were also evaluated in terms of training, validation, testing, and confusion matrices so they could be compared to those achieved based on autonomous feature extraction. Each approach was evaluated based on input data from numerical simulations as well as real-world experiments on laminated composites with delaminations of various sizes. In general, high frequencies excitation and response signals (such as guided waves) are more suitable for detecting and isolating small size delamination defects. However, the delamination detection methods based on high frequencies guided waves suffer from the limitations of

a. The methods are active and require a voltage supply, signal amplifier, function generator, etc., to produce excitation signals within a specific high-frequency range (usually 50–300 kHz).
b. The response to high-frequency excitation should be acquired and stored at a high sampling rate to preserve useful information and keep the required resolution of the signal.

c. For most high-frequency methods, the delamination should be in the path between the excitation source (actuator) and the location for measurement (sensor).

The first requirement demands for more hardware in the damage assessment technology. The second condition requires a large amount of memory and a high-performance data acquisition system. The third condition demands more sensors and subsequent data acquisition and storage systems to cover all the possible paths between the actuator and sensor for damage-prone areas.

Contrary to high-frequency methods, the proposed method employs structural vibration (0–1000 Hz) obtained through smart elements. The structural vibration is more readily available during the service condition of the structure without the need for any specific range of excitation. Though the attributes of structural vibration (natural frequency, modal damping, FRF, etc.) are global and cannot be employed for the localization of delamination, the proposed approach attempts to extract global and local information through deep learning models. The following sections show details of the general workflow used.

2.1. Mathematical Model and Simulated Data

The proposed approach outlined in Figure 1 was carried out using both simulated and experimental data. Improved layerwise theory was employed to simulate delamination in smart composite laminates. A laminated composite plate with 16 plies stacked in a cross-ply symmetric configuration $[0/90]_{4s}$ was modeled in the numerical simulation. A piezoelectric actuator and sensor were modeled on the surface of the laminated plate as shown in Figure 2.

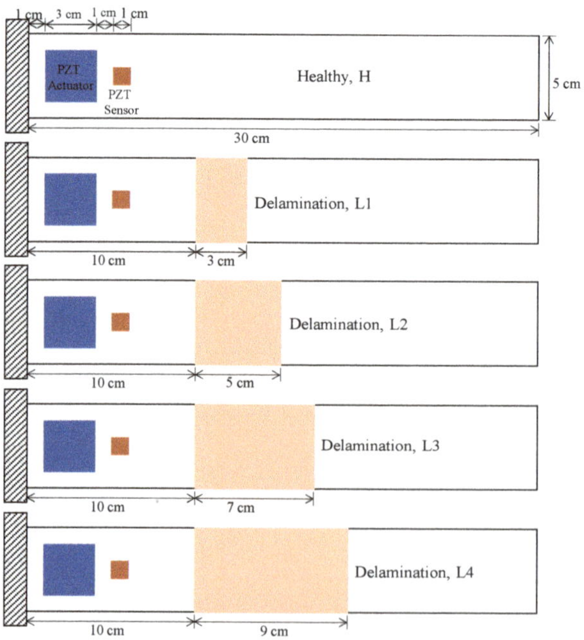

Figure 2. Geometric configuration of laminated composite plate in numerical simulations.

Here, the piezoelectric actuator is attached near the fixed end to excite the plate with low-frequency structural vibrations and the corresponding response is obtained through the piezoelectric sensor. Size of the piezoelectric actuator was 3×3 cm^2 and the dimensions of the piezoelectric sensor were 1×1 cm^2. Four delaminations of different sizes (3 cm–9 cm) were simulated to study the effect the size of the delamination has on the proposed approach. The health states of the composites were denoted by H (for healthy), $L1$ (for 3 cm delamination), $L2$ (for 5 cm delamination), $L3$ (for 7 cm delamination), and $L4$ (for 9 cm delamination). An electromechanically coupled finite element model (shown by Equation (1)) of the smart plate was developed using improved layerwise theory [33,34] and high-order electric potential field [35].

$$\begin{bmatrix} M_{uu} & 0 \\ 0 & 0 \end{bmatrix} \begin{Bmatrix} \ddot{d}_u \\ 0 \end{Bmatrix} + \begin{bmatrix} C_{uu} & 0 \\ 0 & 0 \end{bmatrix} \begin{Bmatrix} \dot{d}_u \\ 0 \end{Bmatrix} + \begin{bmatrix} K_{uu} & K_{u\phi} \\ K_{\phi u} & K_{\phi\phi} \end{bmatrix} \begin{Bmatrix} d_u \\ d_\phi \end{Bmatrix} = \begin{Bmatrix} F_u \\ F_\phi \end{Bmatrix} \tag{1}$$

where d_u and d_ϕ denote the displacement and electrical variables at nodal positions, respectively. M_{uu} is the mass matrix, C_{uu} is the damping matrix, and K_{uu} is the stiffness matrix of the system. The electromechanical coupling of the mechanical and electrical field are denoted by $K_{u\phi}$ and $K_{\phi u}$, respectively, while $K_{\phi\phi}$ is the dielectric stiffness matrix. The terms F_u and F_ϕ denote the respective mechanical force and electric field vectors. A detailed derivation of the finite element model in Equation (1) can be found in references [34,36].

In Equation (1), the electrical excitation, and the corresponding response of the plates in Figure 2 are obtained through the piezoelectric actuator and piezoelectric sensor using the electromechanical coupling matrices, $K_{u\phi}$ and $K_{\phi u}$. These coupling matrices account for the converse and direct piezoelectric effects, where the piezoelectric actuator produces mechanical actuation when a voltage is applied, while the piezoelectric sensor generates a voltage signal when it is mechanically deformed. The matrix form of Equation (1) can be modified through matrix condensation to obtain the governing equation in the following form.

$$M_{uu}\ddot{d}_u + C_{uu}\dot{d}_u + Kd_u = F \tag{2}$$

where the modified stiffness matrix K and force vector F are expressed as follows

$$K = K_{uu} - K_{u\phi}K_{\phi\phi}^{-1}K_{\phi u}, F = F_u - K_{u\phi}K_{\phi\phi}^{-1}F_\phi \tag{3}$$

The thickness of the laminated plate was 0.2 cm and all the delaminations were considered to have occurred in the mid-plane of the laminated composite plate. The material properties of a lamina in the host laminated composite and of the piezoelectric elements used in the numerical simulations are shown in Tables 1 and 2, respectively.

Table 1. Material properties of a lamina in the host laminate.

E_1	E_2, E_3	G_{12}, G_{13}	G_{23}	ρ	ν_{12}, ν_{13}	ν_{23}
372 GPa	4.12 GPa	3.99 GPa	3.6 GPa	1788.5 kg/m^3	0.275	0.42

Table 2. Material properties of the piezoelectric elements.

E	ν	ρ	d_{31}, d_{32}	d_{24}, d_{15}	b_{11}
62 GPa	0.30	7500 kg/m^3	274×10^{-12} m/V	-741×10^{-12} m/V	14.41 nf/m

The numerical model in Equation (2) was implemented using MATLAB by discretizing the piezobonded laminated plate into 60 × 30 elements along the length and width, respectively. Time domain response was obtained from healthy and delaminated smart plates while solving the electromechanically coupled model via Newmark's time integration algorithm [37]. The plates in each of the five health states (H, $L1$, $L2$, $L3$, $L4$) were excited using 10 statistically independent random excitations through the piezoelectric actuator over a period of 2 s, the corresponding responses were obtained through the piezoelectric sensor. The time step chosen for Newmark's time integration was 0.0001 s in order to maintain numerical accuracy. The data from the simulation was comprised of 50 time series signals, 10 for each of the five health states (10 random excitations × 5 health states × 1 sensor for each health state). The essence of the simulated data is that it contains response from the plates with delaminations of different lengths (from 3 cm to 9 cm).

2.2. Experimental Setup and Data

To validate the results of the proposed approach experimentally, healthy and delaminated composite samples were manufactured in a hot press machine. Samples with three different states were produced for the experiments: healthy H, delamination D1, and delamination *D2*. The healthy sample had all of its plies perfectly bonded. The delaminated D1 and D2 samples were seeded in their mid-plane with Teflon film (PTFE film (Model KSC-V1000), thickness: 0.03 mm, heat resistance: up to 280 °C) near the free and fixed end of the cantilever beams. The sizes of both delaminations were chosen to be 10 cm in length and 5 cm in width. T700SC-12k-60E carbon fiber prepreg was cut into 35 cm × 30 cm plies and stacked in a symmetric cross-ply configuration of $[0/90/0/90]_s$ with a total of 8 plies used in each sample. The properties provided by the manufacturer for the fiber are density 1.8 g/cm^3, tensile strength 4900 MPa, Tensile Modulus 230 GPa, elongation 2.1 %, thermal conductivity 9.4 W/m.K, filament diameter 7 μm, and epoxy resin. The healthy and delaminated laminated composite plate samples were cured in the hot press machine for 155 min using the curing cycle suggested by the manufacturer. The cured laminated composite plates were cut to a size of 30 cm × 5 cm with a water jet cutting machine to obtain 5 samples for each health state, as shown in Figure 3.

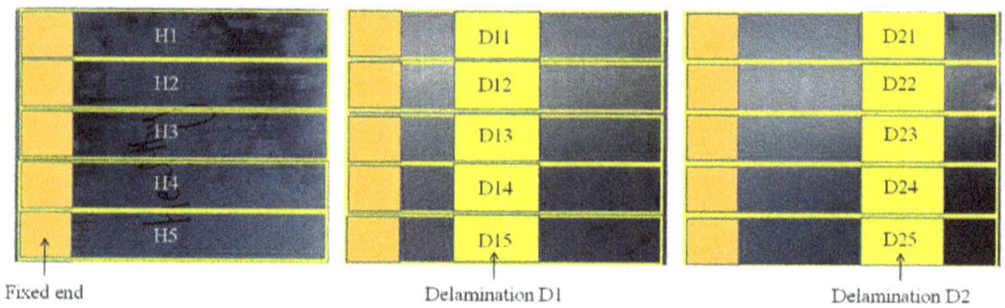

Figure 3. Healthy and delaminated samples representing three health states manufactured using a hot press machine.

Here, the letter H is used to denote a healthy sample while the preceding numbers from 1–5 are used to identify the individual samples. The first digit after the D for the delamination samples D1 and D2 refers to the health state while the second digit identifies the individual samples. Five samples from each health state were used to account for manufacturing and measurement uncertainties. Vibration tests were conducted on the manufactured samples. Figures 4 and 5 show the experimental setup and experimental workflow, respectively, for the random vibration testing.

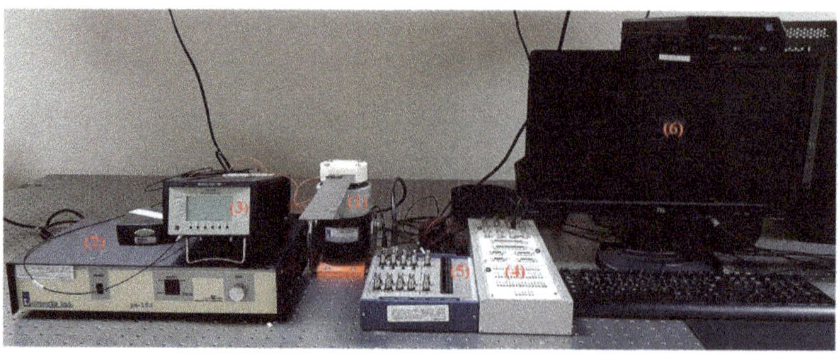

Figure 4. Various apparatus used in the experimental vibration testing; (**1**) shaker; (**2**) amplifier for the shaker; (**3**) amplifier for accelerometer; (**4**) DAQ1; (**5**) DAQ2; (**6**) Lab-view PC.

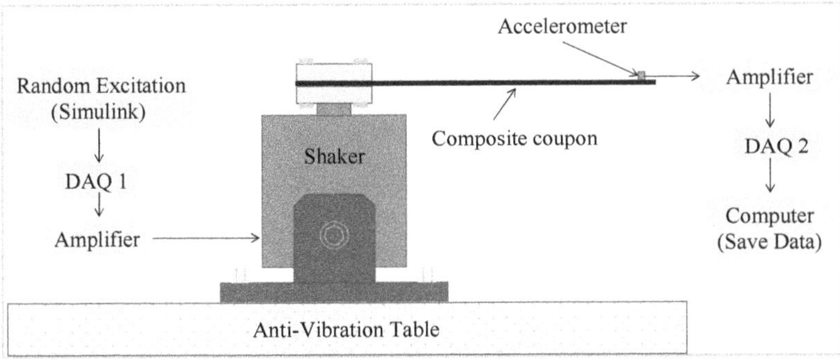

Figure 5. Experimental workflow for vibration testing of composite samples.

The experimental setup consisted of three basic systems: excitation, vibration, and response data acquisition systems. The excitation apparatus consisted of a Lab-view PC that generates random signals using MATLAB Simulink. For the excitation of the laminated composite samples, an acquisition system (DAQ 1 (dSPACE/CLP1104)), a signal amplifier (Labworks/PA-151, Labworks Inc., Costa Mesa, CA, USA) for the shaker, and a shaker (Labworks/ET-126-4, Labworks Inc., Costa Mesa, CA, USA) were used to excite the composite beams with the generated random signals.

The laminated composite beams (H, D1, and D2) were clamped using a 3D printed jig to apply cantilevered boundary conditions to the vibrating system. The response acquisition system was used to acquire data from the vibrating coupon. This consisted of various devices; an accelerometer (Bruel & Kjaer/Type 4517-C, Brüel & Kjær, Nærum, Denmark) that was installed on the vibrating composite beams, an amplifier (Bruel & Kjaer/Type 2692-0S2, Brüel & Kjær, Nærum, Denmark), and a data acquisition system (DAQ 2 (NI/USB-6341)). The responses to the random excitations was measured near the clamped end of the healthy and delaminated samples over 15 s. The experimental data reflects the responses caused by delaminations of the same size but in different locations which could be useful evaluating the possibility of determining the location of the delamination through this kind of data.

3. Results and Discussion

This section discusses the classification results into different health states for the laminated composite samples in the simulations and experiments based on both handcrafted statistical features and autonomously extracted features.

3.1. *Classification Results on Handcrafted Statistical Features*

Time and frequency domain statistical features of the signals have been used to discriminate between different health states of bearings, gearboxes, and rotating machines [38–41]. However, statistical features may not allow us to discriminate between the health states of laminated composites where the dynamic response is dominated by the excitation forces and the response signal may not reveal sufficient information to diagnose the characteristics of the fault present when analyzed in a conventional manner. To show the feasibility of using handcrafted statistical features to discriminate between different health states in the simulated data, time and frequency domain features were extracted from the time series data and analyzed through various machine learning algorithms. A window function of 0.2 s was employed to divide the random response signals over the 2 s period into 10 chunks. The purpose of this division is to look for discriminative features in a smaller portion of the signals instead of looking at the entire signal. The handcrafted time domain features used were the mean, standard deviation, skewness, kurtosis, peak to peak values, root mean squared values, crest factor, shape factor, impulse factor, margin factor, and energy of the signal. The extracted frequency domain features were the spectral kurtosis, spectral mean, spectral standard deviation, spectral skewness, and spectral kurtosis. The fifteen features were extracted from numerical data on the composite laminate samples in all five health states. The data from all samples in the various health states amounted to 1000 instances with 15 features for each instance. The values for the hand-crafted features in each instance was processed with different machine learning algorithms, Table 3 depicts the performance of the different classifiers used to process those values in terms of training/validation accuracy and area under the receiver operating characteristics curve (ROC area).

Table 3. Classification performance of different algorithms on data from simulation using hand-crafted statistical features.

Classifier	Training/Validation Accuracy %	ROC Area
Fine Tree	36.1	0.58
Linear Discriminant	21.7	0.50
Kernel Naïve Bayes	26.4	0.53
Cubic SVM	36.2	0.51
Fine KNN	43.8	0.47

All the classifiers were trained using a 10-fold cross-validation strategy. The results in Table 3 show that the maximum classification accuracy using these statistical features was 43.8% achieved by Fine KNN. However, the ROC area, which reflects the tradeoff between the true positive and false positive rates of the classifiers, indicates that Fine KNN is susceptible to overfitting and will not generalize well to new, unseen data.

Although the classification accuracy is not high enough for practical applications, the training/validation confusion matrix may provide additional insights into the classification performance achieved. Figure 6 shows the training/validation confusion matrix of Cubic SVM.

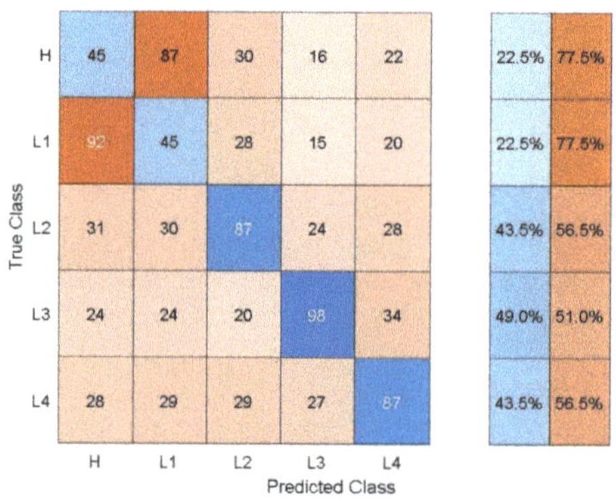

Figure 6. Training/validation confusion matrix of cubic SVM using handcrafted statistical features from simulated data.

In the confusion matrices, the values on the main diagonal denote correctly classified instances and the off-diagonal cells contain the misclassified instances. The blue and red columns on the right side show the recall (true-positive rate) and false-negative rate respectively, for each of the health states. Quantitatively, the recall gives the successful detection and isolation of a given health state while the false negative gives the susceptibility of a model to confusing other health states with the actual health state. From Figure 6, it can be observed that though the classification accuracy is not very high, the classification results are consistent with the physics of the problem. For instance, the classifier can distinguish more severe cases of delamination (L2, L3, L4) with relatively higher classification accuracy and the major loss of accuracy comes from confusion between healthy and less severe cases of delamination i.e., confusion between H and L1.

Experimental data was collected over a period of 15 s, the signal for each sample in each health state was split into 20 chunks using a window of 1875 data points. This created 100 instances for the five samples in each health state, resulting in a total of 300 instances for the three health states. The same statistical features used with the simulated data were extracted from the experimental data and were again processed using various machine learning algorithms. Table 4 shows the performance results for each of the classifiers in terms of training/validation accuracy and area under the receiver operating characteristics curve (ROC area). All classifiers were trained using a 10-fold cross-validation strategy.

Table 4. Classification performance of the various algorithms using handcrafted statistical features from experimental data.

Classifier	Training/Validation Accuracy %	ROC Area
Fine Tree	69.3	0.78
Linear Discriminant	63.0	0.65
Kernel Naïve Bayes	66.7	0.70
Cubic SVM	61.7	0.72
Fine KNN	57.0	0.60

The result in the table shows that delaminations of the same size but at different locations can be correctly classified with a maximum classification accuracy of around 70 %. However, a classifier trained with this accuracy may be susceptible to high misclassification rates when deployed for making predictions on new data.

To give an idea of the per-class misclassification results, Figure 7 depicts the training/validation confusion matrix for the Fine Tree classifier.

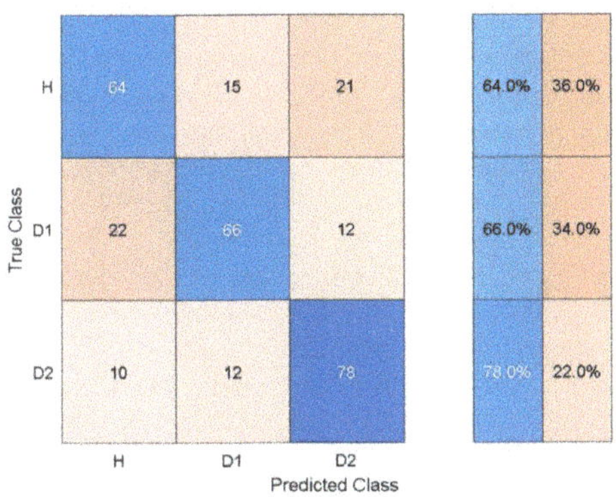

Figure 7. Training/validation confusion matrix for the Fine Tree classifier using handcrafted statistical features from experimental data.

It can be observed that the classifier can distinguish the three health states with 64%, 66%, and 78% classification accuracy. Furthermore, the susceptibility to misclassification was 36%, 34%, and 22% for the three health states.

Although the statistical features provide physically consistent results, the low classification accuracy when using these features hinders their practical application for the detection and isolation of delamination from low-frequency structural vibration in laminated composites. The next section shows the results of using discriminative features that were extracted autonomously via pretrained deep learning models.

3.2. Results from Using Features Extracted Autonomously Using Pretrained Deep Learning Models

In the deep learning framework, pretrained models can be employed for three purposes: to make predictions on new unseen data that is similar to previous data, for feature extraction using the activations of deep layers as features, and for transfer learning based on fine tuning a network that was pretrained on data from a different but related task to work with limited new data [42,43]. In this work, pretrained deep learning models based on AlexNet, GoogleNet, SqueezeNet, ResNet-18, and VGG-16 were employed for autonomous feature extraction from the limited data gathered by simulations and experiments. The reason for choosing different pretrained models for the current problem was to show the effect of the architecture, depth, and the number of pretrained deep learning model parameters for the fault diagnosis in a transfer learning framework. The characteristics features of the adopted pretrained models are shown in Table 5.

Table 5. Comparison of pretrained deep learning models.

Network	Number of Layers	Parameters (Millions)	Memory Size (MB)	Architectural Building Blocks of Hidden Layers
AlexNet	25	61.0	227	Convolution, ReLU, Pooling, Cross Channel Normalization, Grouped Convolution, Dropout, Fully Connected Layers, series connections
GoogleNet	144	7.0	27	Convolution, ReLU, Pooling, Cross Channel Normalization, Inception modules, multi-scale convolutional transformations, convolution layers replaced with small blocks
SqueezeNet	68	1.24	5.2	Convolution, ReLU, Pooling, Depth Concatenation (with squeeze and expand layers), Simple and Complex Bypass
ResNet-18	71	11.7	44	Convolutions, Batch Normalization, ReLU, Pooling, Addition, Residual blocks as structural unit
VGG-16	41	138	515	Convolution, ReLU, Pooling, Drop out, fully connected layers, smaller filters, and series connections

Herein, it is observed that each network has different architectural characteristics, and one can choose a pretrained model based on the initial assessment of the results from different models, depth of the models, memory size, predictive performance, and prediction speed. A detailed description of the architecture of different pretrained deep learning models can be found in the references [44–46].

The mathematical details of autonomous feature extraction via pretrained deep learning models are not discussed here for the sake of brevity but can be found in references [47–49]. The autonomous features were processed via a quadratic support vector machine using 10-fold cross validation and a one-vs-all training strategy for the detection, quantification, and localization of delamination in laminated composites. The mathematical details of SVM for the assessment of discriminative features and classification results can be found in the references [50–52].

Since the existing deep learning models were pretrained on image data, the time domain signals were transformed into time-frequency images using the Synchroextracting Transform (SET) technique introduced by Yu et al. [26]. The reason for choosing SET instead of short-time Fast Fourier Transforms (STFT) or wavelet scalograms was that SET provided better time-frequency resolution. The mathematical details of encoding time series data into an image with SET can be found in references [26,27]. The SET window length is the parameter that affects the time-frequency resolution. In the current work, after trying various lengths of SET windows, a window length equal to the length of the signal was found to provide the best time-frequency resolution for laminated composites. The time series data, windowed into smaller chunks in the same way as for the statistical features, was processed with SET to obtain time-frequency images representing the simulated and experimental data. Some representative time-frequency images of the simulated data obtain through SET are shown in Figure 8.

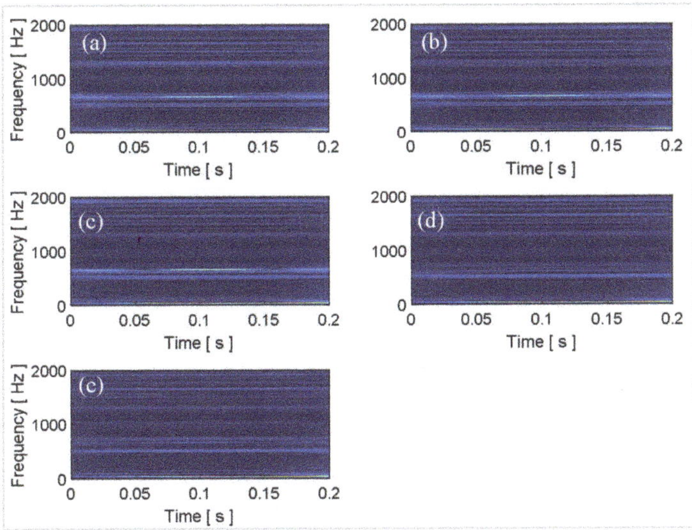

Figure 8. Some representative time-frequency images of the simulated data obtained via the SET technique: (**a**) H; (**b**) L1; (**c**) L2; (**d**) L3; (**e**) L4.

The time-frequency images show different characteristic spectrums for each of the five health states. However, due to the random nature of the response signals, it is difficult to differentiate the health states from their time-frequency images using the naked eye. Moreover, the images in Figure 8 are some representative examples from the 100 images for each health state and only correspond to a single random response signal. The difficulty level is further increased as the number of random excitations and their corresponding responses increase. Owing to their inherent architecture, deep learning models can look for minute differences between images allowing them to differentiate similar looking images autonomously with high accuracy. In the current work, the total number of images is only 500 (100 images for each health state), which is not sufficient to optimize the parameters of a deep learning model which is being developed and trained from scratch. Hence, in this work, off-the-shelf pretrained deep learning models were employed for autonomous feature extraction.

The autonomous features extracted at different layers of the pretrained models have different dimensions and discriminative capabilities. In the current work, features from different layers of the pretrained models are leveraged to discriminate between different health states of the laminated composites based on the simulated and experimental data. To ensure a consistent parametric study, the autonomous features were processed with quadratic SVM, Tables 6–10 show its performance based on the autonomous features extracted from the simulated data in terms of training/validation accuracy, test accuracy, and the number of autonomous features.

Table 6. Classification performance of SVM using the autonomous features from AlexNet.

	Layer Number	Name of Feature Extraction Layer	Training/Validation Accuracy %	Test Accuracy %	Number of Features
AlexNet	23	'fc8'	77.25	75.0	1000
	19	'fc6'	82.75	81.0	4096
	9	'pool2'	72.25	77.0	43,364
	5	'pool1'	70.0	73.0	69,984

Table 7. Classification performance of SVM using the autonomous features from GoogleNet.

	Layer Number	Name of Feature Extraction Layer	Training/Validation Accuracy %	Test Accuracy %	Number of Features
GoogleNet	142	'loss3-classifier'	72.25	75.0	1000
	136	'inception_5b-pool'	64.75	69.0	40,768
	122	'inception_5a-pool'	66.50	77.0	40,768
	93	'inception_4d-pool'	67.25	66.0	100,352

Table 8. Classification performance of SVM using the autonomous features from SqueezeNet.

	Layer Number	Name of Feature Extraction Layer	Training/Validation Accuracy %	Test Accuracy %	Number of Features
SqueezeNet	66	'pool10'	74.75	74.0	1000
	53	'fire8-expand3x3'	70.25	74.0	50,176
	34	'pool5'	70.75	77.0	50,176
	19	'pool3'	71.25	72.0	100,352

Table 9. Classification performance of SVM using the autonomous features from ResNet-18.

	Layer Number	Name of Feature Extraction Layer	Training/Validation Accuracy %	Test Accuracy %	Number of Features
ResNet-18	68	'pool5'	75.75	75.0	512
	59	'res5a_branch1'	72.25	73.0	25088
	45	'res4b_branch2a'	71.75	71.0	50176
	13	'res2b_branch2a'	75.0	79.0	200704

Table 10. Classification performance of SVM using the autonomous features from VGG-16.

	Layer Number	Name of Feature Extraction Layer	Training/Validation Accuracy %	Test Accuracy %	Number of Features
VGG-16	39	'fc8'	77.25	78.0	1000
	33	'fc6'	79.0	76.0	4096
	25	'pool4'	70.25	77.0	100,352
	11	'pool2'	76.50	76.0	401,408

It can be observed that the dimensions of the discriminative features are increasing as the feature extraction layer is moved from the final layers to the inner layers towards the initial layers of the pretrained models. In this problem, the accuracy is not affected much by whether high-level feature (i.e., features from the last layers) or low-level features (i.e., features from the inner and initial layers) are used. However, the dimensions and consequently the computational cost of using the low-level features are higher than for using the high-level features. When compared with the classification performance using the hand-crafted statistical features shown in Table 3, the performance using the autonomous features has substantially increased for all the pretrained models. To get further insight into the classification performance, Figures 9–13 depict the confusion matrices for the SVMs with the highest accuracy seen in Tables 6–10 coming from using the autonomous features of the pretrained models.

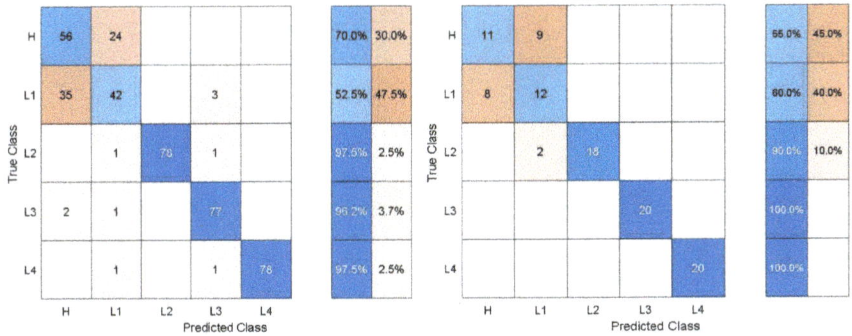

(a) (b)

Figure 9. Classification performance of SVM using the autonomous features from the 23rd layer of AlexNet in terms of its confusion matrices: (**a**) training; (**b**) test.

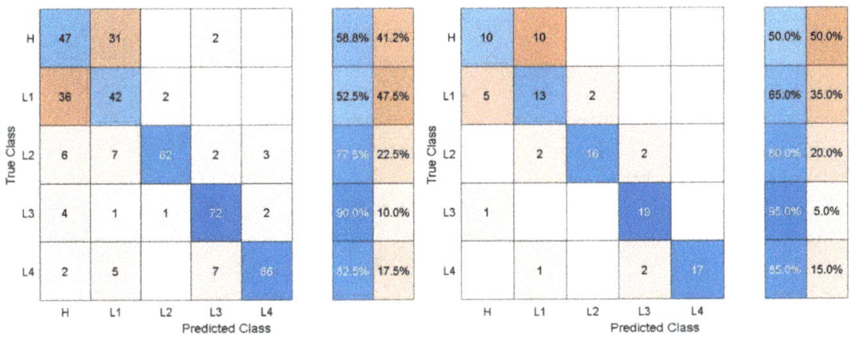

(a) (b)

Figure 10. Classification performance of SVM using the autonomous features from the 142nd layer of GoogleNet in terms of its confusion matrices: (**a**) training; (**b**) test.

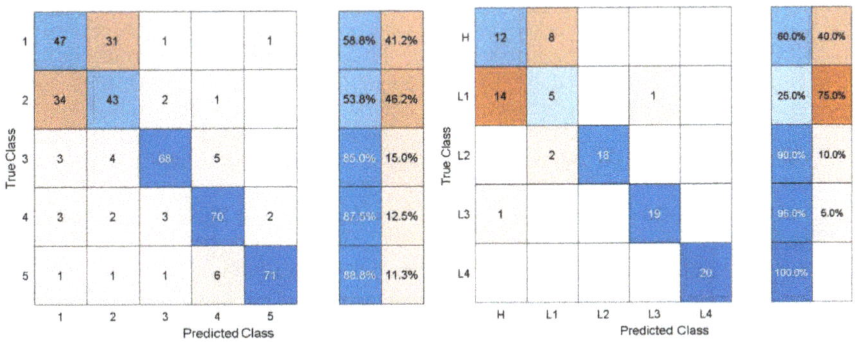

(a) (b)

Figure 11. Classification performance of SVM using the autonomous features from the 66th layer of SqueezeNet in terms of its confusion matrices: (**a**) training; (**b**) test.

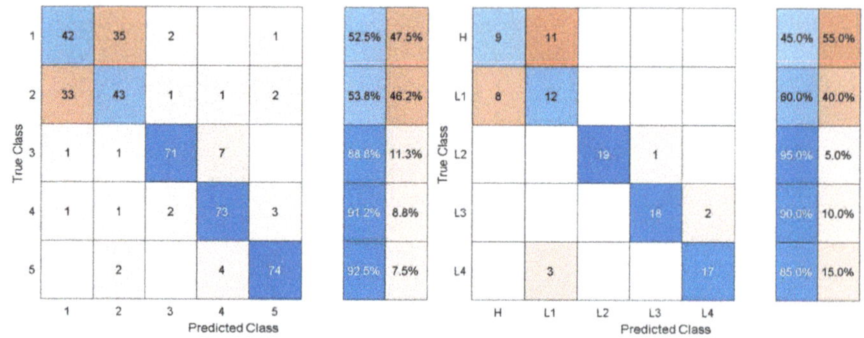

(a) (b)

Figure 12. Classification performance of SVM using the autonomous features form the 68th layer of ResNet-18 in terms its confusion matrices: (**a**) training; (**b**) test.

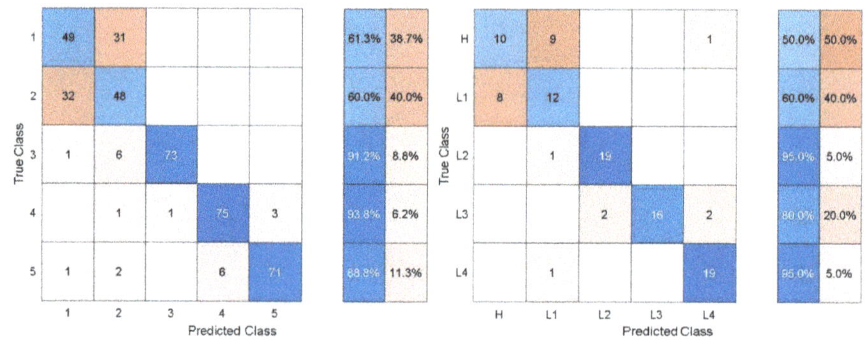

(a) (b)

Figure 13. Classification performance of SVM using the autonomous features from the 39th layer of VGG-16 in terms of its confusion matrices: (**a**) training; (**b**) test.

The dataset from the simulation was split into 80% training data and 20% test data. In the confusion matrices, the values on the main diagonals denote correctly classified instances and the off-diagonal cells contain misclassified instances. The blue and red columns on the right depict the recall (true-positive rate) and false-negative rate, respectively, of the actual classes. Quantitatively, the recall gives the successful detection and isolation of a given health state while the false negative gives the susceptibility of a model to confusing other health states with the actual health state. Specifically from Figure 9, the autonomous features extracted by Alexnet can distinguish healthy cases H from all other health states with 70% accuracy, but it is susceptible to incorrectly classifying other health states as H at a rate of 30%. The values on the main diagonal and in the off-diagonal cells show that the model is susceptible to the incorrect classification of L1 as H.

From all the confusion matrices, it can be observed that the per-class training performance has substantially increased when using the autonomously extracted features from each of the pretrained models when compared with the training performance using the handcrafted statistical features, as shown in Figure 6. Moreover, for all the pretrained models, the major losses of accuracy are associated with confusion between the healthy case H and the least severe case of delamination L1. The physical reason for the confusion of H with L1 is that the 3 cm delamination is only causing a very subtle change in the structural stiffness and consequently in the structural dynamic response when compared with the healthy case. The close resemblance in the characteristic dynamic responses of L1 and H causes the classifier to confuse the two cases. Moreover, from the test confusion matrices, it can be observed that the correct prediction rate from using the pretrained model to autonomously extract features, regardless of the model used, is reasonably high and show similar behavior to that shown by the training confusion matrices. The results for the increasing size of delamination and the ease in its detectability in the current work are supported by the research finding of the articles in the references [50,53].

For the experimental data, the structural vibration data windowed into the same smaller chunks used to extract the statistical features from was encoded to time-frequency images using SET. The image data based on the experimental results was processed using pretrained deep learning models to autonomously extract features. The features were extracted through the last layers of the models and were processed with quadratic SVM using 10-fold cross-validation and a one-vs-all training strategy. The experimental data was randomly split into 80% training and 20% test data. The classification results from the SVM using autonomous features are shown in Table 11.

Table 11. Classification performance of SVM using the autonomous features from various pretrained models.

Pretrained Model	Feature Extraction Layer	Training/Validation Accuracy %	Test Accuracy %	Number of Features
AlexNet	'fc8'	97.92	100	1000
GoogleNet	'loss3-classifier'	93.75	91.67	1000
SqueezeNet	'pool10'	96.25	98.33	1000
ResNet-18	'pool5'	95.83	98.33	512
VGG-16	'fc8'	97.08	98.33	1000

Comparing Table 11 with Table 4 shows that the overall training accuracy has substantially increased when using the autonomous features from all the pretrained models. Among the autonomous features extracted by the various pretrained models, the features extracted by AlexNet achieve the best classification performance, while the worst classification performance was observed when using the features from GoogleNet. For a detailed look at the classification performance using features from AlexNet and GoogleNet, Figures 14 and 15 show the training and test confusion matrices for the SVM trained on the features extracted by AlexNet and GoogleNet, respectively.

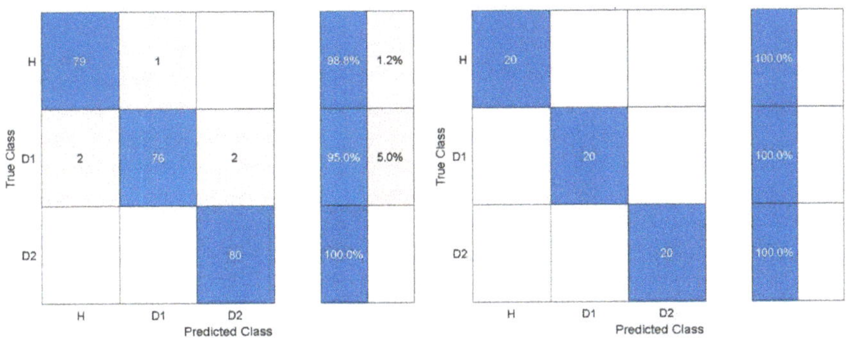

(a) (b)

Figure 14. Classification performance of SVM using the autonomous features form the 23rd layer of AlexNet in terms of its confusion matrices: (**a**) training; (**b**) test.

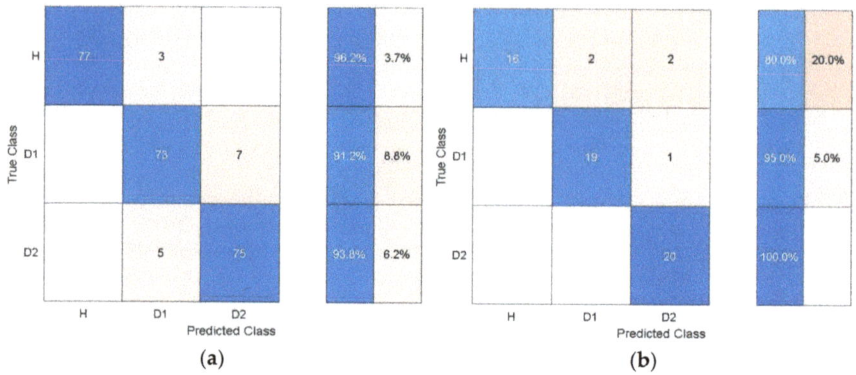

Figure 15. Classification performance of SVM using the autonomous features from the 142nd layer of GoogleNe in terms of its confusion matrices: (**a**) training; (**b**) test.

It can be observed that the SVM can distinguish healthy cases from delaminated cases with 98.8% accuracy while it is only susceptible to incorrectly classify other health states as healthy at a rate o 1.2%. The minimum per-class classification accuracy was 95.0% for D1 delamination samples using th features from AlexNet while the maximum per-class classification accuracy was 100% for D2 delamina tion samples. For the pretrained model with the worst overall classification performance (GoogleNet) the minimum per-class accuracy was 91.2% for D1 delamination samples and the maximum per-clas accuracy was 96.2% for the healthy H samples. In general, the autonomous features from all th pretrained models can be used to distinguish between delaminations of the same size occurring a different locations while they can also be used to distinguish healthy cases H from delaminated case D1 and D2 with higher accuracy. The experimental data has accounted for the uncertainty in th manufacturing of coupons and measurement error by considering five samples of each health stat The existing literature mostly focuses on modal parameters (natural frequencies, modes shapes, mod shape curvature) of structural vibration to assess delamination [54]. However, modal parameters ar global, and the experimental measurement of mode shapes is a difficult task.

The current work showed the feasibility of limited structural vibration data for the detectior localization, and quantification of delamination through classification in a supervised learning frame work. The proposed approach could be employed to assess the size and location of delamination b predicting a label for a given health state which can be interpreted for the size and location of a nev delamination. For instance, the label for a delamination of size 6 cm would be either L2 (delaminatio of 5 cm) or L3 (delamination of 7 cm) due to their similar response characteristics. In general, th delamination of different sizes near the free and clamped ends may have the same dynamics respons characteristics, and the new delamination may be entirely different from the one considered in th pretrained model. Moreover, the labels for the prospective delamination may not be known. The futur extension of the current work will adopt a more generic approach by using more than one senso along the length of the specimen and an unsupervised framework for the detection, localization, an quantification of delamination. These results show our approach could be beneficial for delaminatio localization based on low-frequency structural vibration data.

4. Conclusions

This study investigated the effectiveness of autonomous discriminative features via pretraine deep learning models for assessing delamination in laminated composites. The simulations an experiments were carried out to obtain low-frequency random structural vibration data from th sample in either healthy or delaminated states. The vibration data was encoded into high-resolutio time-frequency images using the Synchroextracting Transform (SET) technique. This image data wa then processed with various pretrained deep learning models (AlexNet, GoogleNet, SqueezeNe ResNet-18, 491 VGG-16) for autonomous discriminative features, and a support vector machine (SVM was employed to assess the health state based on those features. The autonomous features wer found to outperform the handcrafted statistical features for diagnosing delamination in laminate composites. The work also compared the feasibility of low- and high-level features of pretrained models for delamination assessment. The high-level features were found to perform relatively better than the low-level features. The proposed approach eliminates the requirement of extensive training data and labor-intensive process of human-engineered statistical features. The method has the potential for the autonomous diagnosis and prognosis of defects in modern composite structures.

Author Contributions: Conceptualization, H.-S.K., A.K. and J.-W.S.; methodology, A.K. and J.-W.S.; software, A.K., S.K. and I.R.; validation, A.K., H.-S.K. and J.-W.S.; formal analysis, H.-S.K. and J.-W.S.; investigation, A.K., S.K. and I.R.; resources, H.-S.K., J.-W.S.; data curation, A.K., J.-W.S.; writing—original draft preparation, A.K., S.K. and I.R.; writing—review and editing, H.-S.K. and J.-W.S.; supervision, H.-S.K. and J.-W.S.; project administration, H.-S.K. and J.-W.S.; funding acquisition, H.-S.K. and J.-W.S. All authors have read and agreed to the published version of the manuscript.

Funding: This research was supported by the Basic Science Research Program through the National Research Foundation of Korea (NRF-2020R1A2C1006613), funded by the Ministry of Education. This research was also supported by the MSIT (Ministry of Science and ICT), Korea, under the Grand Information Technology Research Center support program (IITP-2021-2020-0-01612) supervised by the IITP (Institute for Information & communications Technology Planning & Evaluation).

Institutional Review Board Statement: Not applicable.

Informed Consent Statement: Not applicable.

Data Availability Statement: The data that support the findings of this study is available from the corresponding author upon request.

Conflicts of Interest: The authors declare no conflict of interest.

References

Xu, Y.; Zhu, J.; Wu, Z.; Cao, Y.; Zhao, Y.; Zhang, W. A Review on the Design of Laminated Composite Structures: Constant and Variable Stiffness Design and Topology Optimization. *Adv. Compos. Hybrid. Mater.* **2018**, *1*, 460–477. [CrossRef]

Saeedifar, M.; Zarouchas, D. Damage Characterization of Laminated Composites Using Acoustic Emission: A Review. *Compos. Part. B Eng.* **2020**, *195*, 108039. [CrossRef]

Hsissou, R.; Seghiri, R.; Benzekri, Z.; Hilali, M.; Rafik, M.; Elharfi, A. Polymer Composite Materials: A Comprehensive Review. *Compos. Struct.* **2021**, *262*, 113640. [CrossRef]

Tabiei, A.; Zhang, W. Composite Laminate Delamination Simulation and Experiment: A Review of Recent Development. *Appl. Mech. Rev.* **2018**, *70*, 030801. [CrossRef]

Garcia, C.; Jurado, A.; Zaba, O.; Beltran, P. Detection and Quantification of Delamination Failures in Marine Composite Bulkheads via Vibration Energy Variations. *Sensors* **2021**, *21*, 2843. [CrossRef] [PubMed]

Garcia, C.; Trendafilova, I. Triboelectric Sensor as a Dual System for Impact Monitoring and Prediction of the Damage in Composite Structures. *Nano Energy* **2019**, *60*, 527–535. [CrossRef]

Kashtalyan, M.; Soutis, C. The Effect of Delaminations Induced by Transverse Cracks and Splits on Stiffness Properties of Composite Laminates. *Compos. Part. A Appl. Sci. Manuf.* **2000**, *31*, 107–119. [CrossRef]

Sahoo, S.S.; Panda, S.K.; Sen, D. Effect of Delamination on Static and Dynamic Behavior of Laminated Composite Plate. *AIAA J.* **2016**, *54*, 2530–2544. [CrossRef]

Gholizadeh, S. A Review of Non-Destructive Testing Methods of Composite Materials. *Procedia Struct. Integr.* **2016**, *1*, 50–57. [CrossRef]

0. Wang, B.; Zhong, S.; Lee, T.-L.; Fancey, K.S.; Mi, J. Non-Destructive Testing and Evaluation of Composite Materials/Structures: A State-of-the-Art Review. *Adv. Mech. Eng.* **2020**, *12*, 1687814020913761. [CrossRef]

1. Ai, L.; Soltangharaei, V.; Bayat, M.; Van Tooren, M.; Ziehl, P. Detection of Impact on Aircraft Composite Structure Using Machine Learning Techniques. *Meas. Sci. Technol.* **2021**, *32*, 084013. [CrossRef]

2. O'Brien, R.J.; Fontana, J.M.; Ponso, N.; Molisani, L. A Pattern Recognition System Based on Acoustic Signals for Fault Detection on Composite Materials. *Eur. J. Mech.* **2017**, *64*, 1–10. [CrossRef]

3. Khan, A.; Ko, D.-K.; Lim, S.C.; Kim, H.S. Structural Vibration-Based Classification and Prediction of Delamination in Smart Composite Laminates Using Deep Learning Neural Network. *Compos. Part B Eng.* **2019**, *161*, 586–594. [CrossRef]

4. Loutas, T.H.; Panopoulou, A.; Roulias, D.; Kostopoulos, V. Intelligent Health Monitoring of Aerospace Composite Structures Based on Dynamic Strain Measurements. *Expert Syst. Appl.* **2012**, *39*, 8412–8422. [CrossRef]

5. Sung, D.-U.; Oh, J.-H.; Kim, C.-G.; Hong, C.-S. Impact Monitoring of Smart Composite Laminates Using Neural Network and Wavelet Analysis. *J. Intell. Mater. Syst. Struct.* **2000**, *11*, 180–190. [CrossRef]

6. Tabian, I.; Fu, H.; Sharif Khodaei, Z. A Convolutional Neural Network for Impact Detection and Characterization of Complex Composite Structures. *Sensors* **2019**, *19*, 4933. [CrossRef] [PubMed]

17. Nash, W.; Drummond, T.; Birbilis, N. A Review of Deep Learning in the Study of Materials Degradation. *NPJ Mater. Degrad* **2018**, *2*, 37. [CrossRef]
18. Toh, G.; Park, J. Review of Vibration-Based Structural Health Monitoring Using Deep Learning. *Appl. Sci.* **2020**, *10*, 1680. [CrossRef]
19. Gong, Y.; Shao, H.; Luo, J.; Li, Z. A Deep Transfer Learning Model for Inclusion Defect Detection of Aeronautics Composite Materials. *Compos. Struct.* **2020**, *252*, 112681. [CrossRef]
20. Ramezankhani, M.; Crawford, B.; Narayan, A.; Voggenreiter, H.; Seethaler, R.; Milani, A.S. Making Costly Manufacturing Smart with Transfer Learning under Limited Data: A Case Study on Composites Autoclave Processing. *J. Manuf. Syst.* **2021**, *59*, 345–354. [CrossRef]
21. Yang, B.; Lei, Y.; Jia, F.; Xing, S. An Intelligent Fault Diagnosis Approach Based on Transfer Learning from Laboratory Bearings to Locomotive Bearings. *Mech. Syst. Signal Process.* **2019**, *122*, 692–706. [CrossRef]
22. Zhang, B.; Hong, X.; Liu, Y. Distribution Adaptation Deep Transfer Learning Method for Cross-Structure Health Monitoring Using Guided Waves. *Struct. Health Monit.* **2021**. [CrossRef]
23. Li, J.; Huang, R.; He, G.; Wang, S.; Li, G.; Li, W. A Deep Adversarial Transfer Learning Network for Machinery Emerging Fault Detection. *IEEE Sens. J.* **2020**, *20*, 8413–8422. [CrossRef]
24. Zhang, R.; Tao, H.; Wu, L.; Guan, Y. Transfer Learning with Neural Networks for Bearing Fault Diagnosis in Changing Working Conditions. *IEEE Access* **2017**, *5*, 14347–14357. [CrossRef]
25. Zheng, H.; Wang, R.; Yang, Y.; Yin, J.; Li, Y.; Li, Y.; Xu, M. Cross-Domain Fault Diagnosis Using Knowledge Transfer Strategy: A Review. *IEEE Access* **2019**, *7*, 129260–129290. [CrossRef]
26. Yu, G.; Yu, M.; Xu, C. Synchroextracting Transform. *IEEE Trans. Ind. Electron.* **2017**, *64*, 8042–8054. [CrossRef]
27. Li, Z.; Gao, J.; Li, H.; Zhang, Z.; Liu, N.; Zhu, X. Synchroextracting Transform: The Theory Analysis and Comparisons with the Synchrosqueezing Transform. *Signal. Process.* **2020**, *166*, 107243. [CrossRef]
28. Krizhevsky, A.; Sutskever, I.; Hinton, G.E. Imagenet Classification with Deep Convolutional Neural Networks. *Adv. Neural. Inf. Process. Syst.* **2012**, *25*, 1097–1105. [CrossRef]
29. Szegedy, C.; Wei, L.; Jia, Y.; Sermanet, P.; Reed, S.; Anguelov, D.; Erhan, D.; Vanhoucke, V.; Rabinovich, A. Going Deeper with Convolutions. In Proceedings of the 2015 IEEE Conference on Computer Vision and Pattern Recognition (CVPR), Boston, MA, USA, 8–10 June 2015; pp. 1–9.
30. Iandola, F.N.; Han, S.; Moskewicz, M.W.; Ashraf, K.; Dally, W.J.; Keutzer, K. SqueezeNet: AlexNet-Level Accuracy with 50× Fewer Parameters and <0.5 MB Model Size. *arXiv* **2016**, arXiv:1602.07360.
31. He, K.; Zhang, X.; Ren, S.; Sun, J. Deep Residual Learning for Image Recognition. In Proceedings of the 2016 IEEE Conference on Computer Vision and Pattern Recognition (CVPR), Las Vegas, NV, USA, 27–30 June 2016; pp. 770–778.
32. Simonyan, K.; Zisserman, A. Very Deep Convolutional Networks for Large-Scale Image Recognition. *arXiv* **2014**, arXiv:1409.1556.
33. Huang, B.; Kim, H.S.; Yoon, G.H. Modeling of a Partially Debonded Piezoelectric Actuator in Smart Composite Laminates. *Smart Mater. Struct.* **2015**, *24*, 075013. [CrossRef]
34. Kim, H.S.; Chattopadhyay, A.; Ghoshal, A. Dynamic Analysis of Composite Laminates with Multiple Delamination Using Improved Layerwise Theory. *AIAA J.* **2003**, *41*, 1771–1779. [CrossRef]
35. Kim, H.S.; Ghoshal, A.; Chattopadhyay, A.; Prosser, W.H. Development of Embedded Sensor Models in Composite Laminates for Structural Health Monitoring. *J. Reinf. Plast. Compos.* **2004**, *23*, 1207–1240. [CrossRef]
36. Huang, B.; Koh, B.-H.; Kim, H.S. PCA-Based Damage Classification of Delaminated Smart Composite Structures Using Improved Layerwise Theory. *Comput. Struct.* **2014**, *141*, 26–35. [CrossRef]
37. Reddy, J.N. *Mechanics of Laminated Composite Plates and Shells: Theory and Analysis*, 2nd ed.; CRC Press: Boca Raton, FL, USA, 2003; ISBN 978-0-203-50280-8.
38. Shukla, S.; Yadav, R.N.; Sharma, J.; Khare, S. Analysis of Statistical Features for Fault Detection in Ball Bearing. In Proceedings of the 2015 IEEE International Conference on Computational Intelligence and Computing Research (ICCIC), Madurai, India, 10–12 December 2015; pp. 1–7.
39. Toma, R.N.; Prosvirin, A.E.; Kim, J.-M. Bearing Fault Diagnosis of Induction Motors Using a Genetic Algorithm and Machine Learning Classifiers. *Sensors* **2020**, *20*, 1884. [CrossRef]
40. Nguyen, C.D.; Prosvirin, A.; Kim, J.-M. A Reliable Fault Diagnosis Method for a Gearbox System with Varying Rotational Speeds. *Sensors* **2020**, *20*, 3105. [CrossRef] [PubMed]
41. Shen, C.; Wang, D.; Kong, F.; Peter, W.T. Fault Diagnosis of Rotating Machinery Based on the Statistical Parameters of Wavelet Packet Paving and a Generic Support Vector Regressive Classifier. *Measurement* **2013**, *46*, 1551–1564. [CrossRef]
42. Farahani, A.; Pourshojae, B.; Rasheed, K.; Arabnia, H.R. A Concise Review of Transfer Learning. In Proceedings of the 2020 International Conference on Computational Science and Computational Intelligence (CSCI), Las Vegas, NV, USA, 16–18 December 2020; pp. 344–351.
43. Zhuang, F.; Qi, Z.; Duan, K.; Xi, D.; Zhu, Y.; Zhu, H.; Xiong, H.; He, Q. A Comprehensive Survey on Transfer Learning. *Proc. IEEE* **2020**, *109*, 43–76. [CrossRef]
44. Khan, A.; Sohail, A.; Zahoora, U.; Qureshi, A.S. A Survey of the Recent Architectures of Deep Convolutional Neural Networks. *Artif. Intell. Rev.* **2020**, *53*, 5455–5516. [CrossRef]

5. Alom, M.Z.; Taha, T.M.; Yakopcic, C.; Westberg, S.; Sidike, P.; Nasrin, M.S.; Hasan, M.; Van Essen, B.C.; Awwal, A.A.; Asari, V.K. A State-of-the-Art Survey on Deep Learning Theory and Architectures. *Electronics* **2019**, *8*, 292. [CrossRef]
6. Bianco, S.; Cadene, R.; Celona, L.; Napoletano, P. Benchmark Analysis of Representative Deep Neural Network Architectures. *IEEE Access* **2018**, *6*, 64270–64277. [CrossRef]
7. Tsiakmaki, M.; Kostopoulos, G.; Kotsiantis, S.; Ragos, O. Transfer Learning from Deep Neural Networks for Predicting Student Performance. *Appl. Sci.* **2020**, *10*, 2145. [CrossRef]
8. Cao, P.; Zhang, S.; Tang, J. Preprocessing-Free Gear Fault Diagnosis Using Small Datasets with Deep Convolutional Neural Network-Based Transfer Learning. *IEEE Access* **2018**, *6*, 26241–26253. [CrossRef]
9. Li, C.; Zhang, S.; Qin, Y.; Estupinan, E. A Systematic Review of Deep Transfer Learning for Machinery Fault Diagnosis. *Neurocomputing* **2020**, *407*, 121–135. [CrossRef]
0. He, M.; Wang, Y.; Ram Ramakrishnan, K.; Zhang, Z. A Comparison of Machine Learning Algorithms for Assessment of Delamination in Fiber-Reinforced Polymer Composite Beams. *Struct. Health Monit.* **2021**, *20*, 1997–2012. [CrossRef]
1. Shyamala, P.; Mondal, S.; Chakraborty, S. Numerical and Experimental Investigation for Damage Detection in FRP Composite Plates Using Support Vector Machine Algorithm. *Struct. Monit. Maint.* **2018**, *5*, 243–260.
2. Widodo, A.; Yang, B.-S. Support Vector Machine in Machine Condition Monitoring and Fault Diagnosis. *Mech. Syst. Signal. Process.* **2007**, *21*, 2560–2574. [CrossRef]
3. Jacobs, E.W.; Yang, C.; Demir, K.G.; Gu, G.X. Vibrational Detection of Delamination in Composites Using a Combined Finite Element Analysis and Machine Learning Approach. *J. Appl. Phys.* **2020**, *128*, 125104. [CrossRef]
4. Montalvao, D. A Review of Vibration-Based Structural Health Monitoring with Special Emphasis on Composite Materials. *Shock Vib. Dig.* **2006**, *38*, 295–324. [CrossRef]

sensors

MDPI

Review

Functionality Analysis of Electric Actuators in Renewable Energy Systems—A Review

Abhijeet Redekar [1], Dipankar Deb [1,*] and Stepan Ozana [2]

[1] Department of Electrical Engineering, Institute of Infrastructure Technology Research and Management (IITRAM), Ahmedabad 380026, India; abhijeet.redekar.21pe@iitram.ac.in

[2] Department of Cybernetics and Biomedical Engineering, Faculty of Electrical Engineering and Computer Science, VSB-Technical University of Ostrava, 17. listopadu 2172/15, 708 00 Ostrava, Czech Republic; stepan.ozana@vsb.cz

* Correspondence: dipankardeb@iitram.ac.in

Abstract: Various mechanical, hydraulic, pneumatic, electrical, and hybrid actuators can alter motion per the requirements of particular applications. However, except for electrical ones, all actuators are restricted due to their size, complex auxiliary equipment, frequent need for maintenance, and sluggish environment in renewable applications. This brief review paper highlights some unique and significant research works on applying electrical actuators to renewable applications. Four renewable energy resources, i.e., solar, wind, bio-energy, and geothermal energy, are considered to review electric actuators applicable to renewable energy systems. This review analyses the types of actuators associated with the mentioned renewable application, their functioning, their motion type, present use, advantages, disadvantages, and operational problems. The information gathered in this paper may open up new ways of optimization opportunities and control challenges in electrical actuators, thereby making more efficient systems. Furthermore, some energy-efficient and cost-effective replacements of convectional actuators with new innovative ones are suggested. This work aims to benefit scientists and new entrants working on actuators in renewable energy systems.

Keywords: electric actuators; linear actuators; rotary actuators; power consumption; spherical actuators; renewable system

1. Introduction

An actuator is a vital component of any physical system enabling movements by converting an energy source into another, primarily electrical, air, or hydraulic energy, into mechanical force [1,2] to modify the current system's state. Some major applications include the automotive industry, furniture-ergonomics, automation, industrial machinery, maritime applications, the medical industry, and renewable energy systems. Various types of actuators are in use (as per the power sources used) such as electrical actuators [3], hydraulic actuators [4], pneumatic actuators [5], mechanical actuators, and a combination of these such as electro-hydraulic [6], electro-pneumatic [7], electro-mechanical [8], self driven thermo-mechanical [9], etc. An electric actuator creates a load movement or an action requiring a force such as clamping, using an electric motor to create the desired force, converting electricity into kinetic energy to automate valves, or damper actions using precise flow control [10]. Electric motors may work on AC or DC supplies depending on the requirements of the application, limit switches, brake mechanisms, resolvers, temperature sensors, etc. The desired force is generated from the motor's torque capability and automates industrial valves, process plants, flow control, thermal power plants, irrigation systems, etc. [11].

Renewable energy plays a pivotal role in tackling climate change by reducing greenhouse gas emissions and fossil fuel use. Energy conversion systems driven by actuators help ensure optimal conversion while transforming renewable sources into valuable forms.

Rotary to linear motion is achievable using a mechanical transmission unit, called linear motion actuators, which are better in accuracy than rotary actuators, compact in size, and easy to use in many applications such as solar tracking systems [12], where solar panels can track the sun's apparent motion; solar furnace systems [13], where the motion required to control the sun-rays by moving shutters; solar heater systems [14], where sun rays are concentrated on oil pipes (called active area) placed over the parabolic solar panels; solar cleaning systems [15,16], where two motions are performed (viz., x-axis and y-axis motion) to clean the whole solar panel; small-scale wind systems [17], where actuators are used as protection devices in high wind speeds; wave systems [1,2,18,19], where actuators are used in the conversion of wave motion to linear and rotary motion for energy generation; and bio-energy systems [20–23], where actuators automate the operation of geologically distributed systems.

Energy optimization is achievable by increasing energy generation or reducing energy consumption. In solar panel cleaning, Williams et al. [15] use piezoceramic actuators for self-cleaning operation. The results show that dust moves away from high to low vibration velocity. The convectional solar panel's mechanical cleaning structure is called a gantry structure [24] and is operated by two rotary actuators. Using telescopic actuators offers the possibility of reducing the use of one rotary actuator for reduced overall cost and energy consumption. In solar furnace systems [25], instead of rotary actuators by applying linear actuators, one may neglect the mechanical dead zone for decreased operation time due to the quick response increased testing quality. For solar tree applications, a 2DOF spherical actuator [26] is more beneficial than a linear actuator. There are two advantages: less space required and a mechanical latch arrangement which uses reduced power while moving. Such an appropriate selection of actuator ensures an optimal system operation.

Actuators are mainly classifiable into active and passive actuators. The active actuators need an electric energy source for functioning. In contrast, passive actuators do not require a source and function based on natural energy, such as thermal expansion material, and energy stored in the spring [27]. Various required motions are circular, rotational, oscillatory (seismic), and rectilinear. Therefore, electric actuators are classified according to their motion, degree of freedom (DOF), and excitation sources for particular functional motion. As per motion, actuators are rotary actuators, linear actuators, and a new one called a spherical actuator with multiple degrees of freedom [26]. Linear motion actuators are split into two categories: rotary-to-linear (hybrid motion) actuators involving mechanical transmission units (also called electromagnetically linear actuators as shown in Figure 1) and direct linear motion ones not requiring mechanical transmissions units.

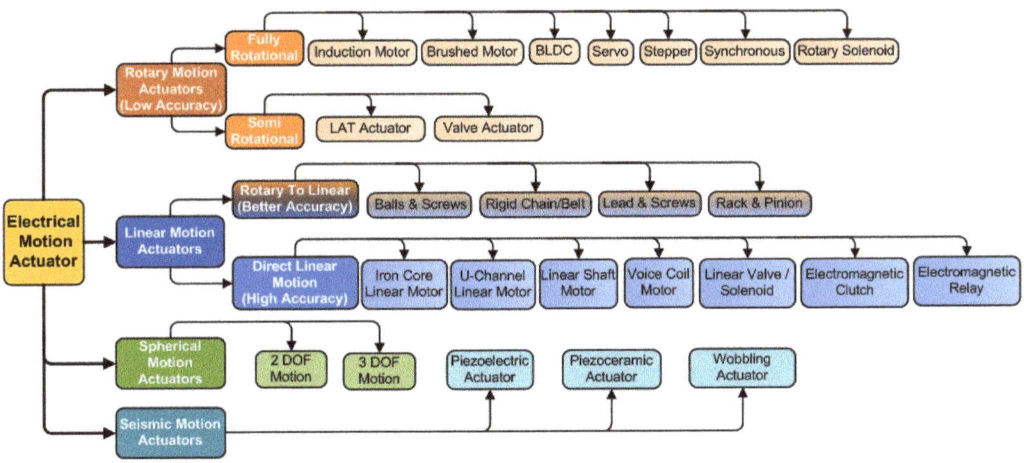

Figure 1. Active electric actuators according to their motion used in the renewable applications.

Actuators are used in renewable energy sources such as solar tracking applications to drive solar panels, solar dishes, heliostats, and solar cookers moving towards the sun throughout the day [28]. Solar water pumping applications use linear actuators in a water pump system [29], and solar tracking, or cooker, applications use actuators (motors or shape memory alloys) to drive the rotation of panels towards the sun throughout the day at a rate of $15^{\circ}/h$ [27,30]. Depending on the temperature, certain materials in a thermo-mechanical actuator help control the collector's angle of inclination to face the sun and provide increased production of about 39% compared to a fixed system. In addition, there are some bioenergy applications, where rotary actuators are used for agricultural machinery and the automation of geographically distributed biogas systems; portable wind energy applications [31] with actuator-controlled over-speeding; wave energy applications, where actuators are used in power-generation purposes [32]; smart grid or microgrid environment [33], where actuators control power generation and consumption; and geothermal power plants with electric valve actuators.

Rotary (also called electromechanical rotary) actuators help move large loads in angular increments. They include electrical motors with gearboxes used to change the direction of rotation or achieve the desired torque and speed with the desired adjustable rotation angle. Examples of the rotary actuators are induction motors, DC motors (brushed and brushless), synchronous motors, and special motors such as stepper and servo motors. Motors' selection factors for actuation are speed, output torque, type of supply required, and motor duty cycle. However, induction motors, which give almost synchronous speed, are effective with high power requirements. Due to the low inertia, servo motor rotors working on AC or DC supply and controlled through pulse width modulation can quickly start and stop and are used to position mechanical parts precisely. Such precise motion control is needed in CNC machines and production assembly lines. In addition, DC servo, DC brushed, and BLDC motors are economical and less complex actuation devices. Unlike servos, a stepper motor that divides the rotary motion into equal small steps can provide 360-degree continuous rotation and is helpful for high torque output and high accuracy at low-cost [34]. However, due to electrical (cogging) and mechanical (dead zone) non-linearities [35], these rotary actuators are relatively less efficient and have lower accuracy. Whenever we may face motor concussion problem [36], accuracy is improved by using special motors such as stepper motors and servo motors, which still face mechanical non-linearities and, thus, low performance. In mechatronic systems, electric rotary actuators find extensive use.

A rotary solenoid is more suitable for small rotations and low torque applications than a costly and complex control circuitry stepper motor. It comprises an electromagnetic coil operated on a DC supply, rotor, and torsion spring. The change in supply polarity is easily handled in the direction of rotation, providing a meager angle rotational movement around a fixed axis (usually $\leqslant\pm360^{\circ}$). Typical rotary solenoids have bi-stable positions that do not require power to hold either end position, minimizing energy consumption. It can drive camera shutters and mirror deflections, fluid/gas valves, the actuation of indicating instruments, application areas such as the packaging industry, medical devices, apparatus construction, and food technology for interlocking purposes. Rotary-to-linear actuators (also called electro-mechanical linear actuators) comprises electric motors and mechanical transmission elements. The rotational motion converts into linear motion using mechanical transmission elements, such as the ball and screw mechanism, the lead screw mechanism, the rack and pinion mechanism, the belt and pulley mechanism, and rigid chains. Such elements slow down electric rotations, thereby increasing torque such that the load on the transmission part moves smoothly. The ball and screw are made out of a cylinder to allow the re-circulation of the balls, causing linear movement, but are used for distinct purposes and are not interchangeable [37].

Unlike rotary to linear ones, direct linear motion actuators have no transmission elements and provide high accuracy, resolution, and speed. There are various types, viz., iron core linear motors, U-channel linear motors, linear shaft motors, voice coil motors, linear motion solenoids, screw-type linear piezoelectric actuators [38], and electromagnetic clutches. The stator and forcer (called rotor), with a permanent magnet and electromagnetic coils, respectively, create linear force. However, the iron core brings cogging and eddy currents, heat, magnetic saturation, and lateral forces. U-channel direct linear motor eliminates such disadvantages, providing increased force through an additional permanent magnet on the opposite side for a double-sided configuration, with electromagnetic coils in an epoxy-mounted non-ferrous forcer aluminum plate. The absence of a magnetic core eliminates cogging, but reduces heat dissipation, and a double-sided configuration increases cost. Although they do not generate the same forces as an iron core motor and U-channel linear motors, they do not have to deal with the additional issues with that design. Voice coil actuators, working on the principle of Lorentz force, are a direct drive mechanism that allows for excellent fine-positioning over small movements.

An electromagnetic device called a "Linear Solenoid" turns electrical energy into a mechanical pushing or pulling force or motion for rapid ON and OFF operation and mainly comprises an electromagnetic coil, a plunger (armature), and a spring. When the coil energizes, the spring is compressed for the valve to open, and when the coil de-energizes, the spring moves downward to close the valve and control fluid/airflow as a solenoid valve. However, the disadvantages are that there are only two available positions, i.e., fully open or closed, generating a limited linear force for electrically opening doors and latches. Electric (also called electromagnetic) clutches have an electric actuation between two rotational gears (drive/input and driven/output) for mechanical transmission of torque through linear motion, i.e., operate electrically and transmit torque mechanically. The passage of a current through the clutch coil makes it an electromagnet producing magnetic lines of flux [39]. Although mainly used in biogas automation [20], clutches are also used in conveyor drives, packaging machinery, and factory automation. Apart from linear and rotary motion, the new spherical motion actuators [26] are multi-degree-of-freedom voice coil devices with a spherical-shaped stator, two coils, a ball-shaped rotor, and multiple permanent magnets (a plurality of magnets) [40] applied to robot eyes, joints, and solar trees.

This comprehensive literature survey focuses exclusively on renewable applications with electrical actuators. Particular actuators find a place in multiple renewable energy systems, while others have only specific or limited usage. Furthermore, there are certain advantages and disadvantages (provided subsequently in each Section) concerning actuator motion. We also suggest some alternative solutions for conventional actuators. However, the overall advantages of electric actuators against non-electrical actuators (such as hydraulic and pneumatic) are given as follows:

(i) Lower response time: Electrical actuators are directly driven without auxiliary accessories, unlike hydraulic actuators, which comprise motors, hydraulic tanks, controlling valve pipes, and cylinders, and hence are fast operating.
(ii) Lightweight: Compared to non-electrical actuators, electric actuators are lightweight and easy to install, with no adverse effect on system performance. The cost of overall installation and material is reduced [15,41].
(iii) Precision control: The low tolerances compared to non-electrical actuators such as gear backlash, slack, and inherent flex provide highly precise positioning [42].
(iv) Compact in size: Electric actuators are compact due to their direct electromagnetic energy conversion system, and they are designed for any space and physical size [26].
(v) Clean: Electrical actuators are a clean source of energy with no chance of leaking or dirty work areas [43].
(vi) IoT compatible: Motion and position are easily measurable due to inbuilt rotation sensors, and encoders and easily transfer to an IoT environment [44,45].

The paper's structure is as follows. Section 2 reviews various types of electric actuators in different solar system applications such as solar tracking, furnace, heater, and solar panel cleaning. Section 3 presents a brief review of electric actuators used in wind applications, especially miniature rooftop wind turbines and wave energy applications. Then, Section 4 reviews actuators used in bioenergy applications, including agriculture and irrigation. Section 5 presents important performance parameters of actuators and evolution. Finally, Section 6 presents the conclusions and future direction.

2. Actuators for Solar System Applications

Solar energy is a free and abundant source available on this planet that is extracted in various ways, such as electric power generation using solar photovoltaic cells or solar-distributed collectors for heating applications, such as a solar water heater, furnace, heater, etc. Next, we present a survey of electrical actuators used in different solar energy applications.

2.1. Actuators for Solar Tracking Applications

Solar panels need to follow the sun's movement to extract the maximum feasible energy. As the sun rises from the east and sets in the west, the tracker system aims to follow the sun's movement or, during the day, keep the solar panel perpendicular to the sun, i.e., to follow the sun's day to day and seasonal movement for maximum power extraction. The tracking system finds use in solar PV panels, in moving heliostat mirrors, parabolic concentrators, and solar collectors. We explore various actuators presently in use for solar tracking. Figure 2 shows individual solar panel tracking and simultaneous-tracking (i.e., group tracking) solar panels. The linear motion of the linear actuator is achieved from the motor's rotary motion. Internally some mechanical motion change arrangements convert the motion, such as ball–screw, lead–screw, etc. This mechanical device also manipulates the torque and speed as required with an appropriate pitch of the screw.

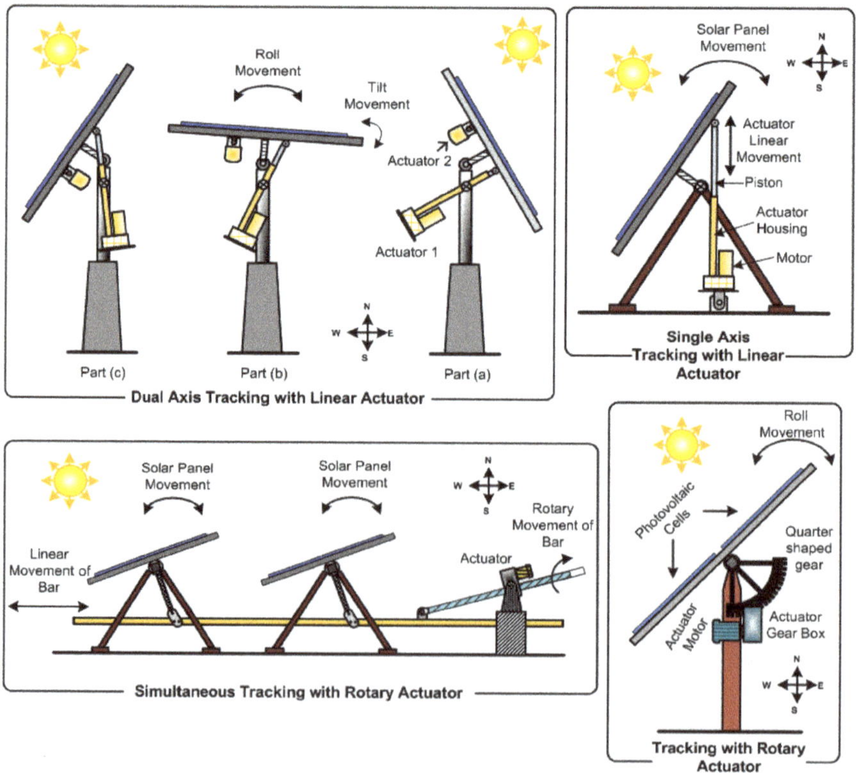

Figure 2. Electric actuators functioning in solar tracking applications.

Linear actuators and rotary actuators achieve solar tracking. Single-axis tracking, where a solar panel tracks the day-to-day movement of the sun (i.e., only east to west movement), is achievable only using one linear actuator [27]. The linear movement of the actuator is also shown through single-axis tracking with linear actuators in Figure 2. In the morning, the solar panel faces the east direction, and the actuators are in the initial position, i.e., actuator pistons are inside the actuator housing. (Note that the figure shows a west-facing panel and the actuator piston is entirely outside the housing.) As the sun's movement starts, the actuator receives signals from the controller, and the motor starts rotating piecewise, which continues until sunset on the western side. At this time, the panel faces the west direction. At night, the solar panel moves back to the initial position through the linear actuator as the motor rotates anti-clockwise and the piston goes inside the housing. The sun changes its azimuth angle position as seasons change, and the panel cannot change the tilted position [12]. Hence, the panel cannot receive perpendicular sun rays, thereby decreasing the yield. When the PV panel is in the vertical position, the panel captures only a few rays and is therefore never used. When the PV panel is horizontal, the panel captures more rays. When moving along with the sun's seasonal movement (i.e., tilt), the panel captures maximum sun rays, increasing the yield.

Moreover, single-axis tracking is achievable through rotary actuators, as shown in Figure 2 which provide roll movement achieved by electric motor gear arrangement. The rotary actuator has a speed reduction gearbox, which decreases speed and increases torque. Although rotary actuators are nothing but motors with a mechanical gear or a rack and pinion, they change the direction of motion with the desired torque. Interaction between the panel and actuator happens through a quarter-shaped gear disc attached to the solar panel base to achieve roll movement of the panel smoothly. This method is quite uneconomical for individual tracking due to the ample torque of rotary actuators. In contrast, such rotary actuators are suitable for simultaneous tracking where ample torque is required [46]. For receiving maximum sun rays, the solar panel should achieve both day-to-day and seasonal movements. So, we need dual-axis tracking achievable using two linear actuators named dual-axis tracking with linear actuators.

Three positions of the solar panel is shown in the dual-axis tracking of Figure 2, which pertain to the time of the day between 9:00 A.M. and 15:00 P.M. [27]. At 9:00 A.M., Actuator 1's initial piston position is inside the housing, as shown in Part (a) in dual-axis tracking, with the panel oriented towards the morning sun. At noon, as shown in Part (b), Actuator 1's piston position is partially outside of the cylindrical block panel oriented towards the afternoon sun. At 3.00 P.M., as shown in Part (c), Actuator 1's piston position is outside the housing such that the panel is oriented directly under the mid-afternoon sun.

Actuator 2 helps account for seasonal changes. In a solar farm, individual panel tracking is non-economical. All solar panels in a particular row need to move simultaneously. Figure 2 also shows simultaneous tracking with rotary actuators. Only east-to-west tracking is possible with simultaneous tracking. Two actuators, used in a model by Hammed et al., is also called a two-axis tracker, where two independent motors (or actuators) ensure motion along an axis: one accounting for the azimuth angle and another for the altitude [47]. Jovanovic et al. use a linear actuator [48] to obtain the east-to-west motion of the solar panel. Weight reduction and cost-effectiveness are the primary considerations in this case and is accomplished by choosing a single linear actuator instead of the usual rotational actuator. Power consumption in the actuator depends on the anchor location. For example, for a moving panel of +45°, more consumption is noted than −45° rotation due to wind pressure during positive angle tracking. The 'tilt' and 'roll' (shown in Part (b) of dual-axis tracking in Figure 2) tracker employs a linear actuator and dampers [12]. Linear movement with a bar mechanism in the 'roll' axis finds use in the concentration of photovoltaic technologies, and two different heliostat-based tracking methods are discussed [49].

The tracking actuator operates in steps in intermittent sun-tracking systems, so the actuator power consumption is more. Ferdaus et al. describe an energy-efficient solar tracking system using a hybrid (continuous as well as static) dual-axis using two fully geared stepper motors for two different axes: 0.6 Watt power consumption in an actuator and a 25.62% energy gain recorded as compared to a static solar panel system [50]. In addition, Batayneh et al. [51] apply three-step intermittent tracking, with 0.4 Watt power consumption in actuator and an around 91–94% energy increment compared to continuous tracking. As we increase the intermittent step by optimization method, there is a saving in actuator power consumption, and almost the same gain is obtained compared to continuous tracking. Still, as per [52], the simulation results at one stage step by step tracking consume considerable power in the actuator because of the overshooting occurring when the motor is turned on/off. The energy consumption for achieving step-by-step motion laws is slightly higher than that for continuous motion. Table 1 shows results that after the fourth step, the actuator consumption is more than the continuous system, and the energy gain is also less.

Table 1. Energy consumption in continuous and step-by-step tracking [52].

Number of Steps	One-Day Energy Production (Wh)	One-Day Energy Consumption in Actuator (Wh)	One-Day Energy Gain (%)
0 (i.e., continuous)	898.46	12.82	39.42
8	893.27	13.32	38.21
6	891.73	12.99	38.02
4	887.32	12.92	37.34
2	868.63	12.88	34.41

Best performance through solar tracking is achievable when the tracker's tilt angle synchronizes with the annual changes of the sun's altitude (i.e., the sun's seasonal motion). However, a single-axis tracker with an electric linear actuator sees higher energy consumption on sunny days and lower at smoggy times, that is, depending on the local weather conditions [53]. Elsewhere, Ponniran et al. evaluate a single-axis solar-tracking system's performance for three different rotary actuator motors used individually (AC induction motor, DC motor, and stepper motor) [46] in comparison with linear actuators. Finally, in terms of economic considerations, Argeseanu et al. evaluate the cost factor in tracking systems by designing a low-cost dual-axis solar tracking structure with motor [54]. Sensor-based tracking is more accurate than azimuth-based tracking [55].

There is a significant benefit of dual-axis tracking [56]. Without tracking, the yield graph looks like a half-sine wave with the peak at noon. With dual-axis tracking, the yield is a trapezoidal shape of a larger area, but there is a need for two actuators, both in the case of linear and rotary actuators. Such actuators with dual-axis tracking are effective in solar farms with several rows of panels placed. The number of actuators required is less in such cases due to simultaneous tracking, causing a significant uptick in power generation. However, such actuators are not suited in the solar tree concept where space is an issue. Two actuators for each panel in dual-axis tracking are needed, group tracking is not feasible, and the number of actuators required is more like a solar street light system.

The advantage of a solar tracking system with one actuator is initial cost reduction, but it decreases yield. In contrast, solar tracking with two actuators increases yield but also the initial cost of the system. To only seek the advantages of both, a new type of actuator called a spherical actuator [26] may be an ideal replacement of the linear actuators in applications such as a solar tree, such that it accomplishes the task of the existing dual-axis tracking with a single spherical actuator. Furthermore, two degrees of freedom (DOF) spherical actuators shown in Figure 3d with mechanical latch can perform tracking on both axes for increased energy extraction and reduced cost. In addition, due to the latching arrangement, the energy is required for the actuator only when motion is required; hence, it can be more energy-efficient than two-axis tracking. The most significant advantage of the spherical actuator is its compactness; it can be used in solar tree application as shown in Figure 3e and street light applications as shown in Figure 3c.

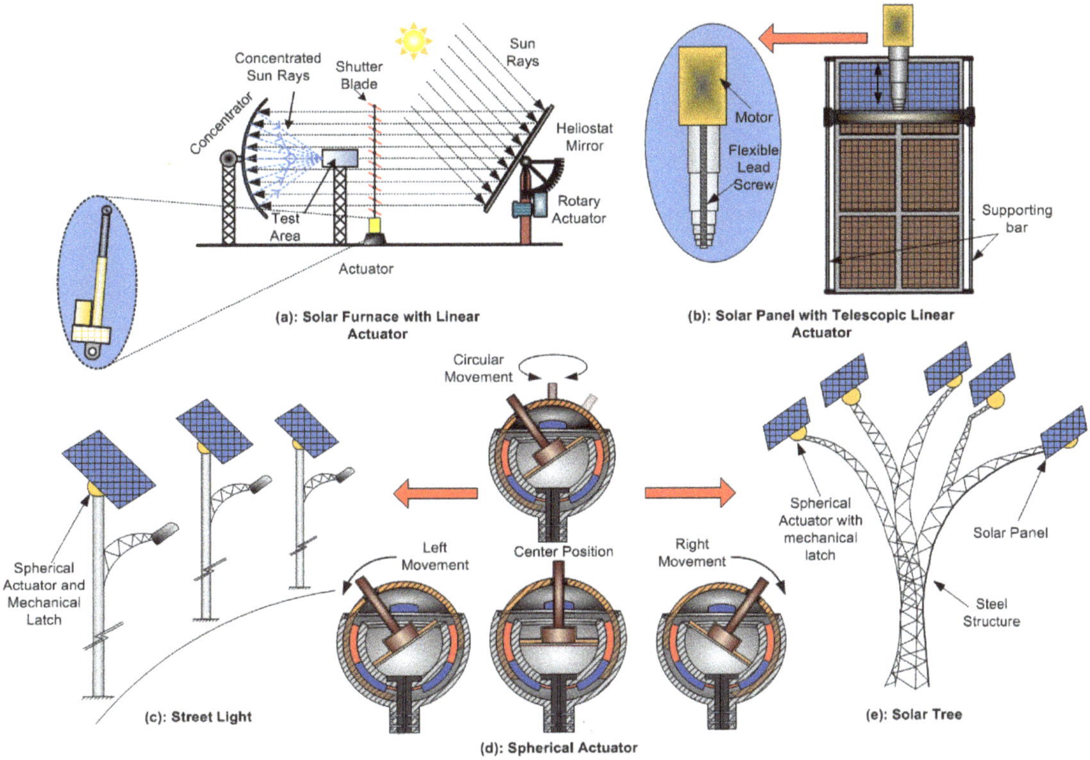

Figure 3. Proposed alternative actuators view: (**a**) solar furnace with linear actuator, (**b**) solar panel with telescopic linear actuator, (**c**) street light application with spherical actuator, (**d**) 2 DOF spherical actuator, (**e**) solar tree application with spherical actuator.

2.2. Actuators for Solar Furnace Applications

A solar furnace is a renewable (green) source of energy. It is a thermodynamic structure [13] that provides concentrated solar radiation, as shown in Figure 4a, using concentrated solar power to produce very clean and high-temperature heat, around 3500 °C (6330 °F) [25].

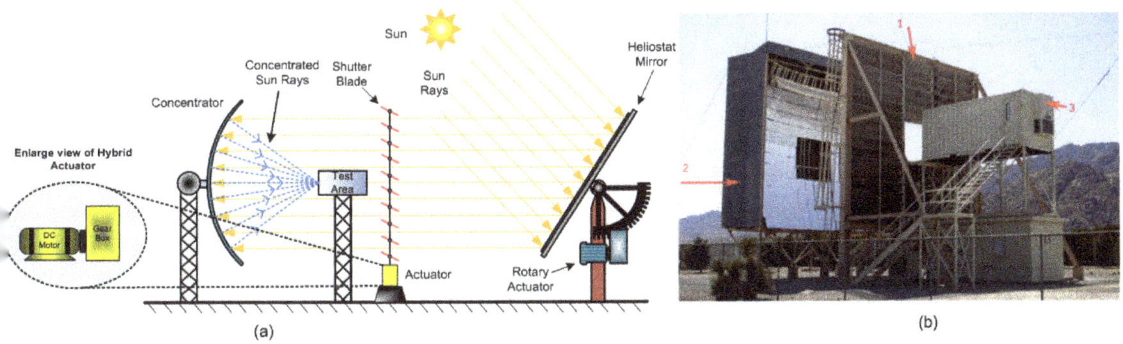

Figure 4. Rotary actuators functioning in solar furnace application (**a**) Schematic Diagram, (**b**) U.S. Army's white sands solar furnace commissioned in 1972 [57,58].

A solar thermal power generation plant commonly employs a solar furnace system different from a solar energy system because it is an indirect use of sunlight reflected and then concentrated on one point of interest. The generated heat is mainly valuable for industrial applications, the heating process of a substance, material testing, charging experiments, melting steel, hydrogen fuel, and foundry applications. The primary solar furnace structure with components such as heliostat mirrors (Not visible in Figure 4b, but shown in Figure 4a), a parabolic concentrator (Number 2, shown in Figure 4b), and mechanical shutters (Number 1, shown in Figure 4b) operated through an electric actuator is standard for all applications. A flat heliostat reflects the solar radiation coming from the sun towards parabolic concentrator mirrors. Rotary actuators or linear actuators placed at the heliostat mirrors track the sun's movement to reflect maximum sunlight, passing through rows of horizontal mounted mechanical shutters whose rotary motion controls and regulates the reflected sun rays as per the heat requirement [59]. The general opening and closing range of shutters are 0–90°, where 0° means shutters are fully open to allow passage of maximum sun rays coming from flat heliostat, and at 90°, the shutters fully close to prevent means sun rays from passing to the concentrator. Rotary and linear actuators achieve such type of motion, but primarily, rotary actuators find preference due to large torque requirements. Generally, AC or DC motors and gearboxes help change the position of the shutters [13]. Finally, the sun rays reaching the parabolic mirrors are concentrated onto a test area (Number 3, shown in Figure 4b) for testing electrical, mechanical, and metallurgical materials under high-temperature heat. Specifically, molten salt is pumped into the furnace test area (also called the tower) in a solar thermal power plant. The molten salt temperature rises from around 300 °C to over 600 °C. Then, high-temperature molten salt is transported to a storage tank and is used for steam generation and eventually electric power generation.

The disadvantage of a rotary actuator in moving shutters is that there will be no quick response of opening and closing shutters due to the backlash nonlinearity of the gearing box. In addition, there is the additional disadvantage of increased space requirement. Instead of a rotary actuator, a low-power linear actuator will be advantageous, as shown in Figure 3a, because there will not be an active load on the shutters, and so the linear actuators perform the shutter's task with low energy consumption and also occupy less space. Adjustment of shutter angle with a rotary actuator enables control of solar flux by controlling the speed of shutter opening, and closing [60]. An aluminum shutter blade operated by a closed-loop serve mechanism helps control sun rays passing from solar mirror to concentrator. Costa et al. also use a solar furnace [13] wherein brushless DC motor and gear mechanism, i.e., rotary actuator, controls the shutter blade movement. Shutters control solar radiation, which helps reduce the concentration so that overheating of the solar furnace is prevented.

For instance, a 1 MW large solar furnace that has been operational since 1968 in Font Romeu Odeillo [43] has seen an upgrade from hydraulic actuators to electric brushless DC motors for efficient operation. These revolutions in the application of the actuator and their power consumption are shown in Table 2.

Table 2. Revolution in application of solar furnace actuator installed at Font Romeu Odeillo.

Generation	Year	Actuator	Power Consumption in Actuator	Maintenance Frequency	Degree of Cleanliness
First generation	1968–1994	Hydraulic jacks	No available data	High (due to hydraulic leaks)	Very less
Second generation	1992–2014	Stepper motors with ball screw jacks	Up to 32 kWe	Moderate	Good
Third generation	2014 onwards	BLDC with ball screw jacks	Up to 6 kWe	Low	Good

The first generation saw the use of hydraulic actuators, but hydraulic leaks and actuator pressure losses harm the effectiveness of solar tracking and cause heliostat oscillations causing rotation axis wear. Therefore, an electric actuator is used in the second generation due to efficient and reliable tracking control. Stepper motor and BLDC motor with ball screw jack obtained linear motion. BLDC has less power consumption than the stepper motor and requires less maintenance.

2.3. Actuators Used for Solar Heating Applications

In the crisis of 1973, oil prices rose dramatically, resulting in the search for sustainable solar heat energy sources for electricity or industrial purposes [14]. Two types of solar heat energy collector mechanisms: (i) a receiver pipe type (active area) and (ii) parabolic trough-based distributed collector system [61], are explored. It is helpful in steam generation for thermal plants, industrial processes, desalination plants, etc. An example of such a plant is the ACUREX parabolic-trough distributed solar field collector located in the Tabernas Desert (Southern Spain), as shown in Figure 5b and their schematic in (a). Trough collector systems comprises a parabolic mirror used to focus sunlight on the active area of the pipe for the oil to absorb thermal energy along the focal line. The system also comprises a heat transfer fluid (HTF) (typically synthetic oil or water), an oil storage tank, a heat exchanger, a electric control valve actuator, an electric pump, etc.

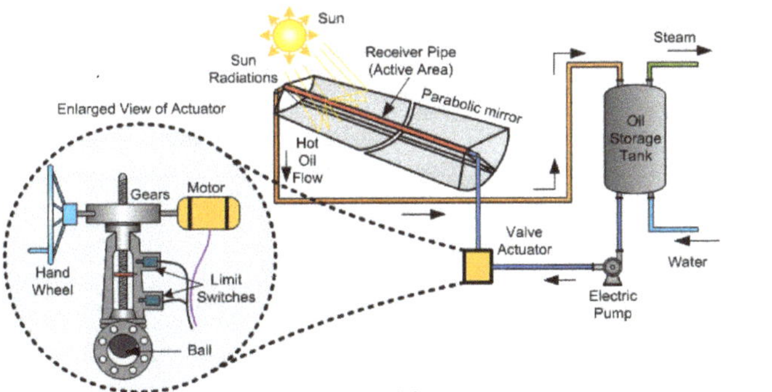

Figure 5. Valve actuator for distributed solar collector field: (**a**) Schematic of plant, (**b**) the ACUREX parabolic-trough-distributed solar field collector [62,63].

The oil pumped from a storage tank (see at the right in Figure 5) passes (at ambient oil temperature) through the electric valve actuator to the field to be heated by concentrated solar radiation and returns to the tank via injection at the top of the storage tank. Once the heat energy is extracted from hot oil, the cold oil is re-injected at the bottom of the storage tank [64]. Suitably adjusting the oil flow rate helps to regulate the outlet oil flow temperature and track a desired temperature reference value using an electric valve actuator (along with the control system) that plays a vital role in temperature maintenance [61]. Electric valves comprise an electric motor, gear mechanism, hand wheel, limit switch, and butterfly valve working at three positions: two end positions and an intermediate one. One of the extreme positions is perpendicular to the oil flow, i.e., valve closed position, the other is parallel to oil flow, i.e., fully open position, and the third (intermediate) position where continuously and smoothly rotates the valve to regulate the oil flow at any flow rate.

To understand the operation of the valve, consider an initial closed position. When the motor starts, it rotates the gear mechanism and changes the direction of rotation. The gear mechanism rotates the vertical shaft at a slow rotation, causing the torque to increase and speed to reduce. The butterfly valve attached to the vertical shaft changes position according to the direction of motor rotation. The oil flow rate is proportional to the valve position. Two limit switches at the upper and lower end operate through a small lever with a vertical shaft for valve automation and protection. When the motor and vertical shaft rotation is clockwise, the valve changes position from closed to open. When fully open, the upper limit switch operates through the lever, then the motor stops rotating. In the anti-clockwise direction of rotation, the valve changes position from open to close. When fully closed, the lower limit switch operates such that the motor stops rotating.

An electric valve actuator is sound in operational control and helpful in protecting from stagnation and overheating. Frank et al. [65] explain the problem associated with the industrial solar heat processing system and different measures to prevent them, stating that weekend shutdowns or company holidays raise the stagnation problem and technical malfunction or a power outage causes overheating. Both problems increase pressure, resulting in the opening of the safety valve. Partial emptying heat transfer fluid (liquid or vapor), with loss of heat, cause additional costs and maintenance. Some passive and active strategies help control stagnation and expansion. In the case of active control strategies, pump-electric valve actuators and appropriate rotary or linear actuators are in use. For handling the stagnation pump-valve system, evacuated tube collectors, and overheating prevention active cooler loop, automatic defocusing of the mirrors is useful. Motlagh et al. propose that pneumatic actuators are helpful to control valves in the energy sector, such as in power plants and the oil and gas industries [66]. The use of electric control-valve actuator technology allows for energy efficiency. Linear actuators control passive structures such as shutters in solar air heaters applications, which are helpful for regulated room heating and reduced electricity bills [67].

2.4. Actuators Used for Solar Panel Cleaning Applications

The performance of solar panel power generation depends on the quantum of sunlight. However, bird poop, dirt (or dust), or pollen on the solar panels prevents light from entering the solar cells, thereby affecting the efficiency of the solar cell, leading to reduced energy production. Maximum solar energy extraction is possible with regular maintenance and cleaning of the solar panels. Using a good cleaning technique recovers up to 95% of the solar cells' power-generating capacity [15]. This section discusses the actuators used for effective cleaning methods specifically include electrical actuators such as electro-mechanical actuators (Figure 6), piezoceramic actuators (Figure 7a), piezoelectric buzzer (Figure 7a), piezoelectric actuator (Figure 7b), and linear-piezoelectric actuator (Figure 8) mechanism.

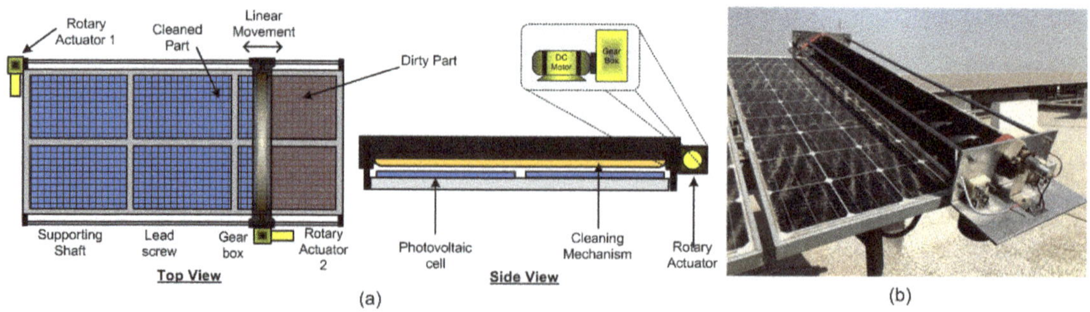

Figure 6. Actuators for solar panel cleaning: (**a**) Hybrid actuators functioning in solar cleaning application, (**b**) prototype model of wheeled automatic cleaning robot [24].

Deb et al. demonstrate a water-free automated cleaning service unit, comprising two DC geared motors, a lead screw, supporting shaft, and timing belt drive mechanism, and the cleaning task is completed using a rolling soft cleaning brush as shown in Figure 6a [24]. These are hybrid actuators that provide rotary-to-linear motion through suitable mechanical transmission arrangements. However, the problem with such mechanisms is that the movable cleaner can become stuck at one side due to there only being one side drive, and so there may be chances of interrupted motion. Moreover, due to the nonlinearity present in the mechanical transmission unit, the performance of the moving system reduces with time.

The solution to this issue is that instead of two rotary actuators, one may use only one actuator called a telescopic actuator, as shown in Figure 3b to push and pull at the center of the cleaner for smooth and symmetric motion. This reduction in an actuator also reduces system nonlinearities, making the system less energy-demanding. Furthermore, such a structure is helpful for solar farms due to its autonomous working. Wheeled semi-automatic cleaning robots having two geared motor actuators connected to two ends of the robot also solve the motion interruption problem [24]. Panat et al. [68] shows the electrostatic dust-removal system with a linear guide slide stage actuator motorized by a stepper motor. Williams et al. [15] present an array of nine piezoceramic actuators attached to the backside of the solar panel surface to handle the dust falling on PV panels by mechanical vibrations (acceleration). Nine actuators are excited by three three-channel piezo amplifiers. Sine sweep excitation of different magnitudes and frequencies (especially significantly higher frequency from 400 Hz to 5000 Hz) shows that high acceleration, i.e., the best dust cleaning motion, efficiently improves the dust-removal process. Such cleaning methods are effective in Martian surface missions or lunar surface environments. Dawson et al. [69] state that instead of piezoelectric actuators, inexpensive and low-power piezoelectric buzzers are helpful for panel cleaning. Incorporate piezoelectric vibrational actuators into the structural supports of photovoltaic panels to occasionally induce vibrations to loosen accumulated dust and are mounted at nodes of a grid of spars, as shown in Figure 7a. Despite being compact, low-power consuming, and inexpensive, the commercially available buzzers, which typically have operating frequencies in kHz, need a redesign for lower photovoltaic panels with lower natural vibrational frequencies.

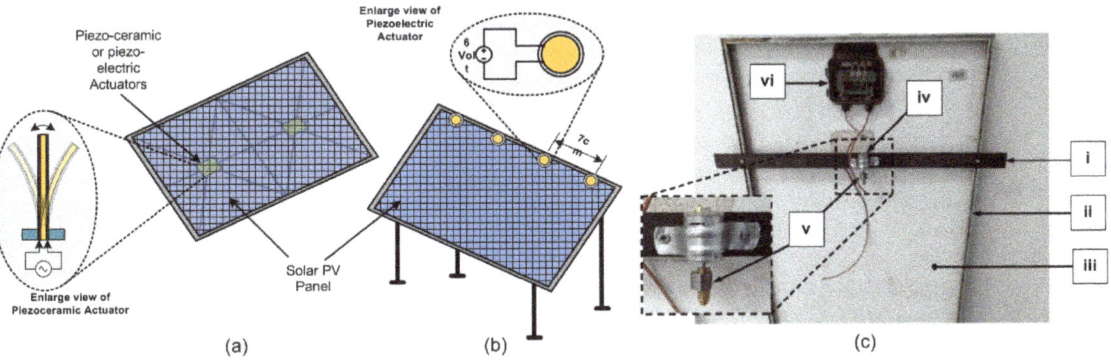

Figure 7. Vibration-based solar panel cleaning applications: (**a**) Piezo-electric actuator located on the back side of the panel; (**b**) Piezo-electric actuator located on the surface of the panel; (**c**) Eccentric Rotating Mass (ERM) motor actuator located on the back side of the solar panel, where (i) iron base for motor, (ii) solar panel frame, (iii) back side of the panel, (iv) ERM motor, (v) ERM, and (vi) connection box [70].

Alagoz et al. [16] demonstrate that the acoustic surface wave (SAW) vibrations, gravitational forces, and panel surface slope play a significant influence on soiling mitigation. Figure 7b shows four piezoelectric actuators attached at the top surface of the panel driven by a pulse generator, generating pulse-type vibrations. Piezoelectric actuator spaced 7 cm apart and working at the resonant frequency. The traveling pulse strikes the surface dust particles, causing acceleration, leading to dust migration from the panel surface. Due to gravity, experiments have shown that bigger particles can easily slide down and away from the panel surface. In addition, a piezoelectric-actuator-based panel cleaning wiper arrangement are used [71,72]. The three actuators, viz., a piezoceramic actuator, a piezoelectric actuator, and a piezoelectric buzzer, are lightweight, so there is no adverse effect on the tracking system. Other important benefits of such actuators are its high torque-to-volume ratio and there being no electromagnetic interference in their operation. However, due to significant vibrations, the solar cell may deteriorate, which poses a research problem of optimization of the location of the piezoelectric actuators. Al et al. [70] generate vibration with motor and Eccentric Rotating Mass (ERM). Such wobbling motion will help to migrate dust particles down as the panel is inclined, as shown in Figure 7c. Lu et al. [41] proposed a prototype of a linear-piezoelectric actuator with an elliptical motion of their feet shown in Figure 8a. It consists of two vibrator plates of piezoelectric material PZT-8H, each plate of size 10 mm × 10 mm × 1 mm. The electrodes and piezoelectric (plates stack not shown here) are bunched between the driving bar and the support as shown in Figure 8b. The application of a sinusoidal signal causes an elliptical motion as well as a linear motion for both the driving feet shown in Figure 8c.

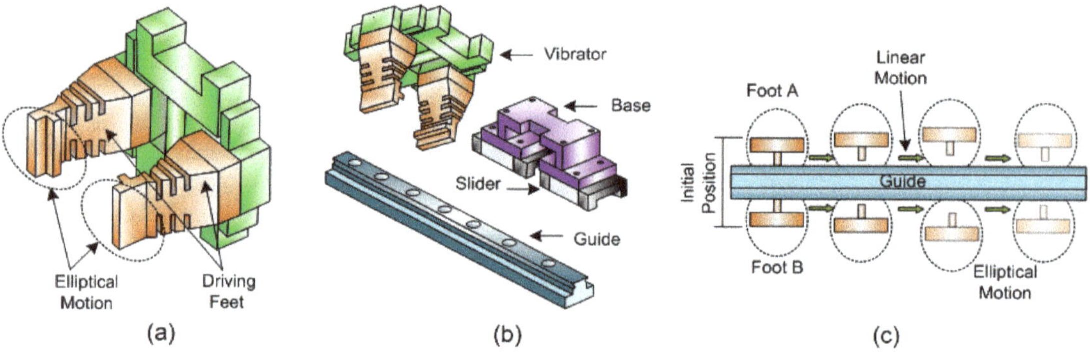

Figure 8. Linear piezo-actuators functioning in solar cleaning applications; (**a**) elliptical motion of their feet, (**b**) the driving bar and the support, and (**c**) the elliptical motion and linear motion for both the driving feet [41].

The entire excitation structure being symmetrical, the actuator moves linearly along the guide fixed over the frame to cover the whole panel area with the attached wiper actuator's base to be driven to aid in the cleaning operation. The advantages of such actuators are five times higher torque-to-volume ratio than electromechanical actuators, smooth operation, high holding position, and lightweight and flexible structure. Examination reveals that energy gain at effective panel width is more significant because $L = 260$ mm > 130 mm. That is, wiping a solar panel in the direction with a shorter wiping distance (or stroke) results in a better energy gain for a given active area. In addition, the speed of cleaning depends on the way of cleaning path. Panel cleaning is also performed with blowing technology. King et al. [73] demonstrated compressed air cleaning with a motor and compressor set. Synthetic jet actuators known for injecting air jet at low power consumption can replace said blowing method for solar panel cleaning application [74,75].

Piezoelectric material creates vibration and also generates electrical energy against vibrations. Hamlehdar et al. [76] review energy harvesting methods by vibration-based techniques and focus on piezoelectric devices in energy harvesting from fluid flow motion. Piezoelectric actuators play a significant role in converting ambient energy into electricity for small and remote devices. In the wind-flow technique, energy harnessing generates electricity through piezoelectric methods such as flutter-induced vibrations (FIV), vortex-induced vibrations (VIV), and galloping. In the water flow technique, a cantilevered transducer having a piezoelectric actuator experiences the vibration due to the converging–diverging channel of water flow. The natural potential of fluid vibration available in ocean waves provides energy density 4–30 times higher than wind energy. This technique uses a piezoelectric beam coupled to the floated buoy to generate electrical energy against wave vibration. Among the various energy-harvesting methods, such as electrostatic, electromagnetic, piezoelectric, and hybrid methods, the vibration-based method has several advantages over other solutions, including high-density power output, high output voltage, and intrinsic cross-conversion capabilities, etc. Han et al. [77] measured the generated power by a PZT-based piezoelectric ceramic actuator, focusing on the deformation that affects power generation. Deformation applied in the size of the piezoelectric ceramic, depth of compression, and speed of compression contribute to the deformation of a single PZT-based piezoelectric element. Nechibvute et al. [78] present a concise review of power generation by piezoelectric microgenerators and nanogenerators for wireless devices such as sensors, microelectromechanical systems, etc. Battery-operated wireless sensor network performance depends on batteries' capacity and needs replacing when discharged. Hazardous sensing environments do not allow for frequent battery replacement, so in such situations, piezoelectric-based continuous generation makes the sensing device autonomous. In addition, such actuators play an essential role in low-power remote sensing devices for optimizing the structure's weight. The actuator's power consumption and net improvement in system power is analyzed and summarized in Table 3. Actuators play a significant role in solar tracking and cleaning, and the efficient use of electric actuators improves system efficiency.

Table 3. Summary of actuator power consumption in different applications.

Application	Axis/Tracking Nature	Actuator Type	Location	Generation Capacity	Actuator Power Consumption	Net Power Improvement	Simulation/ Experiment	Feature
Solar tracking	Single continuous	Linear [53]	Central Vietnam	250 W	2–8%	up to 30.3% (sunny), up to 15.2% (other)	Experiment	Grid connected solar plant
	Single continuous	Linear [12]	Hermosillo	No available data	No available data	No available data	Simulation	Cheaper solar array tracking
	Single continuous	Linear [48]	NA	NA	for +45° more than −45°	NA	Simulation	Optimum actuator location
	Single continuous	Rotary [46]	Zimbabwe	24 W	No available data	up to 25%	Experiment	Affordable tracker
	Dual continuous	Linear [79]	Culiacán, Sinaloa	135.2 W	6.71 W	No available data	Experiment	Maintain power level
	Single, Intermittent (3-step)	Linear [51]	Irbid	0.4 W	Very less	around 91–94%	Experiment	Optimized 3 steps
	Dual, Intermittent	Rotary [50]	Bangladesh	NA	0.6 W	Up to 25.62% (compared to static)	Experiment	Low consumption in actuator
Solar panel cleaning	No tracking	linear piezo-electric [41]	No available data	NA	6.8 W	Increased by 1.29 W	Laboratory testing	Energy gain improvement
	No tracking	Piezoelectric [15]	California	No available data	No available data	up to 95%	Experiment	Creates vibration
	No tracking	Linear [80]	India	20 W	No available data	upto 3.35 w	Experiment	For street lights
	No tracking	Wobbling [70]	Egypt	130 W	No available data	up to 2.912 w after 6 weeks	Experiment	For desert regions
	No tracking	Geared motor [81]	Saudi Arabia	200 W	No available data	up to 4.78%	Experiment	For dry and desert regions
Solar furnace	Single continuous	Stepper motors with ball screw jacks [43]	France	1 MW	Up to 32 kWh	No available data	Experiment	Solar furnace with heliostat tracking
	Single continuous	BLDC with ball screw jack [43]	France	1 MW	Up to 6 kWh	No available data	Experiment	Solar furnace with heliostat tracking
Solar heating	NA	Rotary valve [14]	Portugal	NA	No available data	From 250 °C to 260 °C	Experiment	Only controlling flow
	NA	Rotary valve [63]	Spain	NA	No available data	From 190 °C to 245 °C	Simulation	Flow control

NA: Not applicable.

3. Actuators in Wind Applications

A relatively low wind speed (<5 m/s) makes the deployment of conventional wind turbines in residential areas non-productive. Therefore, new topologies of generators are needed to operate favorably in such low wind speed regimes for adequate energy capture [17]. Usually, wind-direction-oriented turbines are more efficient vis-à-vis turbines operating in winds from any direction. In addition, such small wind turbines favorably orient with a rear rudder as an effective brake for damage mitigation at high speeds. In addition, two actuators protect from higher speeds beyond a threshold voltage via unipolar control—the first is a limited-angle torque (LAT) actuator (called ON-OFF bipolar actuator) used on a 2.5 kW wind turbine with a nominal wind speed of 500 RPM (shown in Figure 9). The second is a stepper motor actuator with a reduction gear (i.e., rotary actuator) [82]. The actuator comprises a permanent magnet rotor mounted on two bearings—the rotor movement is limited by two dead-end pins mounted on one of the steel plates. Stator poles (with two windings connected in series) mounted on supports of the same steel type fixed to the plates' ends together form a rigid assembly.

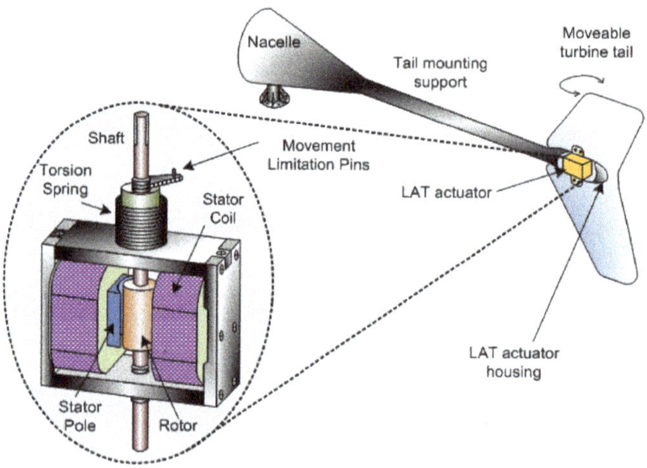

Figure 9. Limited-angle torque actuator (LAT) [42].

The nominal coil operating voltage is $V_{nom} = 200$ V, and the threshold voltage is $V_{min} = 190$ V and $V_{max} = 210$ V. The comparator circuit measures and compares the input voltage to preset values, e.g., $V_{min} = 190$ V and $V_{max} = 210$ V. Because the actuator does not take action between these voltage levels, the system restricts the LAT oscillation at high speed using two comparator values. On reaching the threshold voltage (high generator speed), the rotor moves between two stationary locations, furling the wind turbine. A circular torsion spring placed on the rotor shaft causes the return of the rotor to the initial position after the wind turbine deceleration. The nacelle and tail align with the wind direction for wind speeds below a certain level. As shown in Figure 9, the LAT actuator mounted on a tail is activated when the voltage exceeds the threshold, causing the tail to rotate perpendicular to the rest of the wind turbine [42]. With the nacelle no longer aligned in the wind direction, the blades are not driven efficiently, ensuring speed reduction. Upon the restoration of regular wind speeds, when the windings receive a reverse polarity voltage, the actuator rotor returns from 90° to 0°. The detection levels are adjusted to accommodate different applications and dynamics. The angular displacement solenoid actuators [83] using springs provide torque at a specific desired position despite no electrical current passing through the motor. A solenoid actuator is used as an ON–OFF bipolar actuator in place of LAT. A similar controlling operation has been done with the stepper motor-based actuator. Instead of two coils as LAT, two different comparison and control circuits and one angular moment element play an important role. Table 4 shows a comparison of actuators based on selected parameters. Such actuators are restricted to the system within rated power.

Table 4. Comparison of the wind turbine actuators.

Parameters	LAT Actuator [42]	Stepper-Motor-Based Actuator [82]
Power consumption	high	low
Complexity	low	medium
Weight	high	low
Reliability	high	low
Electronic control complexity	low	medium
Energy harvested at high speed	medium	high

Wave power is a predictable renewable energy source that can create more electricity than wind and solar combined. Wave energy absorption from ocean waves varying in time scales and produced by wind action indirectly from solar energy [1] is a hydrodynamic process resulting in radiation wave phenomena. However, the conversion of wave kinetic energy into smooth electric energy is difficult due to the low frequency of wave motion (~0.1 Hz), random wave, and high-force oscillations [84]. A "smart" oscillating-body device, also called a heaving buoy, that travels freely up and down with the rise and fall of the waves converts wave motion into linear motion for the power-generation system [18]. Linear motion is created by reacting an oscillating buoy against a fixed reference frame called deeply submerged reaction mass, as shown in Figure 10, and it is stationary due to its heavy mass.

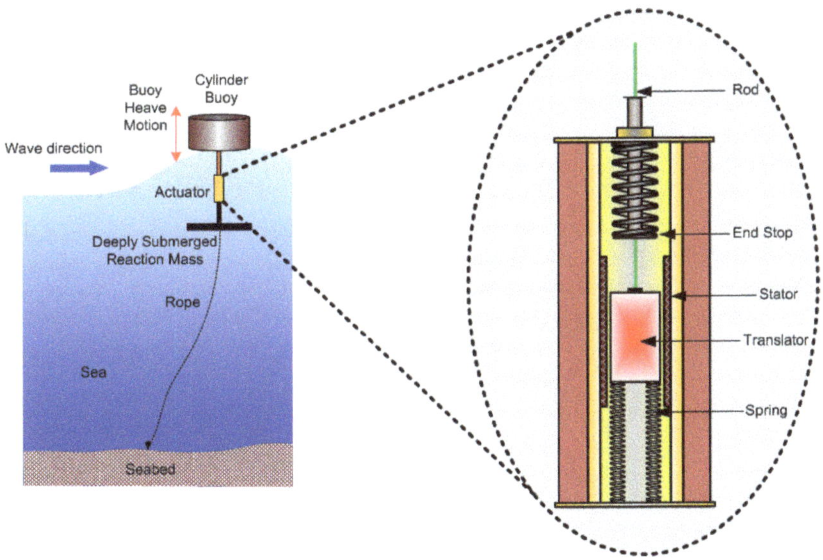

Figure 10. Cylindrical buoy actuator for wave applications.

The reaction mass connected to the seabed through the rope ensures a fixed floated position in the sea. A linear actuator placed between the floated buoy and submerged reaction mass transfers the relative motion of these two through a rod connected between translator and floated buoy. A device called power takeoff (PTO) helps the rod's linear motion (up and down) to convert into usable power [19]. A piston pump device (such as a hydraulic drive or mechanical drive), air or water turbines, rotating electrical generator, or linear electrical generator can serve as a power takeoff device (PTO) [2]. The piston pump supplies high-pressure water to a hydraulic turbine [1], where linear motion converts into rotating motion through water turbines placed onshore. When a rotating electrical generator is available as a PTO, the actuator first converts the linear motion into rotational motion so that the rotating generator generates electricity. For offshore systems, the rotating generator is in the PTO itself. Figure 10 shows the enlarged view of the cross-section of a linear generator that generates electrical power through linear actuation. It has two dead-end stop springs and stator electromagnetic coils placed in the peripheral of the cylindrical body of the actuator. A translator with a solid permanent magnet providing magnetic flux inside the generator can freely move if vertically placed at the center. The relative motion of the buoy and the reaction mass is directly transferred through the rod to the translator for electricity generation through linear motion and the principle of electromagnetic induction.

Actuators used in wave energy conversion (WEC) systems are categorized into oscillating water columns (OWCs) and oscillating body systems or wave-activated body systems. Figure 10 shows the second category along with a linear PTO mechanism. This type of motion conversion system (actuator) and PTO device is essential in the perspective of wave power conversion efficiency. Aderinto et al. [85] explained detailed PTO types, scales, and efficiencies, among which were the efficiencies of five different combinations of actuators and PTO devices, and the results of their numerical simulation with computational fluid dynamics (CFD) are shown in Table 5. The direct-drive linear generator system with an oscillating buoy shows 95% efficiency, which is the highest compared to others. Except for linear generators, all other PTOs have losses in their motion conversion process, but in linear generators, no motion conversion is required and, therefore, comes with higher efficiency.

Table 5. Efficiencies of different PTO.

Actuator and PTO Combination Type	Efficiency (%)
Hydraulic drive with oscillating body	65 [85]
Water turbine with oscillating body	85 [85]
Air turbine with OWCs	55 [85]
Mechanical drive with oscillating body	90 [85]
Piezoelectric device with oscillating body	very low [86]
Linear generator with oscillating buoy	95 [85], 90.03 [86]
Linear generator with OWCs	83.6 [87]

Rahman et al. [86] reported that PTOs, which are also built of piezoelectric actuators, are used to produce a significant quantity of electrical power from the ocean. Therefore, energy converters can be used to power applications that function in the ocean, such as various electrical sensing elements, floating harbors, robotics, etc. Wave energy converters (WECs) can improve their energy conversion efficiency by including several types of piezoelectric materials. Experiments reveal that by straining the material with a pressure of 1.196 kPa at a frequency of 20 Hz, 0.2 W electric power with 2.2 V may be obtained using this technology.

Next, we discuss actuators used in bio energy-related systems such as agriculture, solar irrigation, biogas, biomass, etc.

4. Actuators in Bioenergy Applications

Huo et al. present a distributed control system and communication topology for safe and automatic (start/stop) regulation of biogas driven appliances such as a rice cooker, stove, and water heater, etc., configured with a ZigBee endpoint to control the relevant actuators and transmit the data [20]. For example, a biogas rice cooker typically consists of cooking and warming switches for a three-stage operation: (1) heating with both switches pressed; (2) warm stage with cook switch lifted through a step-motor if cooking is over; and (3) the turn-off stage, with both switches lifted. The step-motor rotates at a fixed angle as commanded to press or elevate the switch. A single standard biogas stove with switches and gears controlled by a Knob's rotation angle comes with a step-motor, an ignition circuit, and an electromagnetic clutch to turn the knob on when the rotational power of the step-motor passes through the clutch to the gear and manually when the clutch is out of power. The mechanical structure of the biogas water heater is similar to the rice cooker. Biomass (grate) combustion being a reasonable alternative to fossil fuels, it is expected to be a significant player in a sustainable future [88] and can fire a wide range of fuels of varying moisture contents, requiring less fuel preparation and handling [89].

Agriculture wastes (most common are wood chips and sugarcane bagasse) are the primary fuel source for biomass furnaces. These biomass grates require actuators to operate an intermittent ash removal system. Biomass-fired power plants are cogeneration facilities producing heat for process heating and electricity generation, driven by electric valve actuators for feedwater treatment, steam generation, turbine injection, flue gas cleaning, and process heating. In addition, electromagnetic actuator systems robustly driven by an induction servo motor drive (rotary actuator) are used in agricultural applications [21]. Sontake et al. [90] present a review of solar PV powered water pumping system with a linear actuator as shown in Figure 11a. The system comprises a piston-type pump, a two-phase variable-reluctance linear stepper motor that works as a linear actuator, a reservoir, a pulley, and a counterweight. The enlarged view of the linear actuator provides the linear vertical motion to the piston-type pump. The solar PV array provides power supply to the control unit, converting it to a two-phase supply using converters and fed to the linear stepper motor. Due to the two-phase supply, electromagnetic coils A and B energized alternately to create downward linear motion.

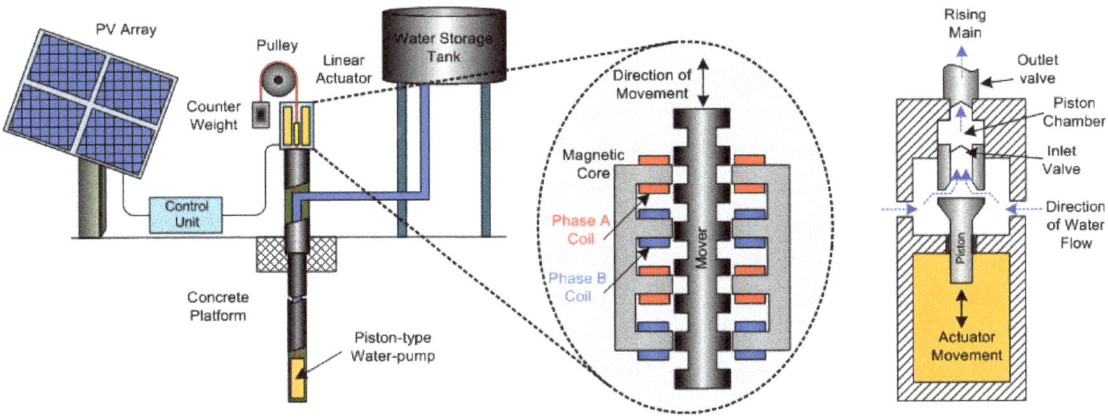

Figure 11. Linear water pump actuators (**a**) variable-reluctance linear stepper motor; (**b**) linear pump [29].

For creating an upward motion, the controller changes the sequence of the excitation supply of the electromagnetic coil. Still, there is a need for a more significant up-stroke to lift water, so a counterweight is helpful. Aliyu et al. [29] present an optimized design for a solar water pumping system, as shown in Figure 11b, which shows a cross-section view of the cylindrical actuator, comprising an electromagnetic coil, a piston, a piston chamber, an inlet valve, an outlet valve, and a piston return spring, which provide restoring force (not shown). The inlet valve is submerged in the water by a water tank. The electromagnetic coil excited by the PV panel supply sets up a solid magnetic attractive force, causing the piston to move upwards. With the coil excitation removed, the piston returns due to the restoring force of a spring. The piston's vertical linear (up–down) movement happens due to the magnetic field and the return-spring. On the upward stroke of the piston, water in the inlet and piston chamber is expelled towards the outlet valve into the rising main. The piston drops downward at the end of the up-stroke, closes the outlet valve, and fills water into the piston chamber via the inlet valve. Due to the non-returning valve arrangements, water flow never returns once it enters the piston chamber, providing an efficient water pumping arrangement.

Talavera et al. discussed agro-industrial and environmental fields where remote actuator devices such as valves, pumps, humidifiers to maintain consistent humidity, relay, and alarms, among others, are in use [22] to optimize the usage of water by scheduling irrigation timings and proportions, fertilizers, and pesticides as per the state of the crop, the growth cycle, and based on available information from weather prediction systems or an on-site wireless network of sensors and actuators over an IoT platform to control the state of the process for irrigation, fertilizers, pesticides, illumination, and access control. Elsewhere, Yoo et al. describe an automated agriculture system based on an interconnected actuator network for monitoring the growth of greenhouses plants, which depend on ambient light, temperature, and humidity, through controlled illumination via a relay actuator [91]. If an agricultural intrusion is detected, the proposed system-generates alarms in the farmer's house [92]. Localized (drip or trickle) irrigation applies water at low pressure (<2 bar) to a small area of the soil surface, sometimes only part of the root zone, and is widely adopted in greenhouse vegetables [93], through hardware components such as pumps, valve actuators, sensors, pipes, tubing, and emitters. A scheme of a localized irrigation system with two-way latch-type solenoid valves with limited stroke length, two ports of water flow (inlet and outlet), and two switching states (open or closed) to start/stop the controlled water flow, which is usually in open condition so that the valve closes on power failure [23,94], is shown in Figure 12.

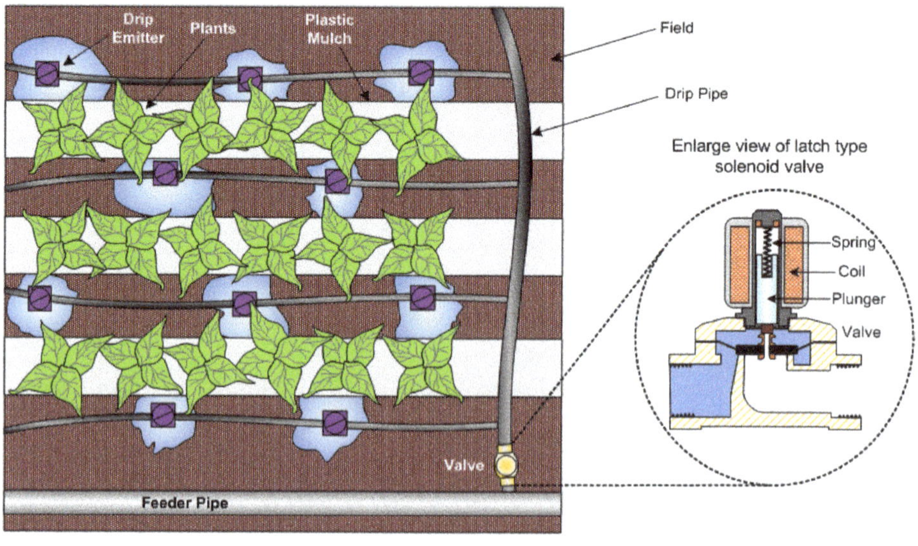

Figure 12. Localized irrigation system and latch-type Solenoids [93].

Generally, latch-type actuators operate on a direct current (DC) supply. The selection criteria for such actuators depend on the pipe connection size, operating voltage, seal material, pressure, temperature, bore size, and more. When the coil receives supply, it becomes energized, and the movable plunger changes its state (ON or OFF).

There are many other renewable energy sources not covered in this manuscript that also deploy electric actuators. For example, the direct utilization of geothermal heat energy extracted in the form of liquid or air using a ground heat pump system and utilized as warm or hot water (in the winter season) and vice versa in the summer season, which is clean and safe, is also helpful for electricity generation and mineral production [95]. The water circulating loops of the ground heat pump system consists of centrifugal pumps and flow controlling actuators [96].

5. Important Performance Parameters of Actuators and Evolution

Some major performance parameters, their specifications, and applications are shown in Table 6, which gives a summary of the actuators' major parameters/specifications. Some of the papers do not mention specifications, so only the parameters are provided. The actuators' drive range determines the size of the feasible solar tracking region [49] (i.e., stroke length). In addition, the maximum rated load, maximum speed, continuous power, and system backlash are all vital performance criteria to consider when selecting linear actuators [97].

Table 6. Summary of Major Performance Parameters, Specifications, and Applications.

Types of Actuator	Major Performance Parameters/Specifications	Applications
Electric rotary actuator (AC motor) [13,60]	Shutter travel time: 0.5 s for opening and closing	Solar furnace shutter operating system
Piezoceramic actuators [15]	Frequency range: 400–5000 Hz	Solar array dust removal system
Electric rotary actuator (DC Motor) [47]	Motor speed and gear ratio	Solar tracker
Piezoelectric actuator [16]	DC voltage: 6 v, panel inclination angle	Solar panel cleaning
Relay [20]	Max. coil voltage, Max contact current capacity	Controlling greenhouse parameters
Electromagnetic clutch [20]	Coil voltage, clutch torque	Bio-gas stove
Induction servo motor (ISM) [21]	Operating voltage, operating speed (in sec/deg), torque (kg/cm or N/m)	Agricultural machines
Latch type solenoids [23]	Operating voltage: 12 V DC, Connection size one-eighth inch, Valve bore: 2.5 mm, Nominal power: 7 VA	Localized irrigation system
AC induction motor [46]	Gear ratio	Biomass grate boiler
Linear actuator [46,48,49,53]	Stroke length: 1005 mm, voltage: 12 V DC, speed: 4 mm/sec, force: 1500 N	Tracking applications
Rotary valve actuator [61]	Range of motion, Valve Stem Stroke Length, Actuation Time, Fail-safe	Solar heating system
Linear piezoelectric actuator [41]	Plates size: 10 mm × 10 mm × 1 mm, piezoelectric material: PZT-8H, operating frequency: 18.8 kHz and operating voltage: 100 V_{o-p}	Solar panel cleaning, astronautic and aeronautic applications
Limited-Angle Torque (LAT) actuator [42]	Threshold coil voltage, rotating angle 0° to 90°	Over-speed protection of wind mill
Solenoid [94]	Stroke length: 10mm, Operating voltage: 24 V DC	For limited stroke on–off position.

A motor, gears, and switches make up an electric rotary actuator. In some applications, the motor speed is an important parameter, e.g., the minimum shutter traveling time is required to operate mechanical shutters of the solar furnace system. There is no significant external load on shutters except for its weight. Some applications require more significant torque than the speed of the operation: in solar tracking applications, the solar panel weight is there; in a biomass grate boiler system, a rotary valve operates against fluid pressure. In such cases, the actuator gear ratio plays a significant role. A gear train is used in electric actuators to increase motor torque and control the actuator's output speed. Therefore, the actuator torque and range of motion should be considered while choosing an electric rotary actuator. In computerized numerical control machines where precise motion control is essential, servo motors are suitable due to low inertia, enabling quick start and stop options. In addition, a servo amplifier can modify torque and speed effortlessly. Figure 5 has already shown a schematic diagram of an electric valve actuator that can control water circulation rate. However, some advantages, disadvantages, and their limitations are given in Table 7.

Overheating protection is available to prevent major breakdowns. Piezoelectric actuators are a new class of actuators for active mechanical system control. Piezoelectric materials have been adapted to applications that require just small quantities of displacement, such as solar panel dust cleaning, despite the magnitudes of piezoelectric voltages, motions, or forces being modest and often requiring amplification. Therefore, the displacement, pressure, resonant frequency, and voltage applied to piezoelectric actuators are vital characteristics. Limited-angle torque actuators are especially used to protect small windmills against over-speeding. They work on a predefined voltage rating of the coil called the threshold actuator coil voltage. Therefore, the voltage and rotation angles are essential.

Table 7. Comparison of electric actuators used in the various renewable applications.

Name of the Actuator	Advantages	Disadvantages	Limitations
Spherical actuator	• Less space required • Low power consumption • Available in desired DOF • Compact	• No counterbalance • Sensitive to vibration	• Suitable for discrete tracking • For moving large weight, latch is required for power saving • Can be used for single panel tracking only
Rotary actuator	• Suitable for group tracking applications	• Backlash nonlinearity effect is more • Quick response cannot obtained due to dead-zone nonlinearity	• Cannot provide precision motion accuracy
Linear actuator	• Backlash nonlinearity effect is less • Compact than rotary actuator	• Large space required in solar tree application	• Space limitation for solar tree applications
Piezo-electric actuator	• Light weight so no adverse effect on solar panel tracking system.	• Surface vibration area is very less • Complex and costly control circuitry	• Need more than one actuator for effective vibration effect
LAT actuator	• Automatic operation based on threshold voltage, no extra control required • Low complex than rotary actuator	• Power consumption is high than rotary actuator	• Applicable only for up to 2.5kW domestic wind turbine speed control application
Solenoid linear valve actuator	• Operate in pressurized fluid flow • Fast response time • Economic feasibility	• Supply voltage deviations affects performance	• Cannot obtain intermediate stage

The electric actuator, initially only a motor with a set of gears and some mechanical torque limit switches enclosed in a box, was later coupled with some auxiliary components such as position feedback switches and over-temperature (thermostat) and over-speed protection, heater, and breather control. After that, electric actuator design improved in various hazardous locations such as oil fields, chemical plants, or gas pipelines, etc. The development of non-intrusive setups combining new and more effective electronic controls with communication protocols was the next step in increasing commissioning capabilities [98]. Wireless actuators are simple to use, implement, and commission these days. They provide both communication and diagnostics while obviating the need for costly wiring and installation [99]. Pneumatic actuators have long been used to control valves in power plants. The use of electric control-valve actuator technology allows for energy efficiency [66]. There are different actuators designed for the power sector, but electric actuators provide solutions for reducing energy waste in renewable energy-producing systems efficiently [100]. Next-generation actuators will develop with new wireless communication technologies such as the Internet of Things (IoT), 5G, and intelligent robots, which are also expected to empower the fourth industrial revolution. For instance, Blanco et al. [101] discusses a wireless actuator network used to create an IoT-based intelligent system, called smart actuators, which accomplishes a reduction in overall energy consumption at a given time by optimizing the functioning of devices and equipment in the IoT [102]. In addition, Llaria et al. [103] apply wireless actuators [104] for intelligent management techniques in a microgrid environment. Junior et al. [105] describe sensors used for management of energy generated by distributed renewable plants.

6. Conclusions and Future Directions

This survey presents a functionality study of significant electric actuators in different renewable applications and the current trends in usage. Traditional nonelectrical actuators are gradually being replaced with electric actuators. Primarily, four significant electric actuators' motion classifications are found in the studied literature: rotary, linear, spherical, and seismic actuators. However, micro-level functionality analysis is necessary to find further energy optimization in renewable applications. Some of the studied conclusive vital points are listed below:

- Use of electric actuators are more advantageous in renewable applications as compared to non electric actuators by their quick response, low wight, precision in control, compact, clean, and more important IoT environment compatible.
- Rotary actuators are suitable wherever accuracy demand is lower but ample torque is needed to drive the load, as preferred in solar group tracking applications and solar furnace heliostat tracking applications. However, electrical cogging and mechanical dead zone nonlinearity increase with time.
- In discrete (step-by-step) tracking, increasing steps are uneconomical due to cumulative intermediate starting current being more than the continuous tracking current.
- In desert regions, net solar power improvement is approximately 3 W to 5 W. Further improvement is not achievable with available actuators; different actuators are required to remove the dust particles efficiently.
- Power consumption in BLDC is low (6 kWe) compared to a stepper motor (32 kWe) in the same capacity tracking application.
- Linear actuated generator with oscillating buoy gives more efficiency than a hydraulic drive, water turbine, or air turbine due to the reduction in power conversion loss in direct actuation.

Additionally, direct linear motion actuators without mechanical transmission elements and iron core free actuators reduce significant problems associated with rotary-to-linear actuators such as cogging, eddy current, heat, magnetic saturation, and lateral forces and hence play an important role nowadays. The spherical actuator is well-suited to solar tree applications due to less space requirement and requires two degrees of freedom. Overall, electric actuators have become more suitable and popular than convectional nonelectrical actuators in renewable applications due to their lower power consumption. Identifying different actuator control problems and challenges of electric actuators such as nonlinearities, saturation problems, and opportunities in optimization and their appropriate control strategies to overcome actuator-related issues for optimal use in renewable applications. The effectiveness of such control methods is part of a possible future work. Major performance parameters are discussed to facilitate actuator selection, which helps find significant parameter-related actuator selection. Next-generation, smart, IoT-based actuators will play an essential role in the energy sector and the chemical and automobile industries due to their two-way communication wireless technologies.

Author Contributions: Conceptualization, A.R. and D.D.; methodology, A.R. and D.D.; software, A.R.; validation, A.R. and D.D.; formal analysis, A.R. and D.D. and S.O.; writing—original draft preparation, A.R. and D.D.; writing—review and editing, D.D. and S.O.; supervision, D.D.; funding acquisition, S.O. All authors have read and agreed to the published version of the manuscript.

Funding: This research was funded by the European Regional Development Fund in the Research Centre of Advanced Mechatronic Systems project, grant number CZ.02.1.01/0.0/0.0/16_019/0000867 within the Operational Programme Research, Development and Education.

Institutional Review Board Statement: Not applicable.

Informed Consent Statement: Not applicable.

Data Availability Statement: Not applicable.

Conflicts of Interest: The authors declared no potential conflict of interest with respect to the research, authorship, and/or publication of this article.

References

1. De Falcão, O.A.F. Wave energy utilization: A review of the technologies. *Renew. Sustain. Energy Rev.* **2010**, *14*, 899–918. [CrossRef
2. Waters, R.; Stålberg, M.; Danielsson, O.; Svensson, O.; Gustafsson, S.; Strömstedt, E.; Eriksson, M.; Sundberg, J.; Leijon, M Experimental results from sea trials of an offshore wave energy system. *Appl. Phys. Lett.* **2007**, *90*, 034105. [CrossRef]
3. Buch, A.; Klöckner, S.; Bernarding, E.; Eid, T. Electrical actuators for modern internal combustion engines. *MTZ Worldw.* **201(** *71*, 26–30. [CrossRef]
4. Hisseine, D. Robust tracking control for a hydraulic actuation system. In Proceedings of the 2005 IEEE Conference on Contro Applications, 2005, CCA 2005, Toronto, ON, Canada, 28–31 August 2005; pp. 422–427. [CrossRef]
5. Prasetya, S.; Adhitya, M.; Budiono, H.; Sumarsono, D. A investigation of braking system actuators for electric shuttle bus. *E3 Web Conf.* **2018**, *67*, 01023. [CrossRef]
6. Quan, Z.; Quan, L.; Zhang, J. Review of energy efficient direct pump controlled cylinder electro-hydraulic technology. *Renew Sustain. Energy Rev.* **2014**, *35*, 336–346. [CrossRef]
7. Koumboulis, F.N.; Skarpetis, M.G.; Tzamtzi, M.P. Robust PI Controllers for Command Following with Application to an Electropneumatic Actuator. In Proceedings of the 2006 14th Mediterranean Conference on Control and Automation, Ancona Italy, 28–30 June 2006; pp. 1–6. [CrossRef]
8. Behbahani, A.R.; Semega, K.J. Control Strategy for Electro-Mechanical Actuators Versus Hydraulic Actuation Systems fo Aerospace Applications. In Proceedings of the Power Systems Conference, SAE Technical Paper Series, Seattle, DC, USA, 11–1 November 2010; p. 15. [CrossRef]
9. Osman, I.S.; Hariri, N.G. Thermal Investigation and Optimized Design of a Novel Solar Self-Driven Thermomechanical Actuato *Sustainability* **2022**, *14*, 5078. [CrossRef]
10. Tisserand, O. How Does an Electric Actuator Work. Available online: https://www.indelac.com/blog/how-does-electric actuator-work (accessed on 14 November 2021).
11. Norgren. What Is an Electric Actuator. Available online: https://www.norgren.com/uk/en/expertise/industrial-automation what-is-an-electric-actuator (accessed on 14 November 2021).
12. Dávila-Peralta, C.; Cabanillas-López, R.; García-Gutiérrez, R.; Rodríguez-Carvajal, R. Design of a Linear Actuator Driven Sola Tracker for High Concentration Photovoltaics Technologies. *J. Clean Energy Technol.* **2015**, *4*, 197–201. [CrossRef]
13. Costa, B.A.; Lemos, J.M.; Rosa, L.G. Temperature control of a solar furnace for material testing. *Int. J. Syst. Sci.* **2011**, *42*, 1253–126 [CrossRef]
14. Silva, R.N.; Lemos, J.M.; Rato, L.M. Variable sampling adaptive control of a distributed collector solar field. *IEEE Trans. Contr Syst. Technol.* **2003**, *11*, 765–772. [CrossRef]
15. Williams, R.B.; Tanimoto, R.; Simonyan, A.; Fuerstenau, S. Vibration characterization of self-cleaning solar panels with piezoce ramic actuation. In Proceedings of the 48th AIAA/ASME/ASCE/AHS/ASC Structures, Structural Dynamics, and Material Conference, Honolulu, HI, USA, 23–26 April 2007; p. 1746. [CrossRef]
16. Alagoz, S.; Apak, Y. Removal of spoiling materials from solar panel surfaces by applying surface acoustic waves. *J. Clean. Pro* **2020**, *253*, 119992. [CrossRef]
17. Ebrahimpour, M.; Shafaghat, R.; Alamian, R.; Shadloo, M.S. Numerical Investigation of the Savonius Vertical Axis Wind Turbin and Evaluation of the Effect of the Overlap Parameter in Both Horizontal and Vertical Directions on Its Performance. *Symmetr* **2019**, *11*, 821. [CrossRef]
18. Lemonis, G. Wave and Tidal Energy Conversion. In *Encyclopedia of Energy*; Cleveland, C.J., Ed.; Elsevier: New York, NY, US 2004; pp. 385–396. [CrossRef]
19. Wilson,D.G.; Robinett, R.D., III; Bacelli, G.; Abdelkhalik, O Non Linear Control Design for Non Linear Wave Energy Converter Available online: https://www.osti.gov/servlets/purl/1648654 (accessed on 11 November 2021).
20. Huo, P.; Yang, F.; Luo, H.; Zhou, M.; Zhang, Y. Distributed monitoring system for precision management of household bioga appliances. *Comput. Electron. Agric.* **2019**, *157*, 359–370. [CrossRef]
21. Chen, D.F.; Chiu, S.P.C.; Cheng, A.B.; Ting, J.C. Electromagnetic Actuator System Using Witty Control System. *Actuators* **202** *10*, 65. [CrossRef]
22. Talavera, J.M.; Tobón, L.E.; Gómez, J.A.; Culman, M.A.; Aranda, J.M.; Parra, D.T.; Quiroz, L.A.; Hoyos, A.; Garreta, L.E. Revie of IoT applications in agro-industrial and environmental fields. *Comput. Electron. Agric.* **2017**, *142*, 283–297. [CrossRef]
23. Lajara, R.; Alberola, J.; Pelegrí-Sebastiá, J. A Solar Energy Powered Autonomous Wireless Actuator Node for Irrigation System *Sensors* **2010**, *11*, 329–340. [CrossRef]
24. Deb, D.; Brahmbhatt, N.L. Review of yield increase of solar panels through soiling prevention, and a proposed water-fre automated cleaning solution. *Renew. Sustain. Energy Rev.* **2018**, *82*, 3306–3313. [CrossRef]
25. Rosa. Solar Heat for Materials Processing: A Review on Recent Achievements and a Prospect on Future Trends. *ChemEngineerin* **2019**, *3*, 83. [CrossRef]
26. Nguyen, K.D. On the adaptive control of spherical actuators. *Trans. Inst. Meas. Control* **2018**, *41*, 816–827. [CrossRef]
27. Hafez, A.; Yousef, A.; Harag, N. Solar tracking systems: Technologies and trackers drive types—A review. *Renew. Sustain. Energ Rev.* **2018**, *91*, 754–782. [CrossRef]
28. Riad, A.; Ben Zohra, M.; Alhamany, A.; Mansouri, M. Bio-sun tracker engineering self-driven by thermo-mechanical actuator fo photovoltaic solar systems. *Case Stud. Therm. Eng.* **2020**, *21*, 100709. [CrossRef]

Aliyu, M.; Hassan, G.; Said, S.A.; Siddiqui, M.U.; Alawami, A.T.; Elamin, I.M. A review of solar-powered water pumping systems. *Renew. Sustain. Energy Rev.* **2018**, *87*, 61–76. [CrossRef]

Almadani, I.K.; Osman, I.S.; Hariri, N.G. In-Depth Assessment and Optimized Actuation Method of a Novel Solar-Driven Thermomechanical Actuator via Shape Memory Alloy. *Energies* **2022**, *15*, 3807. [CrossRef]

Kim, T.; Oh, S.; Yee, K. Improved actuator surface method for wind turbine application. *Renew. Energy* **2015**, *76*, 16–26. [CrossRef]

Mérigaud, A.; Ringwood, J.V. Condition-based maintenance methods for marine renewable energy. *Renew. Sustain. Energy Rev.* **2016**, *66*, 53–78. [CrossRef]

Prasetyo, A.; Asrori; Sipahutar, H.; Wibowo, B.S.C.; Gunawan; Nuryadin, D. Policy and Economy Analysis on the Application of the Smart on Grid Actuator to Public Road Lighting in Magelang City. In Proceedings of the 1st Borobudur International Symposium on Humanities, Economics and Social Sciences (BIS-HESS 2019), Magelang, Indonesia, 16 October 2019; Atlantis Press: Amsterdam, The Netherlands, 2020; pp. 206–210. [CrossRef]

Popovic, M.B.; Lamkin-Kennard, K.A.; Beckerle, P.; Bowers, M.P. 3—Actuators. In *Biomechatronics*; Popovic, M.B., Ed.; Academic Press: Cambridge, MA, USA, 2019; pp. 45–79. [CrossRef]

Nippon Pulse America, I. Linear Motor Systems: Iron Core, U-Channel, and Tubular Linear Motors. 2021. Available online: https://www.nipponpulse.com/learning/white-papers/linear-motor-systems-iron-core-u-channel-and- (accessed on 31 October 2021).

Wang, Y.; Deng, X.; Sun, P.; Qiao, D. A Kind of Electric Actuator Controller Solving The Problem of Motor Concussion. In Proceedings of the 2018 IEEE International Conference on Mechatronics and Automation (ICMA), Changchun, China, 5–8 August 2018; pp. 1015–1019. [CrossRef]

Laskey, D.R. Ball Screw and Nut Linear Actuator Assemblies and Methods of Constructing and Operating Them. U.S. Patent No. 6,101,889, 15 August 2000.

Han, L.; Yu, L.; Pan, C.; Zhao, H.; Jiang, Y. A novel impact rotary–linear motor based on decomposed screw-type motion of piezoelectric actuator. *Appl. Sci.* **2018**, *8*, 2492. [CrossRef]

Gao, B.; Chen, H.; Liu, Q.; Chu, H. Position Control of Electric Clutch Actuator Using a Triple-Step Nonlinear Method. *Ind. Electron. IEEE Trans.* **2014**, *61*, 6995–7003. [CrossRef]

Bandera, P. Multi-Degree of Freedom Spherical Actuator. U.S. Patent No. 9,893,574, 13 February 2018.

Lu, X.; Zhang, Q.; Hu, J. A linear piezoelectric actuator based solar panel cleaning system. *Energy* **2013**, *60*, 401–406. [CrossRef]

Chirca, M.; Dranca, M.; Teodosescu, P.D.; Breban, S. Limited-Angle Electromechanical Actuator for Micro Wind Turbines Overspeed Protection. In Proceedings of the 2019 11th International Symposium on Advanced Topics in Electrical Engineering (ATEE), Bucharest, Romania, 28–30 March 2019; pp. 1–6. [CrossRef]

Guillot, E.; Rodriguez, R.; Boullet, N.; Sans, J.L. Some details about the third rejuvenation of the 1000 kWth solar furnace in Odeillo: Extreme performance heliostats. *AIP Conf. Proc.* **2018**, *2033*, 040016. [CrossRef]

System, A. Top IoT Sensors. 2019. Available online: https://www.avsystem.com/blog/iot-sensors-iot-actuators/ (accessed on 23 May 2022).

Wójcicki, K.; Biegańska, M.; Paliwoda, B.; Górna, J. Internet of Things in Industry: Research Profiling, Application, Challenges and Opportunities—A Review. *Energies* **2022**, *15*, 1806. [CrossRef]

Ponniran, A.; Hashim, A.; Ali Munir, H. A design of single axis sun tracking system. In Proceedings of the 2011 5th International Power Engineering and Optimization Conference, Shah Alam, Malaysia, 6–7 June 2011; pp. 107–110. [CrossRef]

Hammad, B.K.; Fouad, R.H.; Ashhab, M.S.; Nijmeh, S.D.; Mohsen, M.; Tamimi, A. Adaptive control of solar tracking system. *IET Sci. Meas. Technol.* **2014**, *8*, 426–431. [CrossRef]

Jovanovic, V.M.; Ayala, O.; Seek, M.; Marsillac, S. Single axis solar tracker actuator location analysis. In Proceedings of the SoutheastCon 2016, Norfolk, VA, USA, 30 March–3 April 2016; pp. 1–5. [CrossRef]

Arbes, F.; Weinrebe, G.; Wöhrbach, M. Heliostat field cost reduction by 'slope drive' optimization. *AIP Conf. Proc.* **2016**, *1734*, 160002. [CrossRef]

Ferdaus, R.A.; Mohammed, M.A.; Rahman, S.; Salehin, S.; Mannan, M.A. Energy Efficient Hybrid Dual Axis Solar Tracking System. *J. Renew. Energy* **2014**, *2014*, 1–12. [CrossRef]

Batayneh, W.; Bataineh, A.; Soliman, I.; Hafees, S.A. Investigation of a single-axis discrete solar tracking system for reduced actuations and maximum energy collection. *Autom. Constr.* **2019**, *98*, 102–109. [CrossRef]

Alexandru, C. The design and optimization of a photovoltaic tracking mechanism. In Proceedings of the 2009 International Conference on Power Engineering, Energy and Electrical Drives, Lisbon, Portugal, 18–20 March 2009; pp. 436–441. [CrossRef]

Ngo, X.C.; Nguyen, T.H.; Do, N.Y.; Nguyen, D.M.; Vo, D.V.N.; Lam, S.S.; Heo, D.; Shokouhimehr, M.; Nguyen, V.H.; Varma, R.S.; Kim, S.Y.; Le, Q.V. Grid-Connected Photovoltaic Systems with Single-Axis Sun Tracker: Case Study for Central Vietnam. *Energies* **2020**, *13*, 1457. [CrossRef]

Argeseanu, A.; Ritchie, E.; Leban, K. New low cost structure for dual axis mount solar tracking system using adaptive solar sensor. In Proceedings of the 2010 12th International Conference on Optimization of Electrical and Electronic Equipment, Brasov, Romania, 20–22 May 2010; pp. 1109–1114. [CrossRef]

Hariri, N.G.; AlMutawa, M.A.; Osman, I.S.; AlMadani, I.K.; Almahdi, A.M.; Ali, S. Experimental Investigation of Azimuth-and Sensor-Based Control Strategies for a PV Solar Tracking Application. *Appl. Sci.* **2022**, *12*, 4758. [CrossRef]

56. IFC. Utility-Scale Solar Photovoltaic Power Plants, a Project Developer's Guide. Available online: https://www.ifc.org/wps wcm/connect/topics_ext_content/ifc_external_corporate_site/sustainability-at-ifc/publications/publications_utility-scale solar+photovoltaic+power+plants (accessed on 31 October 2021).
57. Ballestrín, J.; Rodríguez-Alonso, M.; Rodriguez, J.; Canadas, I.; Barbero, F.; Langley, L.; Barnes, A. Calibration of high-heat-flu sensors in a solar furnace. *Metrologia* **2006**, *43*, 495. [CrossRef]
58. Egan, D. Giant Magnifier Reaches 5000 Degrees Using Only Sunlight. Available online: https://www.discovermagazine.com technology/giant-magnifier-reaches-5-000-degrees-using-only-sunlight (accessed on 17 May 2022).
59. Ballestrín, J.; Estrada, C.A.; Rodríguez-Alonso, M.; Pérez-Rábago, C.; Langley, L.W.; Barnes, A. High-heat-flux sensor calibratio using calorimetry. *Metrologia* **2004**, *41*, 314–318. [CrossRef]
60. Costa, B.A.; Lemos, J.M. Singular perturbation stability conditions for adaptive control of a solar furnace with actuator dynamic In Proceedings of the 2009 European Control Conference (ECC), Budapest, Hungary, 23–26 August 2009; pp. 1626–163 [CrossRef]
61. Camacho, E.F.; Rubio, F.R.; Berenguel, M.; Valenzuela, L. A survey on control schemes for distributed solar collector fields. Part Modeling and basic control approaches. *Sol. Energy* **2007**, *81*, 1240–1251. [CrossRef]
62. Schiel, W.; Hunt, B.; Keck, T.; Schweitzer, A. Solar Thermal Power Plants For Central or Distributed Electricity Generation. I Proceedings of the POWER-GEN Europe, Barcelona, Spain, 25–27 May 2004.
63. Cirre, C.M.; Berenguel, M.; Valenzuela, L.; Camacho, E.F. Feedback linearization control for a distributed solar collector fiel *Control Eng. Pract.* **2007**, *15*, 1533–1544. [CrossRef]
64. Lemos, J.M. Adaptive control of distributed collector solar fields. *Int. J. Syst. Sci.* **2006**, *37*, 523–533. [CrossRef]
65. Frank, E.; Mauthner, F.; Fischer, S. Overheating prevention and stagnation handling in solar process heat applications. Technic Report A.1.2. *Int. Energy-Agency-Sol. Heat. Cool. Task* **2015**, *49*.[CrossRef]
66. Motlagh , N.H.; Mohammadrezaei, M.; Hunt, J.; Zakeri, B. Internet of Things (IoT) and the Energy Sector. *Energies* **2020**, *13*, 49 [CrossRef]
67. Canaletti, J.L.; Notton, G.; Damian, A.; Colda, I.; Cristofari, C. New concept of solar air heater integrated in the buildin In Proceedings of the 4th IASME/WSEAS International Conference on Energy, Environment, Ecosystems and Sustainab Development, EEESD'08, Algarve, Portugal, 11–13 June 2008; pp. 52–59.
68. Panat, S.; Varanasi, K.K. Electrostatic dust removal using adsorbed moisture–assisted charge induction for sustainable operatic of solar panels. *Sci. Adv.* **2022**, *8*, eabm0078. [CrossRef] [PubMed]
69. Dawson, S.; Nick, M.B.S.; Steve, W. Solar Array Panels with Dust-Removal Capability. Available online: https://www.techbrie com/component/content/article/tb/pub/briefs/mechanics-and-machinery/1553 (accessed on 12 November 2021).
70. Al-Badra, M.; Abd-Elhady, M.; Kandil, H. A novel technique for cleaning PV panels using antistatic coating with a mechanic vibrator. *Energy Rep.* **2020**, *6*, 1633–1637. [CrossRef]
71. Syafiq, A.; Pandey, A.; Adzman, N.; Abd Rahim, N. Advances in approaches and methods for self-cleaning of solar photovolta panels. *Sol. Energy* **2018**, *162*, 597–619. [CrossRef]
72. Altıntaş, M.; Arslan, S. Reduction of Dust on Solar Panels through Unipolar Electrostatic Traveling Wave. *Appl. Sci.* **2021**, *11*, 912 [CrossRef]
73. King, M.; Li, D.; Dooner, M.; Ghosh, S.; Roy, J.N.; Chakraborty, C.; Wang, J. Mathematical Modelling of a System for Solar P Efficiency Improvement Using Compressed Air for Panel Cleaning and Cooling. *Energies* **2021**, *14*, 4072. [CrossRef]
74. Deb, D.; Burkholder, J.; Tao, G. *Adaptive Compensation of Nonlinear Actuators for Flight Control Applications*; Spring Berlin/Heidelberg, Germany, 2021.
75. Deb, D.; Burkholder, J.; Tao, G. Synthetic Jet Actuators and Arrays: Modeling and Control. In *Adaptive Compensation of Nonline Actuators for Flight Control Applications*; Springer: Berlin/Heidelberg, Germany, 2022; pp. 11–41.
76. Hamlehdar, M.; Kasaeian, A.; Safaei, M.R. Energy harvesting from fluid flow using piezoelectrics: A critical review. *Rene Energy* **2019**, *143*, 1826–1838. [CrossRef]
77. Han, H.; Ko, J. Power-Generation Optimization Based on Piezoelectric Ceramic Deformation for Energy Harvesting Applicatic with Renewable Energy. *Energies* **2021**, *14*, 2171. [CrossRef]
78. Nechibvute, A.; Chawanda, A.; Luhanga, P. Piezoelectric Energy Harvesting Devices: An Alternative Energy Source for Wirele Sensors. *Smart Mater. Res.* **2012**, *2012*, 1–13. [CrossRef]
79. Ontiveros, J.J.; Ávalos, C.D.; Loza, F.; Galán, N.D.; Rubio, G.J. Evaluation and design of power controller of two-axis sol tracking by PID and FL for a photovoltaic module. *Int. J. Photoenergy* **2020**, *2020*, 8813732. [CrossRef]
80. Maindad, N.; Gadhave, A.; Satpute, S.; Nanda, B. Automatic Solar Panel Cleaning System. In Proceedings of the 2nd Internation Conference on Communication & Information Processing (ICCIP), Singapore, 26–29 November 2016.
81. Al Qdah, K.S.; Abdulqadir, S.A.; Al Harbi, N.Y.; Soqyyah, A.Z.; Isa, K.J.; Alharbi, M.Y.; Binsaad, N.M. Design and performance PV dust cleaning system in medina region. *J. Power Energy Eng.* **2019**, *7*, 1–14. [CrossRef]
82. Chirca, M.; Dranca, M.; Oprea, C.A.; Teodosescu, P.D.; Pacuraru, A.M.; Neamtu, C.; Breban, S. Electronically Controlled Actuato for a Micro Wind Turbine Furling Mechanism. *Energies* **2020**, *13*, 4207. [CrossRef]
83. Cuches, J.P.; Frazier, R.H. Angular Displacement Solenoid. US Patent US3221191, 30 November 1965.
84. Drew, B.; Plummer, A.R.; Sahinkaya, M.N. A review of wave energy converter technology. *Proc. Inst. Mech. Eng. Part A J. Pow Energy* **2009**, *223*, 887–902. [CrossRef]

Aderinto, T.; Li, H. Review on power performance and efficiency of wave energy converters. *Energies* **2019**, *12*, 4329. [CrossRef]

Rahman, A.; Farrok, O.; Islam, M.R.; Xu, W. Recent progress in electrical generators for oceanic wave energy conversion. *IEEE Access* **2020**, *8*, 138595–138615. [CrossRef]

Sugita, M.R.P.; Wijaya, F.D.; Sarjiya. Design and analysis of tri core permanent magnet linear generator for wave energy conversion in south coast of Java Island. In Proceedings of the 2016 6th International Annual Engineering Seminar (InAES), Yogyakarta, Indonesia, 1–3 August 2016; pp. 256–261.

Schörghuber, C.; Gölles, M.; Reichhartinger, M.; Horn, M. Control of biomass grate boilers using internal model control. *Control Eng. Pract.* **2020**, *96*, 104274. [CrossRef]

Yin, C.; Rosendahl, L.A.; Kær, S.K. Grate-firing of biomass for heat and power production. *Prog. Energy Combust. Sci.* **2008**, *34*, 725–754. [CrossRef]

Sontake, V.C.; Kalamkar, V.R. Solar photovoltaic water pumping system - A comprehensive review. *Renew. Sustain. Energy Rev.* **2016**, *59*, 1038–1067. [CrossRef]

Yoo, S.E..; Kim, J.E..; Kim, T.; Ahn, S.; Sung, J.; Kim, D. A^2S: Automated Agriculture System based on WSN. In Proceedings of the 2007 IEEE International Symposium on Consumer Electronics, Irving, TX, USA, 20–23 June 2007; pp. 1–5. [CrossRef]

Roy, S.K.; Roy, A.; Misra, S.; Raghuwanshi, N.S.; Obaidat, M.S. AID: A prototype for Agricultural Intrusion Detection using Wireless Sensor Network. In Proceedings of the 2015 IEEE International Conference on Communications (ICC), London, UK, 8–12 June 2015; pp. 7059–7064. [CrossRef]

Garcia-Caparros, P.; Contreras, J.; Baeza, R.; Segura, M.; Lao, M. Integral Management of Irrigation Water in Intensive Horticultural Systems of Almería. *Sustainability* **2017**, *9*, 2271. [CrossRef]

Rahman, M.F.; Cheung, N.C.; Lim, K.W. Position estimation in solenoid actuators. *IEEE Trans. Ind. Appl.* **1996**, *32*, 552–559. [CrossRef]

Çetin, G.; Özkaraca, O.; Keçebaş, A. Development of PID based control strategy in maximum exergy efficiency of a geothermal power plant. *Renew. Sustain. Energy Rev.* **2021**, *137*, 110623. [CrossRef]

Sarbu, I.; Sebarchievici, C. Chapter 7—Experimental Ground-Coupled Heat Pump Systems. In *Ground-Source Heat Pumps*; Sarbu, I., Sebarchievici, C., Eds.; Academic Press: Cambridge, MA, USA, 2016; pp. 167–196. [CrossRef]

Zhang, P. (Ed.) Chapter 3—Sensors and actuators. In *Advanced Industrial Control Technology*; William Andrew Publishing: Oxford, UK, 2010; pp. 73–116. [CrossRef]

Tisserand, O. The Evolution of Actuator Design. Available online: https://www.indelac.com/blog/evolution-actuator-design (accessed on 26 March 2022).

Tisserand, O. Wireless Actuator. Available online: https://www.indelac.com/blog/bid/320870/Wireless-Actuator (accessed on 26 March 2022).

Improve Energy Efficiency with Actuator Solutions. Available online: https://www.linak.com/business-areas/energy/ (accessed on 24 March 2022).

Blanco, J.; García, A.; Morenas, J.D.l. Design and Implementation of a Wireless Sensor and Actuator Network to Support the Intelligent Control of Efficient Energy Usage. *Sensors* **2018**, *18*, 1892. [CrossRef]

Bellini, P.; Nesi, P.; Pantaleo, G. IoT-enabled smart cities: A review of concepts, frameworks and key technologies. *Appl. Sci.* **2022**, *12*, 1607. [CrossRef]

Llaria, A.; Terrasson, G.; Curea, O.; Jiménez, J. Application of Wireless Sensor and Actuator Networks to Achieve Intelligent Microgrids: A Promising Approach towards a Global Smart Grid Deployment. *Appl. Sci.* **2016**, *6*, 61. [CrossRef]

Park, B.; Nah, J.; Choi, J.Y.; Yoon, I.J.; Park, P. Robust wireless sensor and actuator networks for networked control systems. *Sensors* **2019**, *19*, 1535. [CrossRef]

Junior, J.; Lima, M.; Balico, L.; Pazzi, R.; Oliveira, H. Routing with Renewable Energy Management in Wireless Sensor Networks. *Sensors* **2021**, *21*, 4376. [CrossRef] [PubMed]

MDPI AG
Grosspeteranlage 5
4052 Basel
Switzerland
Tel.: +41 61 683 77 34

Sensors Editorial Office
E-mail: sensors@mdpi.com
www.mdpi.com/journal/sensors

MDPI

www.ingramcontent.com/pod-product-compliance
Lightning Source LLC
Chambersburg PA
CBHW040317270926
51850CB00007BC/1506
9783725864843